Frank Hoffmann, Mario Köppen, Frank Klawonn, Rajkumar Roy (Eds.)

Soft Computing: Methodologies and Applications

Advances in Soft Computing

Editor-in-chief

Prof. Janusz Kacprzyk
Systems Research Institute
Polish Academy of Sciences
ul. Newelska 6
01-447 Warsaw
Poland
E-mail: kacprzyk@ibspan.waw.pl

Further volumes of this series
can be found on our homepage:
springeronline.com

Henrik Larsen, Janusz Kacprzyk, Sławomir Zadrozny, Troels Andreasen, Henning Christiansen (Eds.)
Flexible Query Answering Systems, 2000
ISBN 3-7908-1347-8

Bernd Reusch, Karl-Heinz Temme (Eds.)
Computational Intelligence in Theory and Practice, 2001
ISBN 3-7908-1357-5

Robert John, Ralph Birkenhead (Eds.)
Developments in Soft Computing, 2001
ISBN 3-7908-1361-3

Mieczysław Kłopotek, Maciej Michalewicz and Sławomir T. Wierzchoń (Eds.)
Intelligent Information Systems, 2001
ISBN 3-7908-1407-5

Frank Hoffmann
Mario Köppen
Frank Klawonn
Rajkumar Roy (Eds.)

Soft Computing: Methodologies and Applications

Springer

Frank Hoffmann
Universität Dortmund
Fakultät für Elektrotechnik
und Informationstechnik
Lehrstuhl für Elektrische
Steuerung und Regelung
Otto-Hahn-Str. 4
44227 Dortmund
Germany

Frank Klawonn
FH Braunschweig/Wolfenbüttel
Salzdahlumer Str. 46/48
38302 Wolfenbüttel
Germany

Mario Kppen
Fraunhofer IPK, Berlin
Institut für Sicherheits-/Prüftechnik
Pascalstr. 8-9
10587 Berlin
Germany

Rajkumar Roy
Enterprise Integration
Cranfield University
Cranfield MK43 0AL
Bedford
United Kingdom

Library of Congress Control Number: 2005925479

ISSN print edition: 1615-3871
ISSN electronic edition: 1860-0794
ISBN-10 3-540-25726-8 Springer Berlin Heidelberg New York
ISBN-13 978-3-540-25726-4 Springer Berlin Heidelberg New York

Springer is a part of Springer Science+Business Media
springeronline.com

Printed in The Netherlands

Typesetting: by the authors and TechBooks using a Springer LaTeX macro package

Printed on acid-free paper SPIN: 11016205 89/TechBooks 5 4 3 2 1 0

Preface

The series of Online World Conferences on Soft Computing (WSC) is organized by the World Federation of Soft Computing (WFSC) and has become an established annual event in the academic calendar and was already held for the 8th time in 2003. Starting as a small workshop held at Nagoya University, Japan in 1994 it has matured to the premier online event on soft computing in industrial applications. It has been hosted by the universities of Granada, Spain, Fraunhofer Gesellschaft, Berlin, Cranfield University, Helsinki University of Technology and Nagoya University. The goal of WFSC is to promote soft computing across the world, by using the internet as a forum for virtual technical discussion and publishing at no cost to authors and participants. The official journal of the World Federation on Soft Computing is the journal *Applied Soft Computing*.

The 8th WSC Conference (WSC8) took place from September 29th to October 10th, 2003. Registered participants had the opportunity to follow and discuss the online presentations of authors from all over the world. Out of more than 60 submissions the program committee had accepted 27 papers for final presentation at WSC8.

What makes WSC conferences unique is the fact that authors and participants disseminate their scientific results and engage in a discussion with colleagues from all over the world at no cost. Participants have immediate access to all contributions, the opportunity to comment the papers and to obtain feedback from the authors. The online discussions result in new contacts within the research community, form the basis for future cooperations and eventually lead to personal meetings with other colleagues in the future.

It has become a tradition of the WSC conferences to publish post-proceedings of selected papers. Authors are encouraged to submit a revised version of their contributions after the conference, taking also the comments from the discussion on their paper into account. This book contains 26 revised contributions from WSC8, covering a wide range of application areas of soft computing like optimization, data analysis and data mining, fault diagnosis, control as well as traffic and transportation systems. The papers collected in this book show how the major soft computing techniques, fuzzy systems, neural networks and evolutionary algorithms and especially hybrid systems combining methods from these fields, lead to successful industrial applications. The reader will find an interesting, inspiring and wide variety of soft computing techniques and applications in this book.

We would like to thank the staff at Springer, in particular Thomas Ditzinger and Heather King, for their professional support and expertise in all phases of this project.

Cyberspace,
May, 2005

Frank Hoffmann
Mario Köppen
Frank Klawonn
Rajkumar Roy

Welcome Message Honorary Chair

As we all know, the genesis of soft computing was motivated by the realization that in science, as in most other realms of human activity; there is a tendency to be nationalistic – to commit oneself to a particular methodology, M, and march under its banner, in the belief that it is M and only M that matters. The well-known Hammer Principle: When the only tool you have is a hammer, everything looks like a nail; and my Vodka Principle: No matter what your problem is, vodka will solve it, are succinct expressions of the one-size-fits-all mentality which underlies nationalism in science. Although it is obvious that one-size-fits-all mentality in science is counterproductive, it is still the case that in many fields of scientific activity, nationalism is the norm rather than an abberation. Given this reality, WSC8 serves an important function by showing that the concept of soft computing has intrinsic validity. I should like to extend my compliments and congratulations to the organizers and participants in WSC8 for contributing so much and in so many ways to the advancement of soft computing and enhancing its visibility and importance in the realms of both theory and applications,

With my warmest regards to all.

Berkeley, California
September, 25th, 2003

Lotfi Zadeh

Welcome Message General Chair

WSC8 has opened its gates. From September 29th to October 10th registered participants will have the opportunity to follow and discuss the online presentations of authors from all over the world. Out of more than 60 submissions the program committee accepted 27 papers for final presentation at WSC8.

The WSC conference has become an established annual event in the academic calendar as it enters its eight round. Starting as a small workshop held at Nagoya University, Japan in 1994 it has matured to the premier online event on soft computing in industrial applications.

The series of WSC conferences has been organized by the World Federation of Soft Computing (WFSC) and has been hosted by the universities of Granada, Spain, Fraunhofer Gesellschaft, Berlin, Cranfield University, Helsinki University of Technology and Nagoya University. The goal of WFSC is to promote soft computing across the world, by using the internet as a forum for virtual technical discussion and publishing at no cost to authors and participants. The official journal of the World Federation on Soft Computing is the journal "Applied Soft Computing", published by Elsevier Press.

What makes WSC conferences unique is the fact that authors and participants disseminate their scientific results and engage in a discussion with colleagues from all over the world at no cost. Participants have immediate access to all contributions, the opportunity to comment the papers and to obtain feedback from the authors. Our hope is that the online discussion will result in new contacts within the research community, forms the basis for future cooperations and eventually lead to a personal meeting with your colleague in the near future. In the end, an online conference can not provide the equivalent social experience equivalent of having a drink or a dinner.

On behalf of all organizers I want to express my thanks to all who have contributed to WSC8, thanks to all the authors who deemed WSC8 to be a suitable place to publish their work, thanks to all who helped to distribute the CFP or organized a session, thanks to the program committee for the review of submitted papers, thanks to the sponsors. Finally, I would like to express my particular gratitude to Mario Köppen, whose help and technical support with setting up the scripts, organizing the review process, answering technical questions was invaluable to me.

We hope that you will enjoy the conference, engage in fruitful discussions, establish novel contacts and draw inspiration from the presentations and discussions.

With my best regards to all participants.

Dortmund, Germany,
September, 29th, 2003

Frank Hoffmann

Organization

The 8th Online World Conference on Soft Computing in Industrial Applications was organized by the World Federation of Soft Computing and the Department of electrical control engineering, Faculty of electrical engineering and information technology, University of Dortmund.

Organizing Committee

Honorary Chair:	Lotfi A. Zadeh, University of California Berkeley, USA
General Chair:	Frank Hoffmann, University of Dortmund, Germany
International Co-Chair:	Takeshi Furuhashi, Nagoya University, Japan Rajkumar Roy, Cranfield University, United Kingdom
Technical Program Chair:	Mario Koeppen, Fraunhofer IPK Berlin, Germany
Web Chair:	Daniel Schauten, University of Dortmund, Germany
Publicity and Online Tutorial Chair:	Seppo Ovaska, Helsinki University of Technology, Finland
Publication Chair:	Frank Klawonn, University of Applied Science Wolfenbüttel, Germany
Industrial Liaison Chair:	Michael Berthold, Tripos, US
Session Chairs:	Erel Avineri, Israel Institute of Technology, Israel Dimiter Driankov, Linköping University, Sweden Eddie Tunstel, NASA, Jet Propulsion Laboratory, California Anders Lansner, Roval Institute of Technology, Sweden Bart Baesens, University of Leuven, Belgium Fernando Gomide, State University of Campinas, Brasil

Program Committee

Ajith Abraham, Korea
Andrea Bonarini, Italy
Andreas König, Germany
Andreas Nürnberger, Germany
Aureli Soria-Frisch, Germany
Avineri Erel, Israel
Bart Baesens, Belgium
Brian Carse, UK
Christophe Marsala, France
Detlef D. Nauck, UK
Dimiter Driankov, Sweden
Edward W. Tunstel, USA
Francisco Herrera, Spain
Frank Klawonn, Germany
Guiseppe di Fatta, Italy
Guy de Tré, Belgium
Hani Abdel Kader Hagras, UK
Janus Abonyi, Hungary
Javier Ruiz-del-Solar,Chile
José M. Benitez, Spain
Katrin Franke, Germany
Leandro dos Santos Coelho, Brazil
Luis Magdalena, Spain
Mauro Dell'Orco, Italy
Renato A. Krohling, Brazil
Sankar Kumar Pal, India
Senén Barro Ameneiro, Spain
Ulrich Bodenhofer, Austria
Vicenc Torra, Spain
Xiao-Zhi Gao, Finland
Yukinori Suzuki, Japan

Sponsoring Institutions

World Federation of Soft Computing
Springer
Elsevier
IEEE Systems, Man, and Cybernetics Society
North American Fuzzy Information Processing Society (NAFIPS)
European Society for Fuzzy Logic and Technology (EUSFLAT)
International Fuzzy Systems Association
University of Dortmund

Contents

Part VI Data Analysis

Part VII Soft Computing for Modeling, Optimization and Information Processing

List of Contributors

Abbattista, F.
Dipartimento di Informatica
Università degli Studi di Bari
fabio@di.uniba.it

Abonyi, Janos
University of Veszprem
P.O.Box 158
H-8201, Veszprem, Hungary
abonyi@fmt.vein.hu

Akad, Murat
Technical University of Istanbul
Faculty of Civil Engineering
Maslak
34469, Istanbul, Turkey
muratakad@ins.itu.edu.tr

Alcalde, Rafael
R&D Department, Cátenon Human Resource Selection
Goya 115
28009, Madrid, Spain
ralcalde@catenon.es

Amato, Paolo
Soft Computing, Si-Optics and Post-Silicon Technologies Corporate R&D STMicroelectronics
Via Remo De Feo 1
80022, Arzano (Na), Italy
paolo.amato@st.com

Avineri, Erel
Centre for Transport and Society
Faculty of the Built Environment
University of the West of England
BS16 1QY
Bristol, UK,
Erel.Avineri@uwe.ac.uk

Barro, S.
Dept. of Electronics and Computer Science, University of Santiago de Compostela
15782 Santiago de Compostela, Spain
senen@dec.usc.es

Bugarin, Alberto
Dept. of Electronics and Computer Science, University of Santiago de Compostela
15782 Santiago de Compostela, Spain
alberto@dec.usc.es

Calixto, Roger
Universidade Tuiuti do Paraná, Campus Torres Faculdades de Ciências Exatas e de Tecnologia
Av. Comendador Franco 1860
CEP 80215-090
Curitiba, PR, Brazil

Callai, T. C.
Federal University of Santa Catarina – Dep. of Automation and Systems
Box 476
88040-900
Florianópolis, SC, Brazil
tomas@das.ufsc.br

Cano, José Ramón
Dept. of Computer Science, Esuela Politécnica Superior of Linares
University of Jaen
23700, Linares, Jaen, Spain
jrcano@ujaen.es

Catucci, G.
Università degli Studi di Bari
Dipartimento di Informatica
gracat@email.it

Celikoglu, H. Berk
Technical University of Istanbul, Faculty of Civil Engineering
Maslak
34469, Istanbul, Turkey
hbcelikoglu@ins.itu.edu.tr

Cesario, Nicola
STMicroelectronics, SST Corporate R&D
Via Remo De Feo 1
80022, Arzano (NA), Italy
nicola.cesario@st.com

Coelho, A.A.R.
Federal University of Santa Catarina – Dep. of Automation and Systems
Box 476
88040-900
Florianópolis, SC, Brazil
aarc@das.ufsc.br

Coelho, Leandro dos Santos
Pontifícia Universidade Católica do Paraná, Grupo Produtrônica Programa de Pós-Graduação em Engenharia de Produção e Sistemas
Rua Imaculada Conceição, 1155
CEP 80215-901
Curitiba, PR, Brazil
leandro.coelho@pucpr.br

Cordón, Oscar
Dept. of Computer Science and A.I.E.T.S. de Ingeniería Informática
University of Granada
18071 Granada (Spain)
ocordon@decsai.ugr.es

De Falco, Ivanhoe
ICAR, National Research Council of Italy
Via P. Castellino 111
80131 Naples, Italy
ivanhoe.defalco@na.icar.cnr.it

Dell'Orco, Mauro
Department of Highways and Transportation Polytechnic
University of Bari
via Orabona, 4
70125 Bari, Italy
dellorco@poliba.it

Della Cioppa, A.
Associato INFM
INFM
Dept. of Computer Science and Electrical Engineering
University of Salerno
Via Ponte don Melillo 1
84084 Fisciano (SA), Italy
adellacioppa@unisa.it

Farina, Marco
STMicroelectronics, SST Corporate R&D
Via C. Olivetti, 2
20041, Agrate(MI), Italy
marco.farina@st.com

Franke, Katrin
Fraunhofer Institute for Production Systems and Design Technology
Department Pattern Recognition
Pascalstr. 8-9
10587 Berlin, Germany
franke@ipk.fhg.de

García-Barriocanal, Elena
Computer Science Department
University of Alcalá,
Ctra. Barcelona km. 33.600
28871, Alcalá de Henares, Madrid, Spain
elena.garciab@uah.es

Herrera, Francisco
Dept. of Computer Science and Artificial Intelligence University of Granada
18071 Granada, Spain
herrera@decsai.ugr.es

Herrera-Viedma, Enrique
Dept. of Computer Science and A.I.E.T.S. de Ingeniería Informática
University of Granada
18071 Granada, Spain
viedma@decsai.ugr.es

Kikuchi, Shinya
Department of Civil and Environmental Engineering, University of Delaware
Newark, DE 19716, USA
kikuchi@ce.udel.edu

Kollias, Stefanos
National Technical University of Athens: Image, Video and Multimedia Systems Laboratory
stefanos@cs.ntua.gr

Köppen, Mario
Fraunhofer Institute for Production Systems and Design Technology,
Department Pattern Recognition
Pascalstr. 8-9
10587 Berlin, Germany
koeppen@ipk.fhg.de

Krohling, Renato
Lehrstuhl Elektrische Steuerung und Regelung, Fakultät für Elektrotechnik und Informationstechnik, Universität Dortmund
D-44221 Dortmund, Germany
krohling@esr.e-technik.uni-dortmund.de

Lamarca, M.
Dipartimento di Informatica, Università degli Studi di Bari
maddriver@libero.it

Lee, Wei-Po
Department of Infromation Management, National University of Kaohsiung
Kaohsiung, Taiwan

Lim, Chee Peng
School of Electrical & Electronic Engineering, University of Science Malaysia, Engineering Campus, Nibong Tebal
14300 Seberang Prai Selatan
Penang, Malaysia
cplim@usm.my

Lozano, Manuel
Dept. of Computer Science and Artificial Intelligence, University of Granada
18071 Granada, Spain
lozano@decsai.ugr.es

Luque, María
Dept. of Computer Science and A.I.E.T.S. de Ingeniería Informática,
University of Granada
18071 Granada (Spain)
mluque@decsai.ugr.es

Madar, Janos
University of Veszprem
P.O.Box 158
H-8201, Veszprem, Hungary
abonyi@fmt.vein.hu

Moreno, D. L.
Dept. of Electronics and Computer Science, University of Santiago de Compostela
15782 Santiago de Compostela, Spain
dave@dec.usc.es

Moya, Felix
Dept. of Information Sciences, Faculty of Information Sciences
University of Granada
18071 Granada, Spain
felix@ugr.es

Mucientes, M.
Dept. of Electronics and Computer Science, University of Santiago de Compostela
15782 Santiago de Compostela, Spain
manuel@dec.usc.es

Mylonas, Phivos
National Technical University of Athens: Image, Video and Multimedia Systems Laboratory
fmylonas@image.ntua.gr

Nagy, Lajos
University of Veszprem
P.O.Box 158
H-8201, Veszprem, Hungary
abonyi@fmt.vein.hu

Oduguwa, Victor
Department of Enterprise Integration
School of Industrial and Manufacturing Science, Cranfield University
Cranfield, Bedford
MK43 0AL, UK
v.oduguwa@cranfield.ac.uk

Přibyl, Ondřej
Czech Technical University in Prague
Department of Applied Mathematics
Na Florenci 25
Praha 1, 110 00, Czech Republic
pribylo@fd.cvut.cz

Puchalski, Bartlomiej
Faculty of Automatic Control, Electronics and Computer Science, Silensian University of Technology
ul. Akademicka 16
41-100 Gliwice, Poland
bartlomiej.puchalski@centertel.pl

Roy, Rajkumar
Department of Enterprise Integration
School of Industrial and Manufacturing Science,
Cranfield University
Cranfield, Bedford
MK43 0AL, UK
r.roy@cranfield.ac.uk

Ruggiero, D.
STMicroelectronics, SST Corporate R&D
Via Remo De Feo 1
80022 Arzano (Na), Italy
davide.ruggiero@st.com

Rutkowski, Jerzy
Faculty of Automatic Control, Electronics and Computer Science, Silensian University of Technology
ul. Akademicka 16
41-100 Gliwice, Poland
jr@boss.iele.polsl.gliwice.pl

Santos, J. E. S.
Federal University of Santa Catarina,
Dept. of Automation and Systems
Box 476 – 88040-900
Florianópolis – SC – Brazil
santos@das.ufsc.br

Sicilia, Miguel-Ángel
Computer Science Department
University of Alcalá
Ctra. Barcelona km. 33.600
28871, Alcalá de Henares, Madrid, Spain
msicilia@uah.es

Sudkamp, Thomas
Department of Computer Science
Wright State University
Dayton OH, USA
tsudkamp@cs.wright.edu

Sumar, R. R.
Federal University of Santa Catarina – Dept. of Automation and Systems
Box 476
88040-900
Florianópolis, SC, Brazil
sumar@das.ufsc.br

Szeifert, Ferenc
University of Veszprem
P.O. Box 158
H-8201, Veszprem, Hungary
abonyi@fmt.vein.hu

Tan, Shing Chiang
Faculty of Information Science & Technology, Multimedia University
Jalan Ayer Keroh Lama
75450 Melaka, Malaysia
sctan@mmu.edu.my

Tarantino, E.
ICAR, National Research Council of Italy
Via P. Castellino 111
80131 Naples, Italy
ernesto.tarantino@na.icar.cnr.it

Tiwari, Ashutosh
Department of Enterprise Integration
School of Industrial and Manufacturing Science, Cranfield University
Cranfield, Bedford
MK43 0AL, UK
a.tiwari@cranfield.ac.uk

Veenhuis, Christian
Fraunhofer Institute for Production Systems and Design Technology,
Department Pattern Recognition
Pascalstr. 8-9
10587 Berlin, Germany
veenhuis@ipk.fhg.de

Wallace, Manolis
National Technical University of Athens: Image, Video and Multimedia Systems Laboratory
wallace@image.ntua.gr

Zambetta, M.
Dipartimento di Informatica, Università degli Studi di Bari
zambetta@di.uniba.it

Zarco, Carmen
PULEVA S.A.
Camino de Rurchil, 66
18001 Granada, Spain
czarco@puleva.es

Zielinski, Lukasz
Faculty of Automatic Control, Electronics and Computer Science, Silensian University of Technology
ul. Akademicka 16
41-100 Gliwice, Poland
lukasz.zielinski@comverse.com

Part I

Keynote Papers

Discovery of Fuzzy Temporal Associations in Multiple Data Streams

Thomas Sudkamp

Department of Computer Science, Wright State University, Dayton OH, USA
tsudkamp@cs.wright.edu

1 Introduction

The need for the rapid analysis of real-time information collected from multiple sensors provides a motivation for the development of algorithms for the identification of temporal relationships among data. The majority of research in knowledge discovery with temporal data has focused on the identification of precise relationships in homogeneous data from a single data stream or database. However, many temporal relationships are inherently imprecise and consider data from multiple sources in which the sole link between the data is the time associated with each item.

We will consider the discovery of relationships among data that include imprecise temporal durations and temporal constraints. An imprecise duration may be described linguistically, for example *a short period* or *approximately 1 hour*, and will be represented by fuzzy intervals. A temporal constraint consists of a duration and a displacement, like *shortly after* or *simultaneously*. Thus associations may have the form

high traffic volume at points *a* and *b*
is *followed shortly* by congestion at point *c*
or
if there is a *large* increase in queries over a *short* period
of time, then the network is under attack

where the italics indicate imprecise or fuzzy quantities. The temporal durations and constraints in the associations will be modeled by fuzzy sets.

To develop a methodology for fuzzy temporal relationships, we begin with a review of data mining for association rules and the assumptions and data representations commonly employed in data mining for temporal patterns in crisp data. We then present the constraints, data representations, and search strategy needed for analyzing fuzzy temporal associations in data from multiple sources.

Thomas Sudkamp: *Discovery of Fuzzy Temporal Associations in Multiple Data Streams*, Advances in Soft Computing **1**, 3–13 (2005)
www.springerlink.com

2 Data Streams, Associations, and Patterns

In this section we examine data sources and representations used in several of the classic works in knowledge discovery. One of the most common and intuitive knowledge discovery models is the market basket representation presented in [1, 2]. The defining metaphor and the source of the terminology comes from considering a single data item to represent the contents of a transaction in a supermarket. The data consist of a set of unrelated transactions and the objective is to determine combinations of items that occur repeatedly within individual transactions. The market basket formulation is a minimally restrictive representation consisting of an unordered set of data from a single source or identical sources.

Employing the market basket metaphor, an itemset I consists of the inventory of a market and a database T of transactions contains of a record of the sales at the market. An association rule represents a correspondence between items in single transaction. The association $A \Rightarrow B$ indicates that transactions containing a set of items A also contain the items B. Two measures, support and confidence, were introduced to assess the validity of an association based on the set of transactions. The support of a set A measures the extent of the occurrence of the set A in the database of transactions. The confidence of the association $A \Rightarrow B$ is the conditional probability that a transaction containing all items in the set A also contains the items in B. The Apriori algorithm [2] computes the support and confidence using support thresholds to limit the search.

Two types of generalizations of the standard association rules are important for the analysis of fuzzy temporal relationships. Generalized association rules [16] were introduced to add linguistic categories and hierarchical relationships between items in the domain. Quantitative association rules [13, 15, 16] describe relationships in linearly ordered domains, such as numerically valued attributes. The ordering is used to partition the domain into a relatively small number of classes to reduce the number of items that must be examined in the search. The only modification required to the basic algorithm is to allow classes as well as individual items in the first level of the search lattice constructed for determining frequent sets. In addition to market basket data, the Apriori algorithm is well suited for discovering associations in tuples in a relational database. Like market basket transactions, tuples are considered independent homogeneous unordered data items and the algorithm discovers relationships between attributes in individual tuples.

Much of the research in temporal data mining has focused on periodic patterns and cyclic behaviour in a single time-ordered data stream. Figure 1 illustrates the relationship between data items A, B, C in a single temporal data stream. The search for sequential patterns or episodes [3, 17] employs the linear ordering between the occurrence of events to describe the pattern, an ordered sequence events recurring in the data stream. For example, the regular expression $\mathbf{A}(\mathbf{B} \cup \mathbf{C})\mathbf{A}$ defines a pattern in which A occurs before either B or C which, in turn, occur before A. This pattern occurs four times in the portion of the sequence shown in Fig. 1.

Mannila et al. [12] define an episode to be "a collection of events occurring frequently close to each other." Temporal information is used to construct a sliding time

⟶ temporal data order ⟶

$\ldots, B, A, A, B, C, A, B, B, \ldots$

Fig. 1. Ordered data stream

window that is used to enforce the "close to each other" requirement for the satisfaction of the episode. Assuming that a single time unit separates every item in the sequence in Fig. 1 and employing a window of duration four time units reduces the number of occurrences of the pattern $\mathbf{A}(\mathbf{B} \cup \mathbf{C})\mathbf{A}$ to two. The support of a pattern is the number of windows in which it occurs divided by the total number of windows. Following the standard Apriori approach, an initial processing determines patterns that exceed a minimum support threshold.

In sequential patterns, the sole role of the time variable is to provide the order on the occurrence of the events and to place a temporal constraint on relevance of one event to another. In cyclic association rules [4, 14] or periodic patterns [8, 9], time is treated as an attribute in the learning process (see Fig. 2) and may be an integral component of the rule. Periodic patterns are also referred to as "calendric association rules" because they describe events that recur cyclicly in time.

In calendric association rules, the antecedent of an association contains a time period. For example, analysis of restaurant orders may produce the relationship "between 6:00 am and 10:00 am, coffee accompanies food orders". The time period used (hours, days, weeks, seasons) defines the granularity of the rules. Cyclic rules may vary with time granules and their assessment may consider only subsets of the data to produce a rule. For example, there may be different support and confidence values for an association *food orders* ⇒ *coffee* when considering all orders or only those that occur between 6:00 am and 10:00 am. The acceptance of a cyclic rule requires "minimum support and confidence at regular time intervals" [14], rather than throughout the entire database.

Each of the preceding knowledge discovery strategies assumed that the data came from a single source or from several sources producing data of the same format. The collection of real-time information from multiple data sources has become

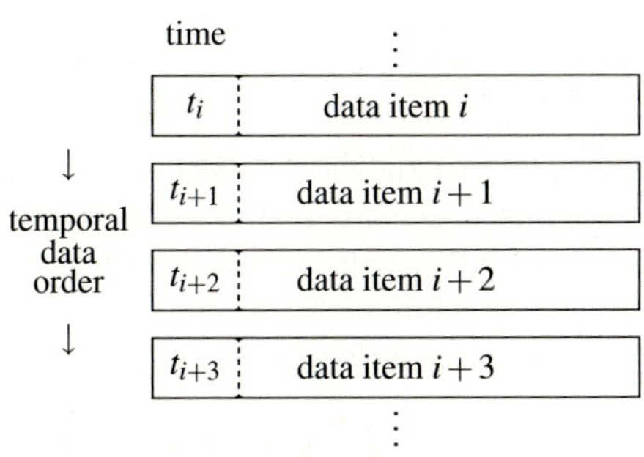

Fig. 2. Calendric data format

increasingly prevalent and provides the need for the development of algorithmic techniques for the identification and evaluation of temporal relationships among data obtained from multiple distributed sources. The type of associations to be examined in distributed data mining are not necessarily temporally cyclic but rather relate the occurrence of one or several events to the occurrence of another with time providing a constraint of temporal relevance. In particular we envision the analysis of historic data from multiple sensors to generate rules suitable for the prediction of future events.

Figure 3 illustrates the scenario under consideration with three information sources $\mathcal{A}, \mathcal{B}$, and $\mathcal{C}$ independently providing data.

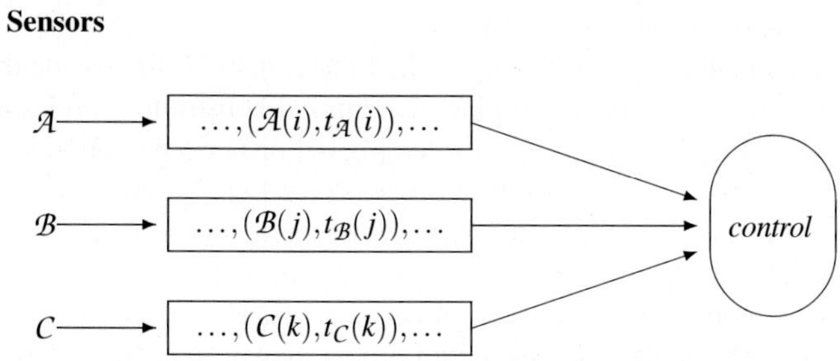

Fig. 3. Multiple data streams

An information source $\mathcal{A}$ consists of a sensor, a set of values $D_{\mathcal{A}}$ that the sensor can observe, and a set of predicates $A_1, \ldots, A_n$ over $D_{\mathcal{A}}$. A predicate may designate a single element, a class or interval, or a fuzzy set. In the latter case, the degree to which a value satisfies the predicate A_i is denoted $A_i(a)$. The ith datum from sensor $\mathcal{A}$ is written as an ordered pair $(\mathcal{A}(i), t_{\mathcal{A}}(i))$, where $\mathcal{A}(i)$ is the value recorded by $\mathcal{A}$ at time $t_{\mathcal{A}}(i)$.

The fuzzy representation of constraints used in the temporal association rules is presented in the following section while the modifications to the Apriori approach for analyzing multiple data streams are given in Sect. 4.

3 Fuzzy Sets in Data Mining

Most applications of fuzzy sets to knowledge discovery have been concerned with partitioning numeric domains in quantitative association rules (see, for example, [11, 6, 10]). The use of fuzzy sets avoids unnatural boundaries between classes and facilitates the linguistic interpretation of the terms. The impact of the extension to fuzzy associations on the efficiency of the Apriori style algorithms has been studied in [7].

Although not a requirement, the predicates associated with a data source frequently form a fuzzy partition of the associated domain to summarize and categorize the values directly recorded by the sensor and to reduce the size of the search space.

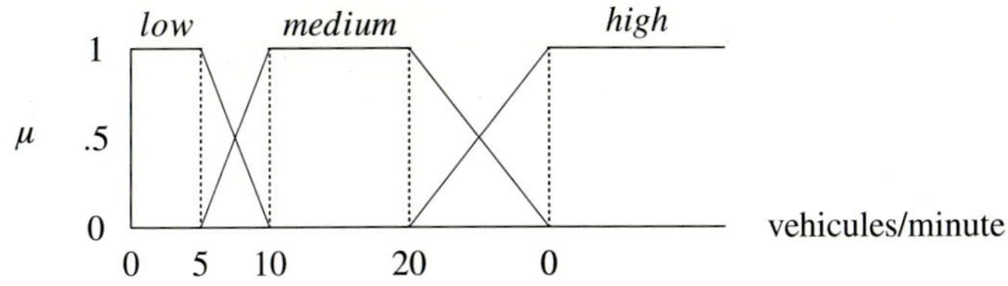

Fig. 4. Fuzzy partition

For example, a fuzzy partition of traffic volume into *low*, *medium*, and *high* volume is given in Fig. 4.

There are two types of temporal properties that must be considered when extending fuzzy representations to temporal data mining: durations and constraints. Durations act as modifiers on predicates. For example, in the term "*low* volume for a *short time*", traffic volume is the domain, *low* is the linguistic term describing the traffic flow, and *short time* indicates the duration. The duration *short* describes a time period in which the volume must be low for the predicate to be satisfied.

A temporal constraint restricts the relevancy of events to a certain fixed, but perhaps imprecise, time period. For example, *follows shortly* and *occurs the next day* both define temporal implicative constraints. The constraint *follows shortly* has a displacement 0 and a duration described by a fuzzy set *shortly*. The duration of *occurs the next day* is 24 hours and the displacement is the length of time from the current time to the beginning of the next day.

An implicative constraint is defined by a fuzzy partition of the interval that specifies the maximum time of relevancy of the satisfaction of the antecedent to the consequent and consists of a temporal duration and a displacement. A fuzzy partition of a 24 hour period representing the linguistic descriptions "immediately after", "a short time after", and "several hours after" given in Fig. 5 form a family of implicative constraints of the time interval 0–24 hours. The fuzzy sets are denoted *ia*, *sta*, *sha*, respectively. Note that only the core of the "several hours after" predicate ends at the window rather than imposing a crisp boundary at 24 hours.

A partition of an implicative constraint provides the ability to assess the associations on various levels of granularity. The granularity effects both the efficiency of the search and the ability to discover relationships. Too fine a granularity precludes having sufficient examples to surpass the frequency threshold for consideration and increases the run time of the algorithm. Too coarse a partition limits the ability to discover associations in which the events are related by time durations less than the partition granularity.

The natural hierarchy associated with a fuzzy partition provides a method of searching several granularities simultaneously. Combining adjacent fuzzy set durations produces a partition of coarser granularity as shown in Fig. 5(b) and (c). Unlike taxonomies in generalized associations, these hierarchical predicates do not need to be incorporated into the search for frequent items; the count of a hierarchical predicate is simply the sum of the counts of the leaf nodes that are descendants of the predicate. The largest granularity, given in Fig. 5(c), may be interpreted as "within a

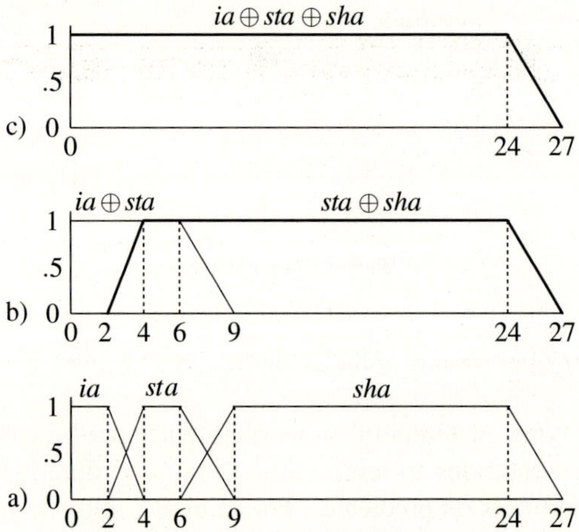

Fig. 5. Temporal sequence hierarchy

day". Note that the hierarchy is not obtained by combining adjacent fuzzy sets using conjunction but rather by making the core of the new fuzzy set contain the cores of the two original fuzzy sets and the region of transition between them.

4 Data Mining with Fuzzy Temporal Constraints

In this section we consider techniques for identifying relationships that are neither cyclic nor calendric, but rather event driven with imprecise implicative temporal constraints. The data are assumed to come from three information sources $\mathcal{A}$, $\mathcal{B}$ and $\mathcal{C}$ as illustrated in Fig. 3.

A sensor produces a data stream consisting of pairs of the form $(\mathcal{A}(i), t_{\mathcal{A}}(i))$. The value $\mathcal{A}(i) \in D_{\mathcal{A}}$ is the ith value recorded by sensor $\mathcal{A}$ and $t_{\mathcal{A}}(i)$ is the time of the occurrence. All that is assumed for analyzing the temporal order is that the data are linearly ordered on the time variable, that is, $t_{\mathcal{A}}(i) < t_{\mathcal{A}}(i+1)$. The following diagram illustrates the data stream notation for a sensor $\mathcal{A}$ with two values a_1, a_2:

time	1	2	3	4	5	6	7	8
$\mathcal{A}$		a_1		a_2	a_1		a_1	

Sequential (clock) time is given in the top row. A blank in the sensor row indicates that no value was recorded at that time. The data stream associated with this is $(\mathcal{A}(1), t_{\mathcal{A}}(1)) = (a_1, 2), (\mathcal{A}(2), t_{\mathcal{A}}(2)) = (a_2, 4), (\mathcal{A}(3), t_{\mathcal{A}}(3)) = (a_1, 5),$
$(\mathcal{A}(4), t_{\mathcal{A}}(4)) = (a_1, 7)$

As in [5, 12], a template is used to define the antecedents, consequents, and implication constraints of the rules to be examined. Letting x_1-after, ..., x_t-after be the partition of the implication constraints, the template

$$C \text{ occurs } x\text{-after } A \text{ and } B$$

describes all rules of the form

$$C_k \text{ occurs } x_r\text{-after } A_i \text{ and } B_j$$

where A_i, B_j, and C_k are predicates concerning sensors $\mathcal{A}, \mathcal{B}$, and $\mathcal{C}$ respectively and x_r is an implication constraint. The template defines a general rule format in which predicates A_i and B_j form the antecedent and C_k the consequent.

The analysis of the data streams requires two time constraints to ensure the relevance of the information; an antecedent time constraint and the implicative constraint. The intent of the temporal constraints is illustrated using the example of traffic flow. Let $\mathcal{A}$ and $\mathcal{B}$ be sensors recording traffic volume at two intersections and let $\mathcal{C}$ contain the sole predicate that a third downstream intersection is congested. The predicates associated with sensors $\mathcal{A}$ and $\mathcal{B}$ are *low*, *medium*, *high* volumes given in Fig. 4. The objective is to determine whether a combination of conditions at $\mathcal{A}$ and $\mathcal{B}$ can be used to predict the conditions at $\mathcal{C}$.

The antecedent constraint is the formalization of the intuitive restriction that only contemporaneous volumes will affect downstream traffic congestion. An antecedent constraint may be an interval or a fuzzy duration. In either case, the constraint requires that the values of $\mathcal{A}$ and $\mathcal{B}$ occur within a specified time of each other. If the constraint is given by an interval $[0,w]$, data $(\mathcal{A}(m), t_{\mathcal{A}}(m))$ and $(\mathcal{B}(n), t_{\mathcal{B}}(n))$ satisfy the antecedent of the template "congestion at $\mathcal{C}$ occurs x-after *high* volume at $\mathcal{A}$ and *med* volume at $\mathcal{B}$" only if $high(\mathcal{A}(m)) > 0, med(\mathcal{B}(n)) > 0$, and $|t_{\mathcal{A}}(m) - t_{\mathcal{B}}(n)| \leq w$. If the constraint is given by fuzzy duration W, the antecedent constraint is satisfied to degree $W(|t_{\mathcal{A}}(m) - t_{\mathcal{B}}(n)|)$.

The implicative constraint enforces a temporal proximity between the satisfaction of the antecedent and that of the consequent; traffic volume at points $\mathcal{A}$ and $\mathcal{B}$ today has little impact on congestion at $\mathcal{C}$ tomorrow.

The rule template "C occurs x-after A and B" describes all possible rules "C_k occurs x_r-after A_i and B_j". A data mining algorithm must examine the validity of all such rules. A rule pattern is used to indicate the particular predicates and implication constraint being evaluated in the determination of the frequency and confidence of the rules.

A pattern for the rule of the form "C occurs x-after A and B" is a quadruple where the first position contains a predicate A_i or a $*$, the second a predicate B_j or a $*$, the third an x-after predicate or a $*$, and the fourth position a C_k or a $*$. A $*$ indicates that there is no condition imposed by the data source or implication constraint corresponding to that position. If A_i is a crisp predicate, the pattern $(A_i, *, *, *)$ is matched by any item $(\mathcal{A}(m), t_{\mathcal{A}}(m))$ when $\mathcal{A}(m) \in A_i$. If A_i is fuzzy, it is matched with degree $A_i(\mathcal{A}(m))$.

The pattern $(A_i, B_j, *, *)$ is matched by data $(\mathcal{A}(m), t_{\mathcal{A}}(m))$ and $(\mathcal{B}(n), t_{\mathcal{B}}(n))$ only if $A_i(\mathcal{A}(m)) > 0$, $B_j(\mathcal{B}(n)) > 0$, and the antecedent constraint W is satisfied. The degree to which these data match is $A_i(\mathcal{A}(m)) \otimes B_j(\mathcal{B}(n)) \otimes W(|t_{\mathcal{A}}(m) - t_{\mathcal{B}}(n)|)$, where $\otimes$ is a T-norm. Similarly, a pattern $(A_i, *, x_r\text{-after}, C_k)$ is matched if there are data items $(\mathcal{A}(m), t_{\mathcal{A}}(m))$ and $(\mathcal{C}(n), t_{\mathcal{C}}(n))$, and $t_{\mathcal{C}}(n)$ is within the time frame x_r-after $t_{\mathcal{A}}(m)$. The degree of the match is $A_i(\mathcal{A}(m)) \otimes C_k(\mathcal{C}(n)) \otimes x_r\text{-after}(t_{\mathcal{C}}(n) - t_{\mathcal{A}}(m))$.

The strategy for discovering temporal relationships follows the Apriori approach of a level-by-level evaluation of the frequency. The learning process is symmetric in the A and B predicates, but not in the consequent C. Due to the conditional probability in the confidence, we are interested in C values only when the A and B predicates have been satisfied. We will outline the steps of the algorithm and provide an example of the computation, but a detailed exposition is not provided due to limitations on space.

Level 1: Items have the form $(A_i,*,*,*)$ and $(*,B_j,*,*)$. For each A_i, data stream $\mathcal{A}$ is traversed and the frequency associated with pattern $(A_i,*,*,*)$ is

$$\sum_{k}^{N_{\mathcal{A}}} A_i(\mathcal{A}(k)),$$

where $N_{\mathcal{A}}$ denotes the number of items in the data stream generated by $\mathcal{A}$. In the same manner, a frequency is obtained for each predicate B_j.

Level 2: Items have the form $(A_i,B_j,*,*),(A_i,*,x_r\text{-after},C_k)$, $(*,B_j,x_r\text{-after},C_k)$. Patterns will be examined for each combination A_i and B_j that surpass the frequency threshold on the preceding level and for each C_k and x_r-after.

Different strategies and constraints are used for the determination of the frequency of the two types of patterns. The approach for computing the frequency of a pattern of the form $(A_i,B_j,*,*)$ is illustrated using the data streams

time	1	2	3	4	5	6	7	8
$\mathcal{A}$		a_2		a_2	a_1		a_1	
$\mathcal{B}$	b_2		b_1	b_2		b_1		b_2

the predicates $A_i = \{a_1\}$ and $B_j = \{b_1\}$. The antecedent time constraint is the crisp duration $[0,3]$.

The data streams are examined synchronously based on time using a time pointer in each stream. The pointers are initialized to 1 and the evaluation begins with $(\mathcal{A}(1),t_{\mathcal{A}}(1))$ and $(\mathcal{B}(1),t_{\mathcal{B}}(1))$. An "earliest time" strategy is employed; the times of the data items $(\mathcal{A}(m),t_{\mathcal{A}}(m))$ and $(\mathcal{B}(n),t_{\mathcal{B}}(n))$ in each stream are compared and the one with the earliest time is processed. If $t_{\mathcal{A}}(m) = t_{\mathcal{B}}(n)$, the order of processing is irrelevant.

Following the earliest time strategy, processing the data in the table examines $\mathcal{B}(1) = b_2, \mathcal{A}(1) = a_2, \mathcal{B}(2) = b_1$ until a element that satisfies either A_i or B_j is encountered. At this point, the other data stream ($\mathcal{A}$ in our example) is traversed until an element is found that satisfies predicate A_i or until the time exceeds the support of the antecedent threshold. In our example, b_1 occurs at time 3, so the search for a matching A_i value must terminate when the time of an element in the $\mathcal{A}$ stream surpasses time 6.

However, the matching data value $(\mathcal{A}(3),t_{\mathcal{A}}(3)) = (a_1,5)$ occurs before the expiration of the antecedent constraint. When a match is found, the frequency of $(A_i,B_j,*,*)$ is incremented by $A_i(a_1) \otimes B_j(b_1)$. The $\mathcal{A}$ and $\mathcal{B}$ time pointers are appropriately incremented and the examination of the streams continues in the earliest first.

In general, the degree of match is also affected by the antecedent constraint. Let W be the antecedent constraint and $(\mathcal{A}(r), t_{\mathcal{A}}(r))$ and $(\mathcal{B}(s), t_{\mathcal{B}}(s))$ be a pair of data items matching A_i and B_j as outlined above. The pattern $(A_i, B_j, *, *)$ is matched by this data to degree $A_i(a_r) \otimes B_j(b_s) \otimes W(|t_{\mathcal{B}}(s) - t_{\mathcal{A}}(r)|)$.

The strategy for determining the frequency of a pattern $(A_i, *, x_r\text{-after}, C_k)$ is simpler than that described above, since the temporal ordering requires the antecedent to be satisfied prior to the satisfaction of the consequent and no antecedent constraint is involved. We will use the data streams

time	1	2	3	4	5	6	7	8
$\mathcal{A}$		a_1		a_2	a_1		a_1	
$\mathcal{C}$	c					c		c

to illustrate the process. Let $[d_1, d_2]$ denote the support of relevance of the implication constraint x_r-after.

The $\mathcal{A}$ stream is scanned until a value is found that satisfies A_i at a time t_0 (in this case, a_1 at time 2). The $\mathcal{C}$ stream is then scanned until either an item $(\mathcal{C}(p), t_{\mathcal{C}}(p))$ with $C_k(\mathcal{C}(p)) > 0$ is encountered whose time is in the interval $[t_0 + d_1, t_0 + d_2]$ or no such $\mathcal{C}$ value is found in that interval. In the former case, the membership of the time $t_{\mathcal{C}}(p)) - t_0$ in "x_r-after" is obtained and $A(a_i) \otimes x_r\text{-after}(t_{\mathcal{C}}(p)) - t_0)$ is added to the frequency of the pattern $(A_i, *, x_r\text{-after}, C_k)$.

Level 3: The patterns $(A_i, B_j, x_r\text{-after}, C_k)$ are examined for every pair A_i, B_j that exceeded the frequency threshold in the level 2 analysis. The strategy utilizes both of the searches employed in level 2. Initially the data streams $\mathcal{A}$ and $\mathcal{B}$ are examined to find data matching A_i and B_j that satisfies the antecedent constraint. When this is discovered, the search for a consequent within the implication constraint is initiated.

Since there are only three levels in our example, the lattice has height three. The frequency data that has been produced yields the confidence for all frequent patterns. For a frequent pattern $(A_i, B_j, x_r\text{-after}, C_k)$, the confidence of the association "C_k occurs x_r-after A_i and B_j" is

$$\frac{freq(A_i, B_j, x_r\text{-after}, C_k)}{freq(A_i, B_j, *, *)},$$

where $freq$ denotes the the sum of the degrees of satisfaction of data elements that satisfy the pattern. In the same manner, the confidence is obtained for all associations with a single antecedent:

$$\frac{freq(A_i, *, x_r\text{-after}, C_k)}{freq(A_i, *, *, *)}$$

is the confidence of "C occurs x_r-after A_i".

Example: The determination of the frequency of the pattern (A_1, B_1, ia, C_1) is illustrated using following crisp predicates and fuzzy constraints:

- $A_1 = \{a_1\}, B_1 = \{b_1\}, C_1 = \{c_1\}$
- implicative constraint: *ia*, immediately-after (see Fig. 5)

- antecedent constraint: $W = \begin{cases} 1, & \text{if } x \in [0,2), \\ -.5(x-4), & \text{if } x \in [2,4), \\ 0, & \text{otherwise.} \end{cases}$

The sensor data and associated times are

time	$\mathcal{A}$	$\mathcal{B}$	$\mathcal{C}$
1	a_1	b_2	–
5	a_1	b_1	c_2
6	–	b_1	c_1
7	a_2	b_1	–
10	a_1	–	c_2
13	–	–	c_1

The analysis of the data streams will produce the following steps

1. $(a_1,1)$-no matching B_1 within antecedent constraint
2. $(a_1,5),(b_1,5)$ antecedent match, $(c_1,6)$ consequent match; satisfaction is $A_1(a_1) \otimes B_1(b_1) \otimes W(0) \otimes C_1(c_1) \otimes ia(1) = 1$
3. $(b_1,6)$-no matching A_1 within antecedent constraint
4. $(b_1,7),(a_1,10)$ antecedent match, $(c_1,13)$ consequent match; satisfaction is $A_1(a_1) \otimes B_1(b_1) \otimes W(3) \otimes C_1(c_1) \otimes ia(3) = .25$

where the product is used for the t-norm $\otimes$ in the determination of the degree of satisfaction of the pattern. □

Note that in the preceding evaluation of frequency, an item in the antecedent is permitted occur in only one antecedent match. The b_1s that occur at times 5 and 6 both match the a_1 occurring at time 5. However, the a_1 at time 5 is paired with the earliest eligible b_1. Depending upon the application, one may wish to count multiple matches for a piece of data.

The frequencies of the x_r-after implication constraints permits the straightforward determination of the frequencies of implication constraints hierarchy obtained by combining elements from the x-after partition as illustrated in Fig. 5(b) and (c). The largest granularity, given in Fig. 5 (c), may be interpreted as "within a day". Using the frequencies generated for the leaf nodes of the hierarchy, the confidence of "C_k occurs within one day after A_i and B_j" is the sum of the frequencies of $(A_i, B_j, ia\text{-after}, C_k), (A_i, B_j, sta\text{-after}, C_k)$, and $(A_i, B_j, sha\text{-after}, C_k)$ divided by $freq(A_i, B_j, *, *)$.

5 Conclusion

Temporal antecedent and implicative constraints are required to ensure the relevancy of events in analyzing temporal data from multiple sources. Fuzzy predicates are used to represent imprecise temporal constraints and durations and a fuzzy partitions provide a hierarchy that allows the analysis of implicative constraints on several levels of granularity. In this paper we have outlined how standard data mining strategies can be adapted to utilize fuzzy representations in the discovery of imprecise temporal relationships between data obtained from multiple sources.

References

1. R. Agrawal, T. Imielinski, and A. Swami. Mining association rules between sets of items in large databases. In Peter Buneman and Sushil Jajodia, editors, *Proceedings of the 1993 ACM SIGMOD International Conference on Management of Data*, pp. 207–216, Washington, D.C., 26–28 1993.
2. R. Agrawal and R. Srikant. Fast algorithms for mining association rules. In Jorge B. Bocca, Matthias Jarke, and Carlo Zaniolo, editors, *Proc. 20th Int. Conf. Very Large Data Bases, VLDB*, pp. 487–499, Santiago, Chile, Sept 1994.
3. R. Agrawal and R. Srikant. Mining sequential patterns. In P. S. Yu and A. S. P. Chen, editors, *Eleventh International Conference on Data Engineering*, pp. 3–14, Taipei, Taiwan, 1995. IEEE Computer Society Press.
4. J. M. Ale and G. J. Rossi. An approach to discovering temporal association rules. In *Proceedings of the 2000 ACM Symposium on Applied Computing*, pp. 294–300, Como, Italy, March 2000.
5. C. Bettini, X. S. Wang, and S. Jajodia. Mining temporal relationships with multiple granularities in time sequences. *Bulletin of the IEEE Computer Society Committee on Data Engineering*, 21(1):32–38, 1998.
6. K. C. C. Chan and W.-H. Au. An effective algorithm for mining interesting quantitative association rules. In *Selected Areas in Cryptography*, pp. 88–90, 1997.
7. M. Delgado, N. Marín, D. Sánchez, and M. A. Vila. Fuzzy association rules: General model and applications. *IEEE Transactions on Fuzzy Systems*, pp. 214–225, 2003.
8. J. Han, G. Dong, and Y. Yin. Efficient mining of partial periodic patterns in time series database. In *Fifteenth International Conference on Data Engineering*, pp. 106–115, Sydney, Australia, 1999. IEEE Computer Society.
9. J. Han, W. Gong, and Y. Yin. Mining segment-wise periodic patterns in time-related databases. In *Proc. Intl. Conf. Knowledge and Data Mining*, pp. 1–12, New York, August 1998.
10. C. M. Kuok, A. W.-C. Fu, and M. H. Wong. Mining fuzzy association rules in databases. *SIGMOD Record*, 27(1):41–46, 1998.
11. J. H. Lee and H. L. Kwang. An extension of association rules using fuzzy sets. In *Proceedings of IFSA 1997*, pp. 399–402, Prague, June 1997.
12. H. Mannila, H. Toivonen, and A. I. Verkamo. Discovering frequent episodes in sequences. In *Proc. Intl. Conf. on Knowledge Discovery and Data Mining*, pp. 210–215, Montreal, August 1995. AAAI Press.
13. R. J. Miller and Y. Yang. Association rules over interval data. In *SIGMOD 1997, Proceedings ACM SIGMOD International Conference on Management of Data*, pp. 452–461, Tuscon, AR, May 1997. ACM Press.
14. B. Ozden, S. Ramaswamy, and A. Silberschatz. Cyclic association rules. In *ICDE*, pp. 412–421, 1998.
15. G. Piatetsky-Shapiro. Discovey, analysis, and presentation of strong rules. In G. Piatetsky-Shapiro and W. Frawley, editors, *Knowledge Discovery in Databases*, pp. 229–248. AAAI/MIT Press, Menlo Park, CA, 1991.
16. R. Srikant and R. Agrawal. Mining quantitative association rules in large relational tables. In H. V. Jagadish and I. S. Mumick, editors, *Proceedings of the 1996 ACM SIGMOD International Conference on Management of Data*, pp. 1–12, Montreal, 1996.
17. R. Srikant and R. Agrawal. Mining sequential patterns: Generalizations and performance improvements. In P. M. G. Apers, M. Bouzeghoub, and G. Gardarin, editors, *Proc. 5th Int. Conf. Extending Database Technology, EDBT*, Vol. 1057, pp. 3–17. Springer-Verlag, 25–29 1996.

Part II

Applications of Soft Computing in Traffic and Transportation Systems

Soft Computing Applications in Traffic and Transport Systems: A Review

Erel Avineri

Centre for Transport and Society, Faculty of the Built Environment, University of the West of England, Bristol, BS16 1QY, UK
Erel.Avineri@uwe.ac.uk

Summary. The use of soft computing methodologies in the field of traffic and transport systems is of particular interest to researchers and practitioners due to their ability to handle quantitative and qualitative measures, and to efficiently solve problems which involve complexity, imprecision and uncertainty. This paper provides a survey of soft computing applications. A classification scheme for soft computing applications is defined. The current frameworks and some future directions of soft computing applications to traffic and transport systems are discussed.

Key words: Traffic, Transport

1 Introduction

Transportation is a wide human-oriented field with diverse and challenging problems waiting to be solved. Characteristics and performances of transport systems – services, costs, infrastructures, vehicles and control systems are usually defined on the basis of quantitative evaluation of their main effects. Most of the transport decisions take place under imprecision, uncertainty and partial truth. Some objectives and constraints are often difficult to be measured by crisp values. Traditional analytical techniques were found to be not-effective when dealing with problems in which the dependencies between variables were too complex or ill-defined. Moreover, hard computing models can not deal effectively with the transport decision-makers' ambiguities and uncertainties.

In order to come up with solutions to some of these problems, much traffic and transport research was done over the recent 25 years in the framework of soft computing. Over the last decade there has been much interest in soft computing applications of traffic and transport systems, leading to some successful applications and implementations.

The use of soft computing methodologies for modeling and analyzing traffic and transport systems is of particular interest to researchers and practitioners due to their

Erel Avineri: *Soft Computing Applications in Traffic and Transport Systems: A Review*, Advances in Soft Computing **1**, 17–25 (2005)
www.springerlink.com

ability to handle quantitative and qualitative measures, and to efficiently solve complex problems which involve imprecision, uncertainty and partial truth. Soft computing can be used to bridge modeling gaps of normative and descriptive decision models in traffic and transport research.

This survey was conducted in an effort to gain a better understanding of how to use soft computing in traffic and transport applications. A summary of the findings is described here. Although this survey is based on many references, it is by no mean a complete bibliography of soft computing applications in traffic and transport systems. The objective of this survey was not to identify every relevant bibliography, but rather to provide the reader with a starting point when investigating the literature dealing with such applications.

2 Classification Scheme for Soft Computing Applications in Traffic and Transport Systems

During the last three decades soft computing approaches have received much attention by the transport research community. Table 1 illustrates a classification scheme for the literature on the applications in traffic and transport systems. Five major categories are defined and the frequency of publications in each category is identified. A total of 1004 citations dealing with applications of soft computing in transport

Table 1. Classification Scheme for Soft Computing Research in Traffic and Transport Systems

#	Research Topics	Number of Citations
1.	Traffic Control and Management	375 (37%)
2.	Transport Planning and Management, Transport Administration, Transport Policy	
	2.1 Modeling of Travel Choice Behavior	83 (8%)
	2.2 Transport Projects Selection	17 (2%)
	2.3 Other Issues of Transport Planning	196 (20%)
	Total	296 (29%)
3.	Logistics	40 (4%)
4.	Design and Construction of Transport Facilities Including Geometric Design, Pavement Management, Construction, Materials Properties	112 (11%)
5.	Other Applications of Traffic and Transport Systems, and Review Papers Including Planning and Operating Public Transport, Operating and Management of Parking Facilities, Maintenance of Traffic and Transport Systems, Airline Network Applications, Airport Planning and Others	181 (18%)
		Total: 1004

systems was found. Traffic control and management resulted in the largest number of publications (375), followed by transport planning and management (296 publications).

Soft computing forms a large collection of methodologies in research and practice of traffic and transport systems. In this survey, methodologies were classified to five main categories: (a) Fuzzy Sets; (b) Neural Networks; (c) Genetic Algorithms; (d) Intelligent Agents; and (e) Other methodologies (such as Data Mining, Bio-inspired Systems, Ant Colonies) and general papers about soft computing applications in transport systems. Table 2 and Fig. 1 provide a breakdown of the number of citations by soft computing methodologies and by the year published.

Not all the papers surveyed here can be classified by single methodology; this is why the left columns in Table 2 do not sum up to the total number on the right column. In many cases a transport system is handled most effectively by using more than one methodology in combination rather than exclusively.

As Table 2 indicates, the survey revealed that soft computing has been applied to a large number of aspects related to traffic and transport systems. We observe that applications of fuzzy sets theory, neural networks and genetic algorithms have been widely published over the last decade. Applications of intelligent agents in transport systems have increased in the last few years. Only little literature on other innovative methodologies of soft computing, such as bio-inspired systems, ant colonies, chaos computing has been published. The peaks during the years 1997, 1999 and 2002 may be explained by the special journal issues and conferences proceedings specialized in applications of soft computing to traffic and transport systems published during these years.

As with living organisms, scientific research and technological developments move through a *life cycle*. A general scheme of such a life cycle is displayed in Fig. 2. Four main phases of a life cycle can be observed: introduction, growth, maturity and decline. Looking at Fig. 1, we observe that applications of soft computing were in the *introductory phase* till the beginning of the 90's, were the soft computing applications were introduced. The dominant methodologies introduced during this stage were based on fuzzy sets theory. The *growth phase* took place till the year 1997, where a record in fuzzy applications to transport systems was observed – 57 citations. Since then we are facing the *maturity phase*, where research and applications areas have been established. Are we about to face the *decline phase* of soft computing applications on the coming years? The answer to that may be depended on the results of some on-going studies. Moreover, applications of some soft computing methodologies have been less explored, thus we may face the growth phase of some methodologies (such as genetic algorithms) and the introductory phase of others (such as various bio-inspired systems).

Table 2. Citation Breakdown by Year and Soft Computing Methodologies (F = Fuzzy Sets, NN = Neural Networks, GA = Genetic Algorithms, Agents = Intelligent Agents, Others = Other Methodologies and Review Papers)

	Soft Computing Methodologies					
Year	F	NN	GA	Agents	Others	Total
1976	1					1 (0.1%)
1977	1					1 (0.1%)
1978						
1979						
1980	1					1 (0.1%)
1981						
1982	1					1 (0.1%)
1983						
1984	2					2 (0.2%)
1985	2					2 (0.2%)
1986						
1987	2					2 (0.2%)
1988	2			1		3 (0.3%)
1989						
1990	5					5 (0.5%)
1991	5	2			1	8 (1%)
1992	10	10	2			21 (2%)
1993	17	10	2		1	30 (3%)
1994	21	21	9		1	46 (5%)
1995	22	30	6			53 (5%)
1996	24	38	6		1	65 (6%)
1997	57	50	7			106 (11%)
1998	37	28	7		1	64 (6%)
1999	42	23	15	2	3	82 (8%)
2000	39	21	18	1		72 (7%)
2001	33	31	23	5	7	94 (9%)
2002	49	36	31	14	10	124 (12%)
2003	45	49	20	3	10	124 (12%)
2004(*)	35	45	29		3	97 (10%)
Total	453 (45%)	394 (39%)	175 (17%)	26 (2.6%)	38 (3.8%)	1004 (100%)

(*) Year 2004 data is updated till July 1, 2004

3 Soft Computing Applications in Traffic and Transport Research

Extensive work has been done on applying soft computing methodologies to applications in traffic and transport systems. By using the classification scheme developed in Sect. 2, research findings in each area of research topics is reviewed in this section.

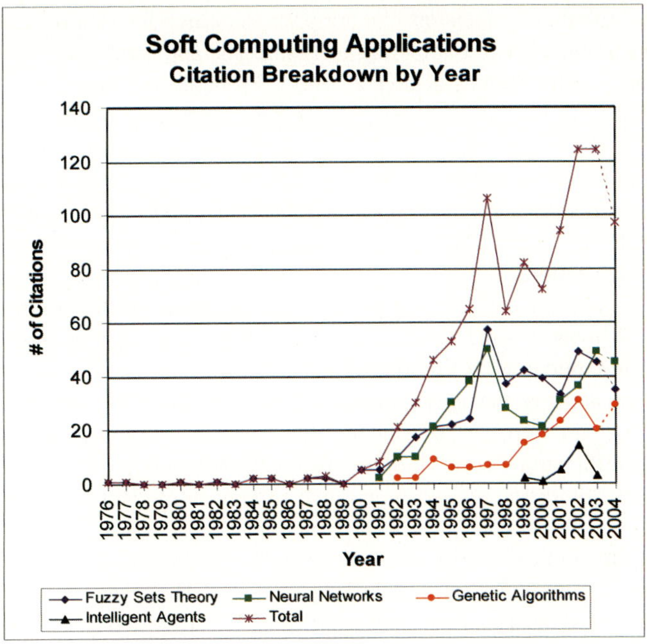

(*) Year 2004 data is updated till July 1, 2004

Fig. 1. Citation Breakdown by the Publication Year(*) and Soft Computing Methodologies

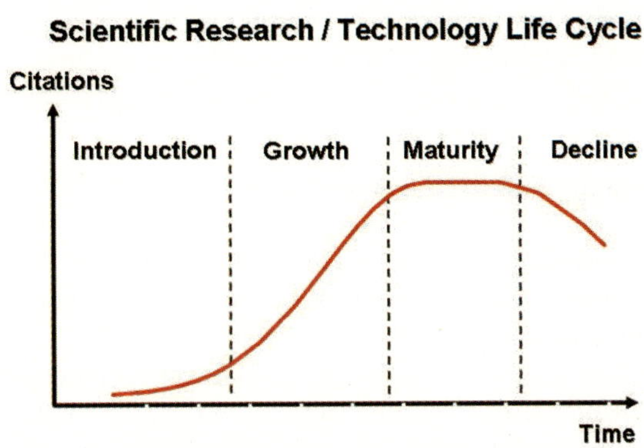

Fig. 2. A General Scheme of Scientific Research/Technology Life Cycle

3.1 Traffic Control and Management

The first known attempt to use fuzzy control in traffic signal control was made by Pappis & Mamdani [1] in their theoretical simulation study of a fuzzy logic controller in an isolated signalized intersection. The strength of fuzzy logic lies in its capability to simulate the human's decision-making process, which is often difficult to be defined using traditional hard computing methods. The application of fuzzy control to traffic signals was the subject of several works. Probably the most famous one is the FUSICO (Fuzzy Signal Control) project [2]. The main goals of this project, started in 1996, were theoretical analysis of fuzzy traffic signal control, generalized fuzzy rules using linguistic variables, validation of fuzzy control principles, calibration of membership functions, and developing of a fuzzy adaptive signal controller. It was suggested that the fuzzy traffic signal control can be the potential control method for signalized intersections.

Although most of the soft computing applications in traffic control are in the framework of fuzzy sets theory [1, 2, 3, 4, 5], some other methodologies were used as well. Papageorgio et al. [6] used a Neural Networks approach to freeway network traffic control. Park & Chang [7] discussed the application of genetic algorithms to adaptive signal control at an isolated intersection. Sha'Aban et al. [8] developed an adaptive traffic signal control system using evolutionary algorithms. Sun [9] described a multi-objective intersection signal timing optimization with evolutionary algorithms.

Most of the applications of soft computing methodologies in traffic control and management are still under research and development. However, real-life traffic control systems, based on soft computing applications, were tested and some of their results look promising. Such applications may be found in Japan, Finland and Poland.

Other studies that were assigned to the "traffic control and management" category include applications of traffic flow forecasting [10, 11]; car-following models [12] and detecting traffic incidents [13]. However, not much literature was found on soft computing applications in other aspects of traffic safety. It is believed by the author that there is much potential in using soft computing methodologies to model the behavior of drivers and pedestrians. Such a research may be inspired by the wide research about travel behavior modeling, which is described in the following section.

3.2 Transport Planning and Management

Transport planning and management deals with many challengeable issues, such as travel demand forecasting, traffic assignment on the transport network and the evaluation of proposed transport policy.

The issue of how to deal with uncertainty has recently become one of the major subjects in transport planning. Common models of travelers' behavior do not offer much insight on the way that travelers make choices. Moreover, the uncertainty presented by common models is from the modeler's point of view; it provides no hypothesis as to how the travelers themselves might consider uncertainty.

First applications of soft computing methods to travel behavior modeling [14, 15] are based on fuzzy rules and on the classical tools of fuzzy control. Those rules are fired simultaneously and conclusions are gathered to extract the traveler decision. Other behavior models are based on the evaluation of possibility theory or with comparison tools that can be demonstrated to be equivalent [16].

Neural networks were suggested to model travel behavior [17]. This approach was based on the assumption that there is a similarity between the process of traveler decision-making and the problem solving approach on neural computing. In principle, however, the neural network approach is less representative of real traveler decision-making than a fuzzy rule-based model. In early example of a hybrid model [18] route-choice decision-making is modeled by a fuzzy rulebase, and a neural network approach is used for calibrating the parameters of the fuzzy model.

Genetic algorithms were recently used in route-choice modeling [19] and in clustering of activity patterns [20].

Soft computing methodologies were applied to other issues of transport planning as well. Xu & Chan [21] applied fuzzy weights for estimating an origin-destination matrix. Tao & Xinmiao [22] used concepts of fuzzy sets theory for urban traffic environment quality evaluation. Kim [23] suggested using Neural Networks for trip generation models. Several researchers used fuzzy sets theory techniques to evaluate transport projects [24].

3.3 Logistics; Freight Transport

Aspects of logistics management in transport systems include freight transport, and physical distribution in the supply chain. The number of citations of soft computing applications to logistics and freight transport is surprisingly low. In order to solve the multicriteria problem of the choice among alternative locations of a logistic center, a method based on fuzzy rating of criteria was suggested [25]. Fuzzy rules were used to describe dispatchers' subjective decision-making process and a neural network was used to tune and adapt the fuzzy system to achieve better performance [26]. The complex Vehicle Routing Problem, with simultaneous pick-ups and deliveries, was recently suggested was recently solved using genetic algorithm [27].

3.4 Design and Construction of Transport Facilities

Many soft computing applications are found in the area of design and construction of transport facilities. The concepts of fuzzy sets theory have been applied to pavement evaluation [28]. Fuzzy logic was used in the process of selecting pile foundation [29]. A neural network-based methodology was suggested for pavement crack detection and classification [30]. Genetic Algorithms were used for pavement management [31]. An evolutionary neural network model was suggested for the selection of pavement maintenance strategy [32]. An application of neural network for the selection of airport rigid pavement maintenance strategies was developed [33].

4 Survey Conclusions

In this paper an extensive survey of soft computing applications in traffic and transport systems was given. It has been observed that (1) soft computing methodologies

13. Guo H, Han [illegible] (199?) [illegible] Fuzzy Logic and Neural Networks in [illegible]
14. [illegible]
15. [illegible]
16. [illegible] Fuzzy Sets and Systems [illegible]
17. Lyons GJ, Hunt JG (1995) [illegible] Neural Networks? The Third International Conference on the Application of Artificial Intelligence to Civil and Structural Engineering, Edinburgh, UK, pp 71–75
18. Vythoulkas PC (1995) An Approach to Travel Behavior Based on the Concepts of Fuzzy Logic and Neural Networks. The 23rd PTRC Transport Forum, Education and Research Services, London, UK, pp 189–205
19. [illegible] (2000) [illegible] Model with Indicated [illegible]
20. [illegible]
21. [illegible] Methodology. Transportation Planning and Technology [illegible]
22. [illegible] Y (199?) Fuzzy [illegible] Assessment [illegible] Clustering Analysis and its Application for Urban [illegible]. Transportation Research [illegible]
23. Kim D (200?) Neural Networks for Trip Generation Model. Journal of the Eastern Asia Society for Transportation Studies 4:201–208
24. [illegible] (2000) [illegible] Process using Fuzzy [illegible]
25. [illegible] (198?) Determining the Best [illegible] on a Motorway. The 9th Mini-EURO Conference, [illegible], Yugoslavia
26. Vukadinovic K, Teodorovic D, Pavkovic G (199?) A Neural Fuzzy Approach to the Vehicle Assignment Problem. The 9th Mini-EURO Conference, Budva, Yugoslavia
27. Jung S, Haghani A (2000) Genetic Algorithm for a Pickup and Delivery Problem with Time Windows. Transportation Research Record 1733
28. Gunaratne M, Chameau JL, Altschaeffl AG (1984) Introduction to Fuzzy Sets in Pavement Evaluation (Abridgment). Transportation Research Record 95?
29. Juang, CH, Ulshafer ML (1990) Development of an Expert System for Preliminary Selection of Pile Foundation. Transportation Research Record 1277
30. Kaseko MS, Ritchie SG (1993) A Neural Network-based Methodology for Pavement Crack Detection and Classification. Transportation Research C [illegible]
31. Fwa TF, Chan WT, Tan CY (1994) Optimal Programming by Genetic Algorithms for Pavement Management. Transportation Research Record 1455
32. Taha MA, Hanna AS (1995) Evolutionary Neural Network Model for the Selection of Pavement Maintenance Strategy. Transportation Research Record 1497
33. Lee C, Chen CF, Huang S-M, Hsu C-J (2002) Application of Neural Network for Selection of Airport Rigid Pavement Maintenance Strategies. The 81st Annual Meeting of the Transportation Research Board, Washington, D.C.

Estimation of Public Transport Trips By Feed Forward Back Propagation Artificial Neural Networks; A Case Study For Istanbul

H. Berk Çelikoğlu[1] and Murat Akad[2]

[1] Faculty of Civil Engineering, Technical University of Istanbul, Maslak, 34469, Istanbul, Turkey
hbcelikoglu@ins.itu.edu.tr

[2] Faculty of Civil Engineering, Technical University of Istanbul, Maslak, 34469, Istanbul, Turkey
muratakad@ins.itu.edu.tr

Summary. Artificial neural networks are one of the recently explored advanced technologies, which show promise in the area of transportation engineering. However, in contrast to the availability of a large number of successful application demonstrations, it is hard to find studies in the literature that provide systematic examinations of the state-of-the-art, the application domains, and the applicability of artificial neural networks to transportation problems. On the other hand, some unseen artificial neural network development has been motivated by transportation engineering objectives. Therefore, this study presents the development of a neural network paradigm for the purpose of daily trip flow forecasting.

In this study, the prediction problem was transformed into the following minimum norm problem: Find a back - propagation neural network such that $||P(U_k) - NN(U_k)||_2$ is minimal. $P(U_k)$ represents the observed values and $NN(U_k)$ represents the neural network generated values. The averages of half-hourly public transport (PT) trip flows using AKBIL (a ticketing system integrated to all PT modes) in Istanbul Metropolitan Area are used as data. As the objective of the study was the forecast of the half-hourly mean trip flows (trips) by using ANN, the analysis process consisted of two steps, which are training and testing respectively. The feed forward back-propagation (FFBP) method was used to train the ANNs. Determination of the number of hidden layers and the number of nodes in the input layer providing the best training results was the initial process of the training procedure. The mean square error (MSE) and the auto-correlation function (ACF) were used as criteria to evaluate the performance of the training simulations. Three hidden layer nodes were found appropriate for the four-nodded input layer. More than three nodes for the hidden layer could not improve the MSE during the testing stage. Following the ANN forecast, a comparison was made by means of a stochastic model. An auto-regressive model of order four was used. The results were then compared. Among the studied models, a significant improvement in the prediction made by the neural network model is noticed, due to its flexibility to adapt to time-series database. Making forecasts by generating negative values after trials is a disadvantage of the algorithm used to train the neural network. It can be said that various kinds of learning algorithms should be assessed during the training process to overcome this drawback.

H. Berk Çelikoğlu and Murat Akad: *Estimation of Public Transport Trips By Feed Forward Back Propagation Artificial Neural Networks; A Case Study For Istanbul*, Advances in Soft Computing **1**, 27–36 (2005)
www.springerlink.com

1 Introduction

Journey is a one way movement from a point of origin to a point of destination. The word "trip" is literally defined as an "outward and return journey, often for a specific purpose" [10]. In transport modelling both terms are used interchangeably. The classification of trips is made upon the trip purpose, person type, and time of the day.

Daily trip time series by different purposes usually vary greatly with the time of the day, which is often classified into peak and off-peak periods. In this study public transport trips are reviewed in respect to the comparison of a stochastic process and an artificial neural network (ANN) used as forecasting methods.

As transportation is one of the major components of the service sector, investments for transportation have a reasonably huge share when the budget of a government is taken into consideration. Because of the infrastructure and the superstructure concept of transportation being generally irreversible, demand analysis for an investment in a system approach is crucial. By defining the demand for transportation as a potential for traffic flow, the importance of the trip amounts is clarified.

Modelling transport generally consists of four steps, which are trip generation, trip distribution, modal-split and flow assignment. In the history, growth factor methods have been widely used to estimate the aggregated zonal trip matrices. The optimisation methods based on different kinds of algorithms have been used to assign the estimated flows to routes. The earlier approach that has been used to determine the mode choice is the diversion analysis. The diversion method has been used both for trip and modal-split and for trip interchanges determined from a distribution model. For the modal-splitting stage, stochastic methods are used. Stochastic methods have been used to be defined as appropriate models, if the inconsistencies in the behaviour of choice makers due to lack of information regarding the attributes of the alternatives available were not taken into consideration. Moreover, for predicting travel behaviour, they provide superior means compared to deterministic models. Due to this, a good model of choice can be set in which the choice function is considered as a random function that takes on different values with certain probabilities. This random function reflects the possibility that given values of the choice function or of any of its attributes are perceived differently by different individuals or by the same individual on different occasions. The perceived utility then becomes a random function, which results in a stochastic choice process, the outcome of which depends on the specific values taken by the random components of the choice function. This postulate of random utility is the most general of many that can be made in order to reflect stochastic choice behaviour, and leads the derivation of most stochastic choice models [3].

Discrete choice models cannot be calibrated in general using standard curve-fitting techniques, because their independent variable "choice proportion" is an unobserved probability. The exceptions to this fact, are models for homogenous groups of individuals, or the situations where the behaviour of every individual is recorded on several occasions. The two most commonly used discrete choice models are the logit and the probit ones.

Most of the transport forecast studies based on stochastic methods have been applied during the modal-split phase. To reflect the behaviour of transportation mode choice [9] used "Monte Carlo Simulation", and [4] used the approximate closed-form solution as multinomial probit (MNP) models. References [5] and [6] developed more efficient algorithms for the estimation of the binary probit models. Reference [3] studied on binary logit model calibration to evaluate the modal-split. References [7] and [8] studied on a comparison between the multinomial logit (MNL) and the multinomial probit (MNP) models in travel choice applications. References [11] and [15] studied on more efficient logit models called hierarchical logit (HL). Recently, [12] and [2] used mixed multinomial logit (MMNL) models to predict the travel behaviour.

Studies considering the daily trip flows in the past, be it time-series models (e.g., [1]) or neural network models (e.g., [14] treat traffic flow as a point process and rarely exploit the spatial dependence of traffic flow in making their predictions.

2 A Neural Network Overview

In the neural network model, the neurone is the basic component. As shown in Fig. 1, the Multilayer Perceptron (MLP) structure is a fully interconnected set of layers of neurones. Each neurone of a layer is connected to each neurone of the next layer so that only forward transmission through the network is possible, from the input layer to the output layer through the hidden layers. In this structure the output Y_i of each neurone of the nth layer is defined by a derivable non-linear function F:

$$Y_i = F\left(\sum_j w_{ji} y_i\right) \tag{1}$$

where F is the non-linear activation function, w_{ji} are the weights of the connection between the neurone N_j and N_i, y_j is the output of the neurone of the $(n-1)^{th}$ layer.

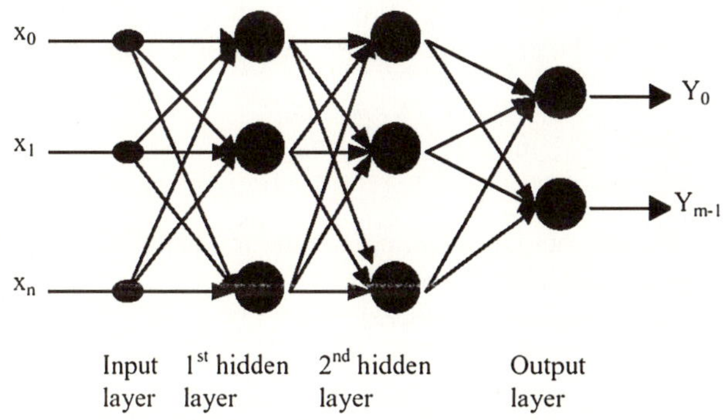

Fig. 1. The configuration of a Multilayer Perceptron

For each input vector presented into the network, the set of connection weights between the different neurones determines the answer of the network in the output layer. Partial specifications of the problem (i.e. input-output pairs) allow us to measure the output error of the network and to adjust its behaviour. An iterative algorithm does the adjustment during the training phase of the neural network. During this phase, a selected set of training patterns is presented to the network. When the network has correctly learned the problem specification (i.e. the error between the network output and the desired output is minimised), the network can be used in a testing phase with test pattern vectors. At this stage, the neural network is described by the optimal weight configuration, which theoretically ensures output error minimisation. The most often used criterion to characterise the best non-linear mapping (F) is the sum square error between the system output d_i and model output y_j [16]:

$$E = \frac{1}{2}\sum_{i=1}^{m}(d_i - y_i)^2 \tag{2}$$

Because one doesn't know the form of the mapping F a priori, an approximation is sought. This can be achieved in a number of ways. For example, one knows that the polynomials and trigonometric series are dense in function space L^2 and can approximate any smooth functions in L^2 on a finite interval $[a,b]$. The development of artificial neural networks offers an alternative to these two function approximators. The back-propagation neural network, which has only a finite number of parameters, can approximate most bounded functions with arbitrary precision (Cybenko 1989), and is used here to approximate F. A multi-layer back-propagation neural network $NN(.) \in R^m$ is a mapping constructed recursively by a vector linear/non-linear function [16]:

$$NN(\Psi, W, \Theta) = \Psi\left[W_n\Psi\left(W_{n-1}\Psi\left(\cdots\Psi\left(W_1U + \theta_1\right)\right) + \theta_{n-1}\right) + \theta_n\right] \tag{3}$$

where n is the number of layers; W_i is the linear operator of proper dimensions where $i = 1,\ldots,n; \theta_i$ is the bias vector, where $i = 1,\ldots,n; U$ is the input vector of the neural network; Ψ is the vector-valued linear/non-linear mapping (transfer function); $W = [W_1,\ldots,W_n]$; and $\Theta = [\theta_i^t,\ldots,\theta_n^t]^t$ (t denotes vector/matrix transpose) [16]. The average system error or the mean square error (MSE), E, for all input patterns is:

$$E = \frac{1}{2N}\sum_{n=1}^{N}\sum_{i=1}^{m}(d_{ni} - y_{ni})^2 \tag{4}$$

where d_{ni} is the system output value, d_i, for the nth pattern and y_{ni} is the neural network output value, y_i, for the nth pattern.

The term NN is a linear neural network if all of its transfer functions $\Psi_{\alpha,\beta}(x)$ are linear:

$$\Psi_{\alpha,\beta}(x) = \alpha x + \beta \tag{5}$$

and a non-linear neural network if at least some of its transfer functions are non-linear. A typical non-linear transfer function is the sigmoid function:

$$\Psi_{\alpha,\beta}(x) = \frac{1}{1 - e^{-\alpha x + \beta}} \tag{6}$$

The weight matrices $W_1, W_2, \ldots, W_n$ and bias vectors $\theta_1, \theta_2, \ldots, \theta_n$ are adjusted by a learning rule called back-propagation, which is a gradient descent method to minimise output error with respect to the weights and thresholds [13].

3 Neural Network Prediction Model

The prediction problem was transformed into the following minimum norm problem:

Find a back-propagation neural network such that $||P(U_k) - NN(U_k)||_2$ is minimal. $P(U_k)$ represents the observed values and $NN(U_k)$ represents the neural network generated values.

The solution to the above minimum norm problem involves a number of steps. The first one is the choice of U_k-the model inputs; the second is the selection of layers and Ψ, and the final step is the attainment of parameters that minimise the 2-norm. Because the second step is largely a trial-and-error process and experiments involving neural networks with input layers more than 4 and hidden units more than 3 did not show any sizeable improvement in prediction accuracy, a two-layer, 3 hidden units back-propagation neural network with the sigmoid function $\Psi_{\alpha,\beta}(x)$, as it is a non-linear transfer function, is selected.

4 Analysis of Data

In this study, the averages of half-hourly public transport (PT) trip flows using AKBIL (a ticketing system integrated to all PT modes) in Istanbul Metropolitan Area are used. The data relevant to PT vary within public bus, tram, LRT, metro, funicular system, seabus, and urban marine lines modes. The collected data are obtained from the observations made during the first eleven months of the year 2002, starting on January 1st and ending on November 30th. Table 1 gives a brief description of the data set.

Table 1. Information for the analysed data

Type of the Data	Region	$x_{\min}$	$x_{\max}$	Mean	Standard Deviation	Skewness	Observation Period
Trip flow (Lag: 1/2 hour.; unit: trips)	Istanbul Metropolitan Area	15	2367412	521128	451776	0.654	from 01.01.2002 to 30.11.2002

5 Ann Application

A code was written to run the neural network toolbox in the MATLAB software for the simulation process.

As the objective of the study was the forecast of the half-hourly mean trip flows (trips) by using ANN, the analysis process consisted of two steps, which are training and testing respectively. The feed forward back-propagation (FFBP) method was used to train the ANNs. Determination of the number of hidden layers and the number of nodes in the input layer providing the best training results was the initial process of the training procedure. The mean square error (MSE) and the auto-correlation function (ACF) were used as criteria to evaluate the performance of the training simulations.

The number of input nodes to the neural network model is dependent on the number of time lags that is determined by the analysis of the ACF belonging to our time series data set. Four time lags were found sufficient; hence, the input layer consisted of four nodes.

The values of the training inputs and outputs were scaled between 0 and 1 as in (7).

$$\frac{x_i - x_{min}}{x_{max} - x_{min}} \tag{7}$$

After the completion of the training stage, testing stage took place. The optimum learning, η, and the momentum, α, rates were determined after trials observing the MSE produced at the end of the testing stage. It is seen that η and α should be decreased if the number of the input and output layers are increased. On the other hand, the iteration number increases by decreasing η and α values.

5.1 Forecasts of Data

The last 348 half-hourly mean PT trip flow values (the interval starts on April 1st and ends on November 30th in the year 2002) were analysed during the training stage of the neural networks. The first 180 values (the interval starts on January 1st and ends on March 30th in the year 2002) were then used for the testing phase. As mentioned above, also with respect to the ACF of the trip flow time-series data, the last four consecutive trip flow values set the input layer. The MSE, which was calculated by scaled values, for the testing period was equal to 0.000277. The MSEs and the determination coefficients for different orders of time series depending on the optimum hidden unit number for each order is shown in Table 2.

Three hidden layer nodes were found appropriate for the four-nodded input layer. More than three nodes for the hidden layer couldn't improve the MSE during the testing stage. The values forecasted by the ANN seemed to be closer to the observed values as shown in Fig. 2.

Following the ANN forecast, a comparison was made by means of a stochastic model. An auto-regressive model of order four was used. The auto-regressive (AR) process $\{X_t\}$ or order p has a stochastic structure, if there exist constants δ, ϕ_1, ϕ_2, ..., ϕ_q and a white noise $\{\varepsilon\}$ such that (8) is satisfied.

Table 2. The MSE (calculated by scaled values) and the determination coefficient values of ANN predictions

Order	MSE	R^2
1	0.003028	0.906
2	0.000374	0.986
3	0.000435	0.987
4	0.000277	0.991
5	0.000421	0.988

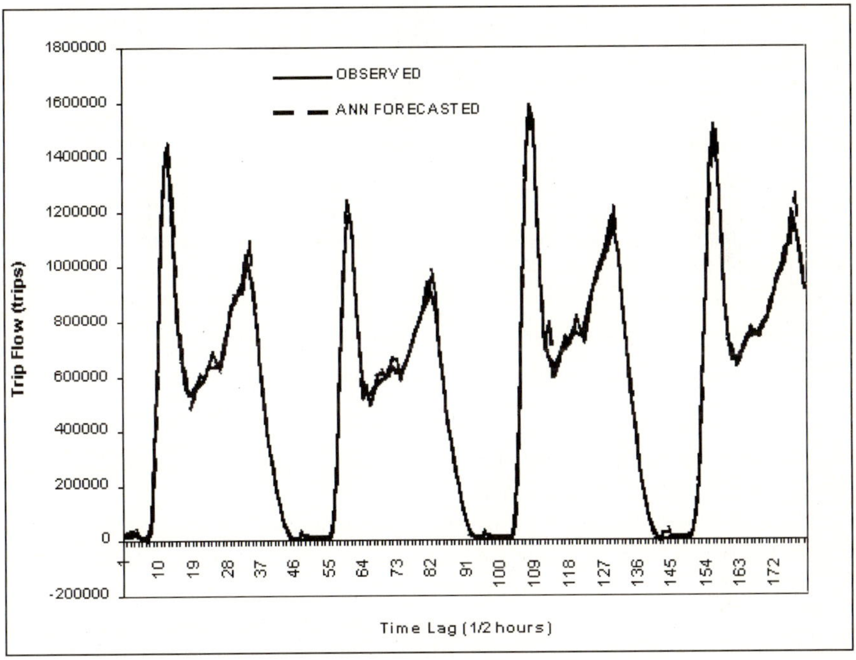

Fig. 2. Comparison of the observed half-hourly mean PT trip flows with ANN forecasts

$$y_t = \delta + \varphi_1 y_{t-1} + \varphi_2 y_{t-2} + \cdots + \varphi_p y_{t-p} + \varepsilon_t \tag{8}$$

The data period between April 1st and November 30th in the year 2002 (training period of ANNs) was used to obtain the AR(4) model. Again the data period between January 1st and March 30th in the year 2002 (testing period of ANNs) was used to make a forecast with the AR(4) model. The results are presented in Fig. 3. The AR(4) is indicated with (9).

$$y_t = 2.05121 y_{t-1} - 1.52353 y_{t-2} + 0.41155 y_{t-3} + 0.04237 y_{t-4} \tag{9}$$

The MSE obtained from the stochastic model was equal to 0.03025, which is higher than the value (0.000277) obtained after the ANN application. The results

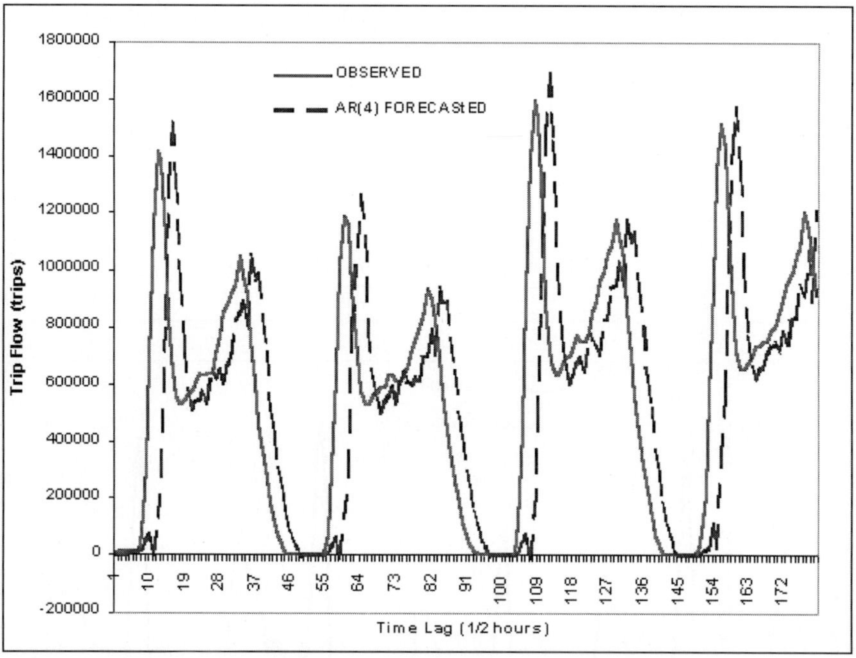

Fig. 3. Comparison of the observed half-hourly mean PT trip flows with AR(4) forecasts

point out that the ANN forecasts for the observed values are closer to the original ones compared with the AR(4) model.

To present the difference between two forecasting methods clearly, the first 180 forecasts obtained from each method were plotted with the corresponding observed ones in Fig. 4.

Both in the peak and off-peak hour periods, the ANNs provided estimates are too close to the observed values where the AR(4) model fails by over and underestimating.

6 Conclusions

Prediction of the future PT trip flows has a key role for developing precautions before the forthcoming investments for an urban area. In this study, a trip flow-estimating model based on neural networks has been developed and has been compared with a stochastic model. Among the studied models, a significant improvement in the prediction made by the neural network model is noticed, due to its flexibility to adapt to time-series database. Another advantage of the application of neural networks is the fact that it is trained with measurements, hence; the included propagation effects are more realistic. Making forecasts by generating negative values after trials is a disadvantage of the algorithm used to train the neural network. It can be said that

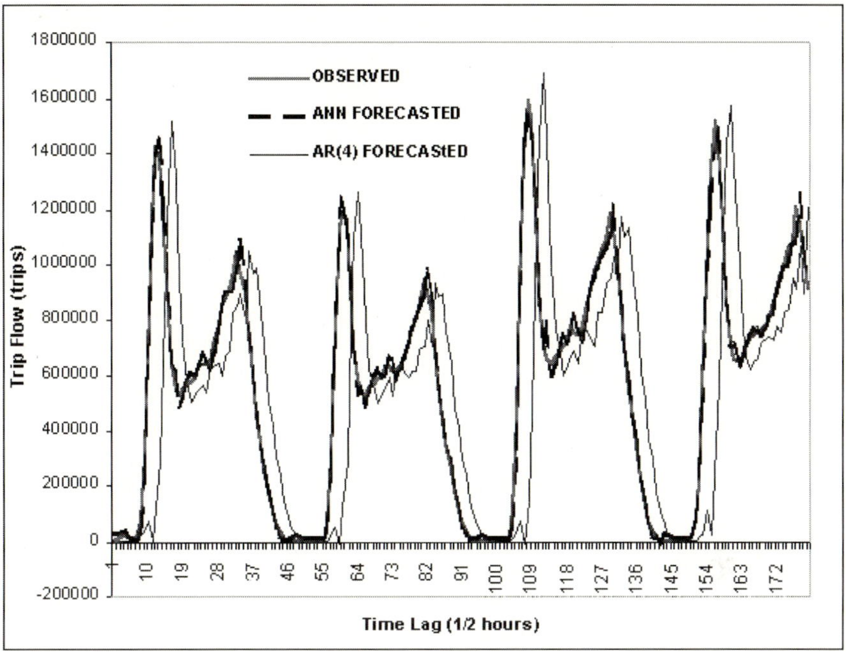

Fig. 4. Comparison of the observed half-hourly mean PT trip flows with ANN and AR(4) forecasts

various kinds of learning algorithms should be assessed during the training process to overcome this drawback.

References

1. Ahmed, S.A., and Cook, A.R., (1979). Analysis of Freeway Traffic Time Series Data by Using Box-Jenkins Techniques. Transportation Research Record 722, Transportation Research Board, Washington, D.C., 1–9.
2. Bhat, C.R., (2000). Quasi-Random Maximum Simulated Likelihood Estimation of the Mixed Multinomial Logit. Bhat C.R. Department of Civil Engineering, University of Texas, Austin.
3. Celikoglu, H.B., (2002). Logit Modal-Split Model Calibration: An Evaluation for Istanbul, Unpublished MSc Thesis, INS UZ1734 CEL, ITU, Istanbul.
4. Clark, C.E., (1961). The Greatest of a Finite Set of Random Variables. Operations Research, 9, 145–162.
5. Daganzo, C.F., Bouthelier, F. and Sheffi, Y., (1977). Multinomial Probit and Qualitative Choice: A Computationally Efficient Algorithm. Transportation Science, vol. 11, no. 4, 338–358.
6. Daganzo, C.F. and Bouthelier, F., (1979). Aggregation with Multinomial Probit and Estimation of Disaggregate Models with Aggregate Data: A New Methodological Approach. Transportation Research, 13B, 133–146.

7. De Donnea, F.X., (1971). The Determinants of Transport Mode Choice in Dutch Cities. Rotterdam University Press.
8. Kanafani, A., (1983). Transportation Demand Analysis. McGraw-Hill.
9. Lerman, S.R. and Manski, C.F., (1977). An Estimator for the Generalized Multinomial Probit Model, Transportation Research Board Meeting.
10. McLeod, W.T. and Hanks, P. (eds.) (1986). The New Collins Concise Dictionary of the English Language. William Collins, Sons & Co., Glasgow.
11. Ortuzar, J. de D., (1980b). Modelling parknride and kissnride as submodal choices: a comment. Transportation, 11(4), 383–385.
12. Revelt, D. and Train, K., (1997). Mixed Logit with Repeated Choice: Households Choices of Appliance Efficiency Level. Review of Economics and Statistics, July.
13. Rumelhart, D.E., Hinton, G.E. and Williams, J.E., (1986). Learning Internal Representations by Error Propagation. Parallel Distrubuted Processing-Explorations in the Microstructure of Cognition, volume 1: Foundations. D.E. Rummelhart and J.L. McClelland, eds., MIT Press, Cambridge, Massachussets.
14. Smith, B.L. and Demetsky, M.J., (1994). Short-term Traffic Flow Prediction: Neural Network Approach. Transportation Research Record 1453, Transportation Research Board, Washington, D.C., 98–104.
15. Sobel, K.L., (1980). Travel demand forecasting by using the nested multinomial logit model. Transportation Research Record 775, 48–55.
16. Zhang, H.M., (2000). Recursive Prediction of Traffic Conditions With Neural Network Models. Journal of Transportation Engineering, Vol. 126, No. 6, 472–481.

Clustering of Activity Patterns Using Genetic Algorithms

Ondřej Přibyl

Czech Technical University in Prague, Department of Applied Matematics, Na Florenci 25, Praha 1, 110 00, Czech Republic
pribylo@fd.cvut.cz

Abstract. Finding groups of individuals with similar activity patterns (a sequence of activities within a given time period, usually 24 hours) has become an important issue in models of activity-based approaches to travel demand analysis. This knowledge is critical to many activity-based models, and it aids our understanding of activity/travel behavior. This paper aims to develop a methodology for the clustering of these patterns. There is a large number of well-known clustering algorithms, such as hierarchical clustering, or k-means clustering (which belongs to the class of partitioning algorithm). However, these algorithms cannot be used to cluster categorical data, so they do not suit the problem of clustering of activity patterns well. Several other heuristics have been developed to overcome this problem. The k-medoids algorithm, described in this paper, is a modification of the k-means algorithm with respect to categorical data. However, similar to the k-means algorithm, the k-medoids algorithm can converge to local optima. This paper approaches the medoids-based formulation of clustering problem using genetic algorithms (GAs), a probabilistic search algorithm that simulates natural evolution. The main objective of this paper is to develop a robust algorithm that suits the problem of clustering of activity patterns and to demonstrate and discuss its properties.

Key words: Cluster analysis, genetic algorithms, activity behavior

1 Introduction

Finding groups of individuals with similar activity patterns (a sequence of activities within a given time period, usually 24 hours) has become an important issue in models of activity-based approaches to travel demand analysis. This knowledge is critical to many activity-based models, and it aids our understanding of activity/travel behavior. Previous research [1, 2] gave us evidence that people with similar socio-demographic characteristics have also often similar schedules. One direction of research focuses on the microsimulation of the observed activity patterns [3]. An essential step in their modeling effort is to find an average or typical schedule that represents the activity/travel behavior of these individuals. The ultimate goal of this project, then, is to find a relatively small set of activity patterns that would represent all observed patterns in given data set.

Ondřej Přibyl: *Clustering of Activity Patterns Using Genetic Algorithms*, Advances in Soft Computing **1**, 37–52 (2005)
www.springerlink.com

The problem as stated belongs to so-called subset selection problem that is interpreted as a special case of cluster analysis. The number of different partitions of N objects into K clusters, $p(N,K)$, is extremely large. Everitt et al. [4] provides an equation of how to compute $p(N,K)$. To illustrate the complexity, let's look at the following examples: $p(10,3) = 9330$, and $p(50,4) \approx 5.3 \cdot 10^{67}$. In our case, the size of the data set, N, as well as the number of clusters, K, is expected to be even higher. For this reason we are not able to evaluate all possible combinations.

The data used in this paper are of categorical type (as will be shown in section two). Many well-known clustering algorithms, such as hierarchical clustering and k-means clustering [4], are not suitable for this type of data. They are based on Euclidean distances, and they need to find an average of several objects. In our case, however, there is not a simple and realistic way to find the average of several schedules. Several heuristics have been developed to overcome this problem. For example, Kaufman and Rousseeuw [5] described a modification of the k-means algorithm to suit categorical data. This algorithm is called *medoid-based clustering*, and it is described in greater detail in this paper. Similar to the k-means algorithm, the medoid-based algorithm can converge to local optima [4]. For this reason this paper uses an alternate heuristic for the same formulation of the problem. It is based on the principles of genetic algorithms (GAs), probabilistic search algorithms that simulate natural evolution.

The main objective of this paper is to develop a robust algorithm that suits the problem of clustering activity patterns, and to demonstrate its properties. The algorithm should suit relatively large data sets to enable its application to real-world problems. Approaches based on genetic algorithms have usually many parameters that need to be set, such as the size of population, number of runs, probability of mutation, and others. The performance of the algorithms is often very sensitive to the setting of these parameters, but at the same time no general guidance about the parameter setting exists, since it is problem specific. This paper shows the performance of the developed algorithm for different settings of parameters. The objective is to provide recommendations concerning parameter settings that would enable the broader use of this tool without the need for further in-depth study.

2 The Data

Knowledge of the data set's structure would help to evaluate the algorithm and also it would enable more truthful discussion of its properties. For this reason, artificially generated data rather than observed data are used in this preliminary phase of the project. Before we describe the data set used for this analysis, the representation of activity patterns used in this paper must be discussed.

2.1 Representation of Activity Patterns

The proposed algorithm will be evaluated on its ability to cluster activity patterns. An essential feature for forming the activity patterns is the type of each activity. Though there are many different activity types that an individual can perform, only a limited number of activities will be considered in this work, for simplicity. The activities considered are: **H**ome (everything that is an in-home activity), **W**ork (work, work related or school), **M**aintenance (dining out, shopping and others), and **D**iscretionary (social events, recreation, visiting friends and others)[1].

Within this project we use a discretized representation of the activity patterns. For each individual in the sample, we look at her/his activity participation in regular time intervals (10 minutes in this project) and record it. It means that the whole 24 hour long activity pattern consists of 144 values. Each of the values corresponds to an activity type. Integer values from the range 1 to 4 are used to represent given activity types; Home (1),Work (2), Maintenance (3), and Discretionary (4). It must be clearly stated that these values do not have real numerical meaning. The activity types are clearly of categorical nature. For example, we can see ten activity patterns in Fig. 1. They correspond to different types of activity patterns as described in the next chapter. The *x-axis* in this figure represents time per day (hours) and the *y-axis* represents the four activity types as described above. For example, the first pattern shows the schedule of an individual who left from home to work at about 9:30 am, he went for lunch at about 11 am (maintenance activity) and returned back to work at about noon. This individual returned back home shortly before 6 pm. The other patterns in Fig. 1 can be understood similarly.

2.2 Synthetic Data

The data used in this paper are generated artificially so their structure is known, but at the same time they should realistically represent the observed patterns. In order to meet these objectives, five different groups of schedules, using results reported by Kulkarni and McNally [3] were generated in this paper. Examples of two schedules from each group are shown in Fig. 1.

The first group is called *standard work* (SW). As the name implies, people in this group work for about 8 hours. Their work starts sometimes in the morning (between 7 am and 9 am). Some of these individuals go for lunch at about noon (maintenance activity up to one hour long). The second group, called *power work* (PW), contains of people whose schedules are similar to those in the previous group, only their work activity is in average two hours longer, for example from 8 am to 6 pm. The next group is called *late work* (LW). People in this group also have, on average, an 8 hour-long working day (similar to SW), but their work starts in the afternoon, around 2 pm. The fourth group is called *work-discretionary* (WD). People in this group have standard work starting in the morning, but they participate in some discretionary activity after work. Some people in this group go to the discretionary activity directly from work, some people first stop at home. The last group is called *various short*

[1] The algorithm developed in this project should be general enough to enable later extension to more activity types. We might want to distinguish if the four activities were undertaken alone or joint with some other family member. This step will increase the number of activity types up to ten.

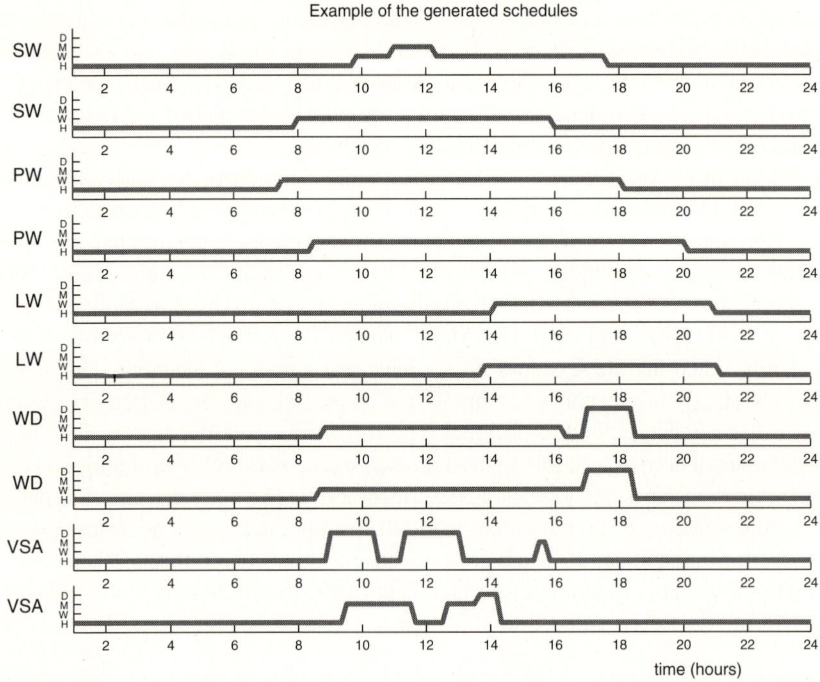

Fig. 1. Examples of generated activity patterns. There are two patterns of each type; standard work (SW), power work (PW), late work (LW), work discretionary (WD), and various short activities (VSA), as described in Sect. 2.2

activities (VSA). People in this group participate in several activities (up to five per day). Their type varies randomly among work, maintenance and discretionary type.

The code enables the user to set the number of schedules generated in each group, as well as set all probabilities (for example probability to go for lunch for the first two groups) and the starting and ending times for each activity. The data, as generated, are rather similar to the observed schedules of individuals. However, this brings also a challenge to the algorithm. The groups (or clusters) of the data are overlapping. It is not true that schedules in one generated group are the closest to each other based on the used dissimilarity measure. Unfortunately, this is also the case of any observed data.

3 Medoid Based Clustering

The method in this paper is based on so called *k-medoid clustering* [5, 6]. Consider a set $S = \{S_1, \ldots, S_N\}$ of N objects (in our case activity patterns). Each object S_i is a vector, containing of L integer values describing an activity type at particular time instant. Our objective is to find K objects, $m_1, \ldots, m_K$, which represent all objects in the data set. The remaining objects are then assigned to the nearest representative object, using a given dissimilarity measure.

A simple example of 12 two-dimensional objects is shown in Fig. 2. The arrows in this figure point to objects that represent each of the clusters. These representative objects are called *medoids*, because we expect them to be located around the geometrical center of each cluster. Each object in the data set belongs exactly to one medoid. All objects belonging to the same medoid form a *cluster*. The objects in each cluster are clearly more similar (based on given dissimilarity measure) to each other than to objects in any other group.

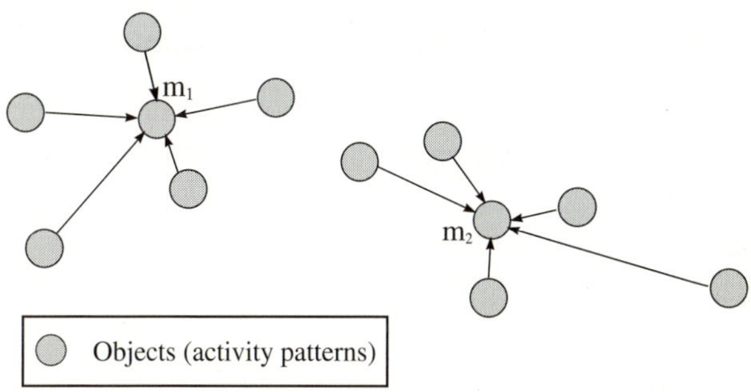

Fig. 2. Suggested principle of k-medoid clustering approach

The developed algorithm requires the number of clusters, K, to be known. In order to find the optimal number of clusters, the algorithm will be repeated for different numbers of clusters. The optimal number will be chosen off-line, after the clustering problem for different K will be performed. For example, silhouette width defined in [5] can be used.

Mathematically, we are looking for a set of K objects, $K \ll N$, that minimize the following objective function, F_K:

$$F_K = \sum_{i=1}^{N} \min_{t=1,\dots,K} d(S_i, m_t) \,. \tag{1}$$

In this equation, F_K is the sum of dissimilarities of all objects S_i to their nearest medoid. In case the arrows in Fig. 2 represent the dissimilarity among objects, the objective function F_K can be understood as sum of the lengths of all the arrows. We are seeking the medoids so that this sum is minimal. An advantage of this method is also in the fact that it uses only a dissimilarity matrix and not the original data. This implies that this method can be used for any type of data as long as we know how to measure dissimilarities among objects in the dataset.

3.1 Dissimilarity Measures

A very important issue in any clustering algorithm is definition of dissimilarity between two objects. We cannot use common Euclidean distance, because we deal with objects of a categorical type (activity types). The distance measure used in this paper is defined in [6] as follows: The dissimilarity measure between two objects S_1, and S_2 reflects the total number of mismatches of activity types at a corresponding time index. This can be written as:

$$d(S_1, S_2) = \sum_{j=1}^{|S_i|} \delta(S_1(j), S_2(j)),$$

$$\delta(S_1(j), S_2(j)) = \begin{cases} 0, & S_1(j) = S_2(j) \\ 1, & S_1(j) \neq S_2(j) \end{cases} \tag{2}$$

and $|S_i|$ denotes the length of a schedule (in our case it equals to 144). Therefore, the value of the dissimilarity measure between two patterns can range in our case from 0 (the two patterns are the same) to 144 (totally different).

The dissimilarities between particular objects i and j from the DATA 50, d_{ij}, are shown in Fig. 3 (a similar visualization was introduced in [5]). Only a triangular matrix is depicted since the dissimilarity matrix is symmetric $(d_{ij} = d_{ji})$. As the color chart implies, darker colors depict objects that are closer to each other (small dissimilarities). The objects are sorted by the five groups of origin. For example, the dissimilarities for the fourth group, work discretionary (WD), are shown in the

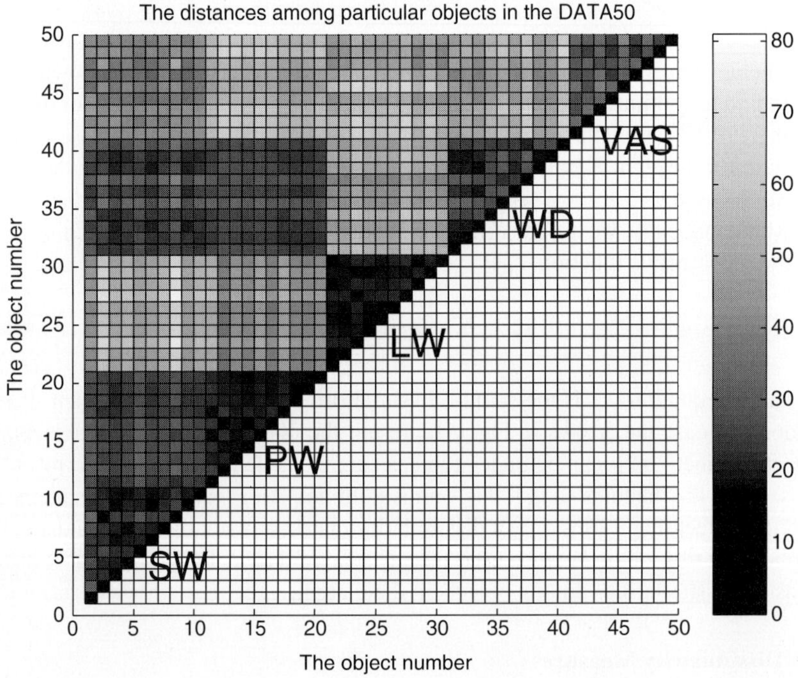

Fig. 3. The dissimilarities among objects for a data set containing 50 objects. These objects are sorted by groups of origin

triangle that ranges from 30 and 40 on both axes. If the groups did not overlap, the values of dissimilarity would be the smallest among objects in the same group. This is true for some objects and some groups. For example, the objects in the third group, late work (LW), are closer to each other than to any other object in the data set. On the other hand, some objects from the work discretionary group are closer to other objects in standard work (SW) group (range from 1 to 10 in the figure) than to each other.

4 GA Based Approach to K-Medoid Clustering

Several different representations have been used in the literature (for an overview see [7], or [8, 9, 10], and others). The most suitable representation for the given problem (considering the size of the problem and also its categorical character) can be adapted, for example, from [11].

Each chromosome is a vector of the length K (number of clusters), and every element is obtained from a uniform distribution in the range $(1, N)$. The ith value is an index of the ith median. For example, if K equals 4, and the medoids are patterns with numbers 4, 23, 54, and 72, the result chromosome can be expressed as the following vector $\{4, 72, 23, 54\}$ (there is more than just one representation of the same solution). The objective function of each chromosome is then computed based on (1). One advantage of integer representation is that we do not have to convert chromosomes into phenotypes before computing value of the objective function, and so we save computational time. Two different algorithms have been developed in this paper. Model 1 is an integer coded GA with $(m+l)$ binary tournament selection (the symbols m and l are common in the GA related literature and represent the parent and children population respectively). Model 2 is a modification of model 1. It modifies the selection operator to maintain more diversity in the population. The principle of the selection operator is briefly described in this paragraph and also shown in the provided pseudo-code. Given two randomly selected parents, we apply a recombination and mutation operator and so produce two offspring. The selection operator is applied to the set of both parents as well as both offspring. Two chromosomes with the lowest fitness functions are deterministically selected from this set. The pseudo-code for both models is provided in the section below.

Notation:

```
GN    ... number of generations
NP    ... size of population
pi    ... parent i
ci    ... offspring i
PmB   ... probability of in-built mutation
P(t)  ... population at time t
```

Pseudo-Code for Model 1:

```
for t = 1: GN,
     P(t) <-- shuffle P(t-1)
    for j = 1:NP/2-1,
      p1 = P2j+1(t)
      p2 = P2j+2(t)
      {c1,c2}   <-- recombine (p1,p2)
      {c1',c2'} <-- mutate (c1,c2)
      append {c1',c2'} to matrix C
    end
 Create matrix Q that combines both, parents and
 offsprings {P;C} Binary tournament selection (matrix Q)
end
```

Pseudo-Code for Model 2:

```
for t = 1: GN,
   P(t) <-- shuffle P(t-1)
 for j = 1:NP/2-1,
   p1 = P2j+1(t)
   p2 = P2j+2(t)
   {c1,c2}   <-- recombine (p1,p2)
   {c1',c2'} <-- mutate(c1,c2)
   select 2 best individuals from the
   set {p1, p2, c1',c2'} and place them to the
   new population P(t+1)
 end
end
```

4.1 Population Management

In order to improve the performance of the algorithm, a method of multiple population approach similar to [12] is used. The algorithm performs several iterations. The population size is doubled in every iteration, by adding a randomly initiated population. The inserted population has the same size as the current population. This doubling of population aims to decrease the computational time, it helps to prevent premature convergence, and also it aims to decrease the sensitivity of the algorithm to the setting of its parameters, such as size of population, number of runs, or the random number seed used. Figure 4 shows an example of progress of the best and average fitness function for each generation. In this example there are 300 objects in the data set (N). The algorithm performs forty runs (GN) for each population size ($NP = 10, 20, 40$, and 80).

The points of injection (every $GN = 40$ populations) show a large increase in the average fitness function. This reflects the fact that the diversity in the population increases by inserting random individuals. This also often leads to improvement of the best fitness function shortly after insertion of new random individuals.

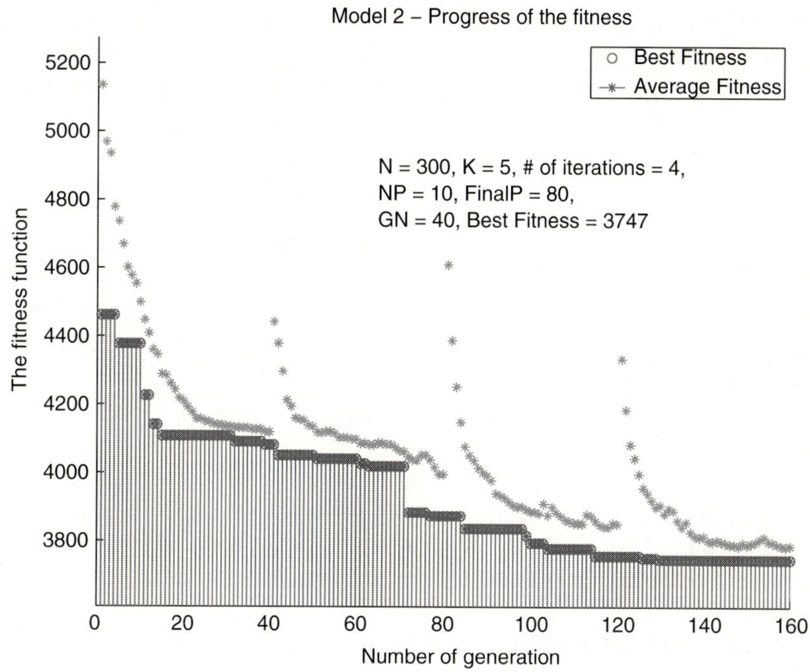

Fig. 4. The progress of the best and average fitness for DATA 300 and four iterations (3 injections)

4.2 Detailed Description of the Algorithm

The whole initialization phase, as well as the recombination and mutation operator, is the same for both developed models. The main difference is in the management of the population and in the selection process. The following paragraphs explain the elementary building blocks of the algorithm in more detail.

Create Initial Population

The size of each population is denoted *NP*. The initial population is created randomly. Each value is an integer number in the range of 1 to *N* (number of objects in the data set). It is ensured that the values within each chromosome are unique (not repeated).

D_MX Recombination with In-Built Mutation

Several recombination operators were considered in this project. The use of the standard operators, such as uniform crossover, is problematic, because they produce infeasible offspring (the values in each chromosome cannot be repeated). The Random Respectful Recombination (R3) as described by Estivill-Castro and Murray [8] was also considered. Its principle is based on the theory of building blocks [13]. If a certain value is present in both parents, it must be present also in both children. The rest of the algorithm is the same as for uniform crossover (however, feasibility of offspring is ensured by this algorithm). The author decided not to use this operator, because it might decrease the diversity of offsprings and lead to premature convergence to a suboptimal solution. This is true especially in the case of clustering, when the length of each chromosome is short. The recombination operator, D_MX (*Mixed Subset Recombination*), used in this project was modified from [11]. The "*D*" in its names reflects the fact that it is applied "directly" to the phenotype, rather than to the binary representation (genotype). The operator, when applied to two parent strings (P1, P2) produces two offspring (C1, C2), and works as follows:

1. Mix $P1$ and $P2$
   (a) Append copy of $P1$ at the end of $P2$, and form a vector Q.
   (b) Randomly scramble the elements in Q.
2. Apply the built-in mutation to all elements in Q (this step is different from [11], who applies the built-in mutation only to the first k elements). If an element is chosen (with probability *PmB*), its value is replaced by some other random value, that points to some other medoid (random integer number in the range 1 to number of objects in the data set).
3. Randomly scramble elements in Q again.
4. Create first offspring, $C1$, by copying k elements from Q into $C1$, starting at the leftmost element. Elements that are already in $C1$ are skipped (to ensure the feasibility of offspring).
5. Similarly, create second offspring, $C2$, by copying k elements from Q into $C2$, this time starting from the rightmost element.

With a certain level of simplification, the operator can be considered to be a uniform crossover with in-built mutation that ensures generating of feasible offsprings (feasibility for this application was discussed above).

Selection **Model 1**

In the previous literature, most of the researchers used roulette wheel selection [11]. In order to improve its performance, a linear scaling had to be applied to the objective function. For the purpose of this project, we decided to apply the standard binary tournament selection. It is well supported by theory and previous work has shown its superiority to roulette wheel selection. For example, a "super solution" – solution with much higher fitness function than the rest of the population – would occupy most of the roulette wheel area, so the next generation would be occupied mostly by this solution (for more detailed comparison of different selection operators see, for example, [13]).

Selection **Model 2**

The two best solutions are deterministically selected from a set of parents and their offspring $\{p1, p2, c1, c2\}$. The application of the selection operator only on a small subset of the population and not on the entire population should again increase diversity in the population and prevent preconvergence. In order to evaluate each solution, the objective function, D_K, for each chromosome (solution) in the population is computed using equation 2. We use the dissimilarities among objects as stored in the dissimilarity matrix D.

5 Results

All results were obtained using an original code developed in the software environment MATLAB [14]. This paper focuses on showing properties of the algorithm for different sizes of the data sets (from one hundred to one thousand). Also, it aims to provide recommendations concerning the setting of its parameters and so enable its usage for application. Because of the limited space available, the number of clusters is set to five. For future use, this number would not be known in advance and the results for different number of clusters would be compared.

Approaches based on genetic algorithms are stochastic in nature. We do not obtain the same result every time we run the algorithm. For this reason we performed 10 runs for each setting of the algorithm. All tables in this chapter show the average values of these ten runs. Figure 5 shows the average objective function of the algorithm on DATA300 in case we did not use any injections. We can see results for both models (model 1 and model 2), different sizes of initial population ($NP = 50, 100$, and 150), and different number of generations ($GN = 50, 100, 150$ and 200).

The figure suggests that the performance is rather sensitive to the parameters of the algorithm. Also, the obtained results depend on the random seed used. In case we use a multiple population approach, the effect of random seed number, as well the

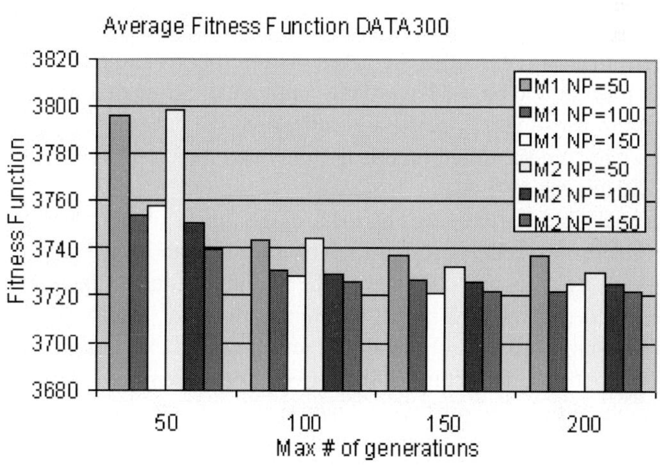

Fig. 5. Fitness function (average of 10 runs) for both models (M1, M2), and different settings (NP, GN) on the DATA300

sensitivity to particular parameters is reduced. Results supporting this argument are provided in the following section.

5.1 Performance for Various Sizes of the Data Sets

In this part, I will demonstrate the performance of the algorithm for different sizes of the data sets. Table 1 summarizes the results for the following data sets: `DATA100`, `DATA300`, `DATA600`, and `DATA1000`. Each number in the name of the data sets represents the size of particular data set.

Notation Used in Table 1:

Model	*Type of Model (1 or 2)*
NP	*Initial size of population*
FinalP	*Size of population at the end of run*
Iter	*Number of iterations*
GN	*Number of generations for each iteration*
PmB	*Probability of in-built mutation*

Evaluation:

Fitness	*Average fitness function*
Time	*Computational time of the algorithm*
NumMin	*How many times out of 10 runs did the algorithm reach the minimum fitness function*

The value of computational time provided in the results is only informative. It is clearly a function of population size and number of generations for each iteration. In some cases, the final solution can be found already in an initial phase of the computation, although the algorithm continues to run without any improvement of the solution. Termination criteria that are better than simply reaching the maximal number of generation should be used in order to improve the computational performance of the algorithm. However, since the algorithm does not have to perform in real time applications, the computational time is not the most important issue and its current values are satisfactory.

All parts of Table 1 are ordered by the average fitness function (the lowest value of fitness function first). In the top of each of the tables above we provide the name of each data set used, as well as value of the minimum fitness function obtained. It is not guaranteed that this is a global minimum. However, it indicates the minimal value obtained out of all runs of the algorithm. We could expect that this number is close to global optimal. This belief is supported also by the results depicted in Fig. 6.

An example of the resulting "typical patterns" (medoids) found by the algorithm is shown in Fig. 6. Since the objective function does not have a very clear meaning, these medoids can serve as an additional evaluation of results. Clearly, the medoids really represent the five original groups: *SW*, *PW*, *LW*, *WD*, and *VAS*. Similar results were obtained in every run of the algorithm.

Table 1. The obtained results for different sizes of data sets (100, 300, 600, and 1000) and for two types of models, and different parameter settings

DATA 100	Min. Fitness = 1200							
Model	NP	FinalP	Iter	GN	PmB	Fitness	Time (sec)	NumMin
2	20	320	5	60	0.05	1222	119	10
2	20	320	5	60	0.1	1222	115	10
2	10	640	7	50	0.05	1222	200	10
2	20	320	5	60	0.01	1224	124	8
2	10	160	5	50	0.05	1223.2	49	7
DATA 300	Min. Fitness = 3721							
Model	NP	FinalP	Iter	GN	PmB	Fitness	Time (sec)	NumMin
2	20	320	5	60	0.05	3721	135	10
1	30	1920	7	60	0.05	3721	1001	10
2	20	320	5	60	0.1	3721.7	123	7
2	20	320	5	60	0.1	3724	124	8
2	20	320	5	60	0.05	3724.2	127	8
DATA 600	Min. Fitness = 7239							
Model	NP	FinalP	Iter	GN	PmB	Fitness	Time (sec)	NumMin
1	20	1280	7	60	0.05	7239.6	657	9
2	20	1280	7	60	0.05	7240.7	620	9
2	20	640	6	60	0.1	7242.5	308	9
2	20	640	6	60	0.01	7243.1	323	7
2	30	480	5	40	0.1	7243.4	145	4
DATA 1000	Min. Fitness = 11780							
Model	NP	FinalP	Iter	GN	PmB	Fitness	Time (sec)	NumMin
2	30	480	5	40	0.1	11780	189	10
2	20	640	6	60	0.05	11794	414	9
1	20	640	6	60	0.05	11794	421	9
1	30	480	5	40	0.05	11802	184	4
1	30	480	5	40	0.1	11804	177	6

5.2 Recommended Parameter Setting

There are several suggestions that can be implied from Table 1. First we can look at performance of particular models. The majority of the best few solutions was obtained using model 2. Also, the computational time of the second model outperform the first model for the same settings. The table also implies that the probability of in-built mutation, *PmB*, should be set in the range from 0.05 to 0.1, and values closer to 0.05 can be preferred. It is not surprising that the algorithm performs better with a larger population size, higher number of generations, and also higher number of iterations. All these parameters increase the size of space being searched. On the other hand, increasing these parameters also increases the number of function evaluations, and so leads to an increase in computation time. Finding the equilibrium between these two features is desirable. It will clearly depend on the desired precision.

In general it is possible to recommend setting the initial population size, *NP*, equal to 20 or 30. The number of iterations, *Iter*, can be set in the range from 5 to

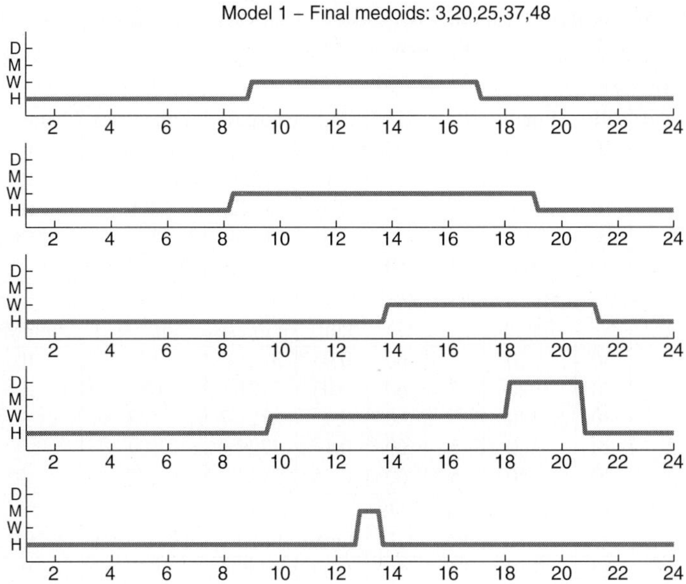

Fig. 6. An example of the resulting medoids for DATA 50, model 1, and 4 iterations

7, and the number of generations for each iteration, *GN*, can be set to 50 or 60. All these parameters should have higher values for data sets of a larger size.

6 Conclusion and Future Research

In this paper, algorithms to cluster activity patterns were developed. These algorithms, as designed, can be used for clustering large data sets of categorical objects. A dissimilarity measure that suits categorical data was applied here. In general, the algorithm performs well even for large data sets. The modification of the selection operator used in model 2 leads to improvement in the model's results and it can be recommended for future applications. The successive insertion of random individuals also improved the performance of the algorithm. The algorithm, as developed, was designed not to be overly sensitive to the exact setting of its parameters. It performs reasonably well for a wide range of these parameters. This feature is of an important concern of all approaches based on genetic algorithms. The recommended setting of the parameters is provided in the previous part of this paper.

In order to be able to evaluate performance of the developed algorithm with more precision, it should be compared to other tools used for solving of the same problem. For example, the development of other heuristic for the k-medoid algorithm, such as PAM [5], could be considered for future research. Also, a more detailed study should be dedicated to the problem of dissimilarity among activity patterns. For example, the use of multidimensional sequence alignment method as described in [15], or the use of multivariate logic (instead of Boolean logic) for comparison of two activity patterns should be studied. Also, the author would like to use genetic algorithms to solve a multi-objective clustering problem [16]. The two objectives that we want to minimize are the number of medoids (clusters) as well as the resulting objective function. In this case, we would like to develop an algorithm that finds all pareto-optimal (nondominated) solutions. Such an algorithm would be superior to the developed algorithm, because we could directly determine the optimal number of clusters.

Acknowledgement

The author would like to thank to Dr. Patrick Reed from the Pennsylvania State University for providing many valuable comments and additional references concerning this paper.

References

1. Pas, E.I. "A Flexible and Integrated Methodology for Analytical Classification of Daily Travel-Activity Behavior." Transportation Science (Operations Research Society of America) 17, 4 (1983): 405-429.
2. Ma, June. "An Activity-Based Approach and Micro-simulated Travel Forecasting System: A Pragmatic Synthetic Scheduling Approach." PhD Thesis, The Pennsylvania State University, Department of Civil and Environmental Engineering, University Park, Pennsylvania (1997).
3. Kulkarni, A.A., and M.G. McNally. "An Activity-Based Travel Pattern Generation Model.," Institute of Transportation Studies, University of California, Irvine, December, 2000. UCI-ITS-AS-WP-00-6.
4. Everitt, B.S., S. Landau, and M. Leese. (2001) Cluster Analysis. Fourth Edition: Arnold, A member of the Hodder Headline Group, London, 2001.
5. Kaufman, L., and P.J. Rousseeuw. Finding Groups in Data, An Introduction to Cluster Analysis. John Willey & Sons, Inc., 1990.
6. Huang, Z. "Clustering Large Data Sets with Mixed Numeric and Categorical Values." In Proceeding of the First Pacific-Asia Conference on Knowledge Discovery and Data Minning (Singapore, World Scientific), 1997.
7. Lozano, J.A., and P. Larranaga. (1996) "Using Genetic Algorithms to Get the Classes and Their Number in a Partitional Cluster Analysis of Large Data Sets." http://citeseer.nj.nec.com/457425.html: NEC Research Institute.
8. Estivill-Castro, V. and A.T. Murray. (1997) "Spatial Clustering for Data Mining with Genetic Algorithms." http://citeseer.nj.nec.com/estivill-castro97spatial.html: NEC Research Institute.
9. Moraczewski, I.R., W. Borowski, and A. Kierzek. "Clustering Geobotanical Data with the Use of a Genetic Algorithm." COENOSES (C.E.T.A., Gorizla, Italy) 10, 1 (1995): 17-28.
10. Maulik, U., S. Bandyopadhyay. "Genetic Algorithm-Based Clustering Technique." Pattern Recognition 33 (2000): 1455-1465.
11. Lucasius, C.B., A.D. Dane, and G. Kateman. "On k-Medoid Clustering of Large Data Sets with the Aid of a Genetic Algorithm: Background, Feasibility and Comparison." Analytica Chimica Acta (Elsevier Science Ltd) 282 (1993): 647-669.

travel time on each path, they could easily plan the travel choosing the best path from their point of view. However, complete and precise information about network conditions is hardly ever available, so that uncertainty causes anxiety and stress in decision makers.

Several researchers have dealt with conceptual models of drivers' behavior – see for example [2, 5, 16]. In these studies, uncertainty has been usually modelled through random utility models. The basic idea in these models is that each user or service provider behaves rationally: he compares the costs of several alternatives and chooses, among them, the best one from his standpoint. Since the knowledge of alternatives is rarely perfect, uncertainty affects single person's decision. Randomness is then used to represent uncertainty and therefore the potential for different choices. In such a case, the probability of a choice can be calculated. This approach, when applied to the transportation assignment problem, leads to a stochastic equilibrium [10].

Very complex structures of random utility models have been analyzed to obtain both day-to-day and within-day decision processes (see for example [7] or [8]). Additionally, a simulation system for dynamic traffic assignment (DYNASMART) was developed [15] to take into account various information strategies. For this kind of models, the unavailability of complete and reliable data limits the capabilities of calibrating model parameters: if incomplete data are used, estimates will contain significant levels of uncertainties. On the other hand, traditional choice models do not provide neither with levels of uncertainty imbedded in the estimates, nor easily with a perspective on the confidence on the estimates.

On the contrary, the concepts of approximate reasoning can be helpful in treatment of uncertainty.

A modelling framework based on Evidence Theory represents the uncertainties in the perception of travel attributes and measures the *Belief* rather than the *probability* of a choice. Belief and its dual Plausibility establish a system of uncertainty measures when dealing with incomplete data. More explanations about Evidence Theory will be given in the Sect. 4.

Recently, approaches based on Fuzzy Theory have been followed to obtain choice models [3] or control strategies [20]. The correspondence of fuzzy models with generalized logit models has been studied [12]. Also users' behaviour in presence of information has been modelled through a set of *if*.... .*then* rules [18].

In this paper, the influence of information provision on drivers' behavior is examined in detail, according to Uncertainty-based Information Theory. In particular, Evidence Theory is used to obtain, by means of basic probability assignment functions, a relation between released and actually received information sets. Since several studies point up that the drivers' trust in information services is a function of Uncertainty, a relation between uncertainty and trust level has been developed; finally, the effects of information are illustrated through a numerical example.

2 Decision Making Mechanism and Processes: A Brief Overview

First of all, a distinction between static and dynamic choice has to be made: in the static case, mode, route and time departure are decided in pre-trip planning; decision is influenced by historical experiences and by information provided before the trip. In the dynamic case, en-route switching choice is made according to current perception of dynamic conditions of network.

In both cases, travelers can be influenced in making choices both by day-to-day variations in travel time and by their capabilities of learning from personal experience [6], as well as from available information. Three sources of information are in general used to make routing decision [1]:

- historical experiences;
- current perceptions of the network conditions;
- information acquired via an informative system.

It is worth noting that within-day decision process is influenced by values of traffic and network characteristics at a given instant in a given day. Models of within-day pre-trip route selection and en-route path switching have been developed by [19], and [2].

Evolution of drivers' spatial knowledge over time and in particular day-to-day is better described by dynamic formulation; nevertheless, there are some issues regarding spatial cognition and path choice:

- path choice between two points in a network depends on mental representation of the network, that is, on how travelers imagine streets are linked to each other. Usually there are maps providing this kind of information. Missing these maps, the knowledge of a network can be obtained by making repeated trips through the network and developing a mental map. Notion of cognitive maps was discussed by [24] with respect to decision making and design of informative system;
- day-to-day and within-day variations of traffic patterns are experienced by users. Drivers repeating trips through the network are able to understand these variations. Travelers without experience have to learn traffic dynamics through informative systems;
- an experience increase can change behavioural tendencies. Over time, experienced people mature and some attitudes with respect to routing behaviour may change.

3 Information as a Counterpart of Uncertainty

Given a generic finite set S of messages, the measure of uncertainty related to this set was developed at first by Hartley [11] and had the form:

$$U(S) = a \cdot \log b|S| \tag{1}$$

where:

- $|S|$ is the cardinality of S;
- a and b are positive constants ($a > 0,\ b > 1$) that determine the uncertainty unit of measure.

Assuming $a = 1$ and $b = 2$,

$$U(S) = \log 2|S| \tag{2}$$

and total uncertainty is expressed in bits.

Consider now two states A and B and the corresponding uncertainties $U(A)$ and $U(B)$ with $U(A) > U(B)$; information $I(A,B)$ which produces the transition from A to B is defined as:

$$I(A,B) = U(A) - U(B)$$

or, in terms of Hartley function:

$$I(A,B) = \log 2(|A|/|B|) \tag{3}$$

If information eliminates all alternatives except one, $|B| = 1$ and

$$I(A,B) = \log_2(|A|) = U(A) \tag{4}$$

Then, from the relation (4) we can derive that Information and Uncertainty are two facets of the same problem, and they can be handled through the same mathematical tools.

Also another classical measure of Uncertainty developed by Shannon in 1948, calculated Uncertainty as Information associated with a message x_k:

$$I_k = -\log_2 P\{x_k\}\ ,$$

where $P\{x_k\}$ is the probability associated with the selection of message x_k. It was solely a probabilistic approach, since from a significant standpoint the actual message is one selected from a set of possible messages. The average Information (Uncertainty) is:

$$I = -\sum_{k=1}^{n} P\{x_k\} \log_2 P\{x_k\} \tag{5}$$

This measure, so-called Shannon entropy, has been applied in any case as long as Uncertainty was faced only by Probability Theory. Afterwards, in consequence of wider studies, it became clear that Uncertainty is a multidimensional concept, and different mathematical frameworks were added to this type of measure to deal with different types of Uncertainty. In particular, the usefulness of Uncertainty-based Information Theory has been demonstrated [4, 9], and a modelling framework based on concepts of approximate reasoning was developed in the 1970s.

4 Handling Uncertainty Through Evidence Theory

The approximate reasoning is based on the hypothesis that, given the proposition "x is S", the evidences (information) about x support only partially the predicate S. The problem in this model is to measure the truth of the proposition, that is how much evidences support the alternative expressed in the predicate.

Basically, there are three patterns in dealing with this problem (Fig. 1):

- each element of evidence supports one and only one alternative;
- evidence supports more than one alternative (nested alternatives);
- combination of cases 1 and 2.

In the first case, Probability Theory is the appropriate mathematical framework to measure Uncertainty; in the second case, Possibility Theory; in the third one, Evidence Theory (Dempster – Shafer Theory). Evidence Theory subsumes both Probability and Possibility Theory. It provides the most general definition of the Uncertainty measure [22].

Let:

- X be a universal set;
- $P(X)$ be its the power set;
- A and B be two generic subsets of X.

A Belief measure is a function

$$\text{Bel: } P(X) \rightarrow [0,1]$$

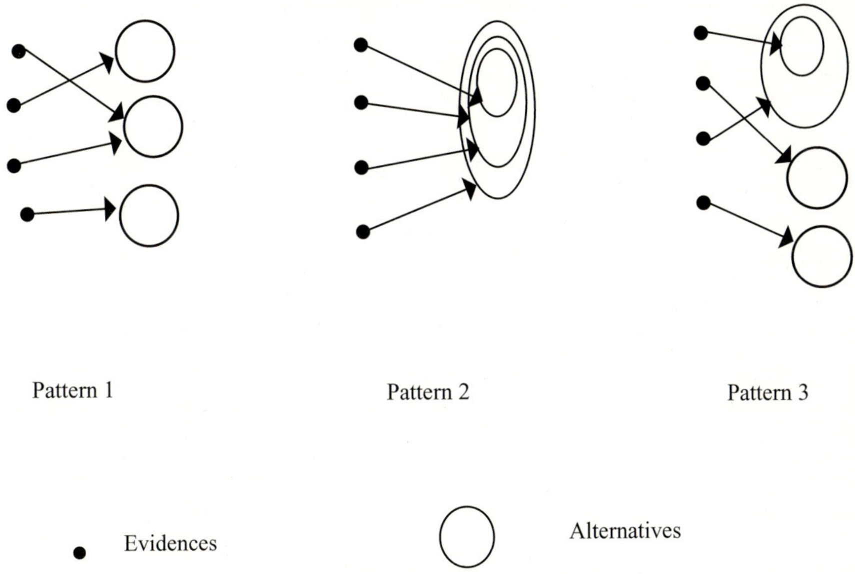

Fig. 1. Evidences Alternative Relationship

$$\text{Bel} : P(X) \to [0,1]$$

satisfying the following axioms:

boundary conditions: $\text{Bel}(\emptyset) = 0$, $\text{Bel}(X) = 1$

monotonicity:

$$A \subseteq B \Rightarrow \text{Bel}(A) \leq \text{Bel}(B)\ \forall\, A, B \in P(X) \tag{6}$$

continuity: $\lim_{i\to\infty}(A_i) = \text{Bel}(\lim_{i\to\infty} A_i)\ A_i \in P(X)$ and the additional axiom:

$$\begin{aligned}\text{Bel}(A_1 \cup A_2 \cup \ldots A_n) \geq \Sigma_i\, \text{Bel}(A_i) - \Sigma_{i<j}\, \text{Bel}(A_i \cap A_j) + \cdots . \\ + (-1)^{n+1}\, \text{Bel}(A_1 \cap A_2 \cap \ldots . A_n)\end{aligned} \tag{7}$$

The first axiom states that an element definitely does not belong to empty set and definitely belongs to universal set, which contains all elements under consideration in each particular context.

The second one requires the Belief that an element belongs to a set must be at least as great as the Belief that same element belongs to any subset of that set. The third axiom is valid only for infinite sets and requires that Belief is a continuous function. Finally, the axiom (7) is a weaker version of the additivity axiom of Probability Theory. It represents the main difference between these two theories. In fact, in Probability Theory axiom (7) is:

$$P(A_1 \cup A_2 \cup \ldots A_n) = P(A_1) + P(A_2) + \cdots P(A_n)\ ,$$

because in this case each element of evidence supports one and only one alternative. Belief and Plausibility merge in the unique measure of Probability.

With Belief, associated measure of Plausibility can be defined as:

$$\text{Pl}(A) = 1 - \text{Bel}(\sim A)\ \forall\, A \in P(X)$$

This measure is a function:

$$\text{Pl}: P(X) \to [0,1]$$

satisfying axioms (6) and the additional one:

$$\begin{aligned}\text{Pl}(A_1 \cap A_2 \cap \ldots A_n) \leq \Sigma_i\, \text{Pl}(Ai) - \Sigma_{i<j}\text{Pl}(A_i \cup A_j) + \cdots . \\ + (-1)^{n+1}\, \text{Pl}(A_1 \cup A_2 \cup \ldots . A_n)\end{aligned} \tag{8}$$

Belief and Plausibility measures can be expressed in terms of a function:

$$m : P(X) \to [0,1]$$

such that

$$\begin{aligned} m(\emptyset) &= 0 \\ \sum_{A\in P(X)} m(A) &= 1 \end{aligned} \tag{9}$$

According to (9), $m(A)$ can be thought as the degree of evidence, or the degree to which we believe, that an element of X belongs to the set A but not to any special subset of A.

Since (9) is similar to equation of Probability distribution, the function m is usually called "basic probability assignment" and expresses the proportion to which an element of X belongs to a generic subset A of $P(X)$. Note that, because of ignorance or incomplete information, $m(A)$ pertains only to the set A and does not regard subsets of A. Sets $A \in P(X)$ for which $m(A) \neq 0$ are called "focal elements".

Given a basic probability assignment, Belief and Plausibility measures are obtained by the formulas:

$$\mathrm{Bel}(A) = \sum_{B \subseteq A} m(B) \tag{10}$$

$$\mathrm{Pl}(A) = \sum_{B \cap A \neq \emptyset} m(B) \tag{11}$$

When subsets are nested like in the pattern 2 of Fig. 1, the degrees of evidence allocated to focal elements should not conflict with each other; the associated Belief and Plausibility measures are called consonant, and have the following properties:

$$\mathrm{Bel}(A_1 \cap A_2) = \min[\mathrm{Bel}(A_1), \mathrm{Bel}(A_2)]\ \forall A_1, A_2 \in P(X)$$

$$\mathrm{Pl}(A_1 \cup A_2) = \max[\mathrm{Pl}(A_1), \mathrm{Pl}(A_2)]\ \forall A_1, A_2 \in P(X)$$

Consonant Belief and Plausibility measures are called Necessity and Possibility measures, respectively, then:

$$\mathrm{Nec}(A_1 \cap A_2) = \min[\mathrm{Nec}(A_1),\ \mathrm{Nec}(A_2)]$$

$$\mathrm{Poss}(A_1 \cup A_2) = \max[\mathrm{Poss}(A_1),\ \mathrm{Poss}(A_2)]\ .$$

Between Necessity and Possibility measures there is the dual relationship:

$$\mathrm{Nec}(A) = 1 - \mathrm{Poss}(\sim A)\ .$$

Now, let

X, Y be two universal sets;
A, B be two sets defined in X and Y, respectively;

a joint basic probability assignment is defined:

$$m : P(X \times Y) \to [0,1]$$

Let R be a relation on $X \times Y$; projecting it on X and Y, sets RX and RY are obtained as follows:

$$R_X = \{x \in X | (x,y) \in R \quad \text{for some } y \in Y\}$$
$$R_Y = \{y \in Y | (x,y) \in R \quad \text{for some } x \in X\}$$

In consequence, from sets R_X, R_Y and given a joint probability assignment m, marginal basic probability assignments m_X and m_Y can be calculated:

$$m_X(A) = \sum_{R|A=R_X} m(R) \ \forall A \in P(X) \tag{12}$$

$$m_Y(B) = \sum_{R|B=R_X} m(R) \ \forall B \in P(Y) \tag{13}$$

For example, consider two sets X and Y, whose elements are, respectively, released and perceived information. Note that, although released information can be crisp or vague, human mind always revises them in an approximate way; therefore, a relation between released and perceived information is always a fuzzy one, as represented in Fig. 2. In the same figure, the meaning of R_X and R_Y is represented as well.

The 45°-straight line represents the information released and correctly perceived. In this simple communication model, we call $RX \cap RY$ "information transmitted".

Assume that $X = \{a,b,c\}$ and $Y = \{\alpha,\beta,\gamma\}$, and that the matrix of joint basic probability assignments is the following one:

Table 1. Joint basic assignments

	$X \times Y$									
	αa	αb	αc	βa	βb	βc	γa	γb	γc	$m(R_i)$
$R_1 =$	0	0	0	0	1	1	0	1	1	0.0625
$R_2 =$	0	0	0	1	0	0	1	0	0	0.225
$R_3 =$	0	0	0	1	1	1	1	1	1	0.125
$R_4 =$	0	1	1	0	0	0	1	1	0	0.250
$R_5 =$	0	1	1	0	1	1	0	0	0	0.125
$R_6 =$	0	1	1	0	1	1	0	1	1	0.0375
$R_7 =$	1	0	1	1	1	0	0	1	0	0.175

Then, marginal basic probability assignments are:

$m_X(\{\alpha,\beta\}) = m(R_5) = 0.125;$
$m_X(\{\alpha,\gamma\}) = m(R_4) = 0.25;$
$m_X(\{\beta,\gamma\}) = m(R_1) + m(R_2) + m(R_3) = 0.4125;$
$m_Y(\{a\}) = m(R_2) = 0.225;$
$m_Y(\{b,c\}) = m(R_1) + m(R_5) + m(R_6) = 0.225.$

Given the information A, due to the fuzzy revision of human mind, the individual user perceives the information B|A, whose possible marginal basic probability assignment is represented in Fig. 3.

Now, let:

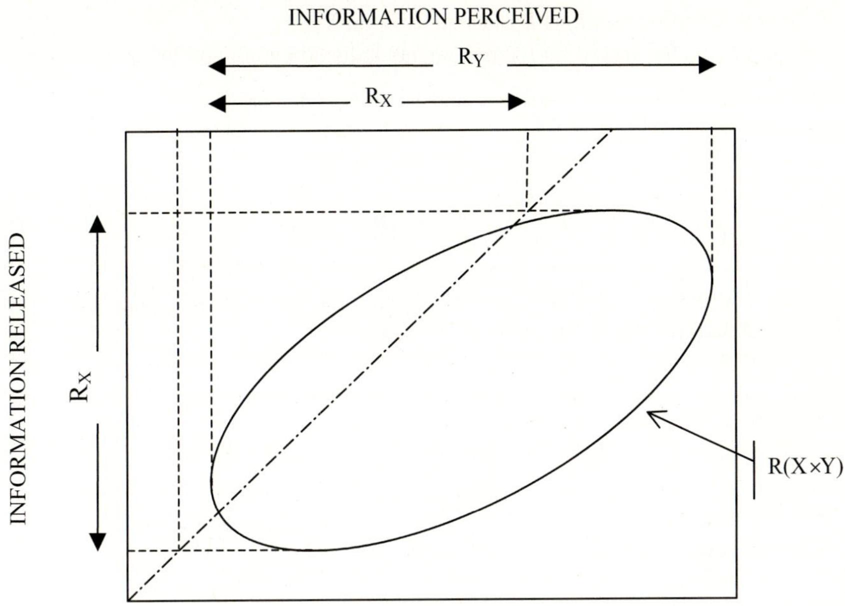

Fig. 2. Relationship between released and perceived information

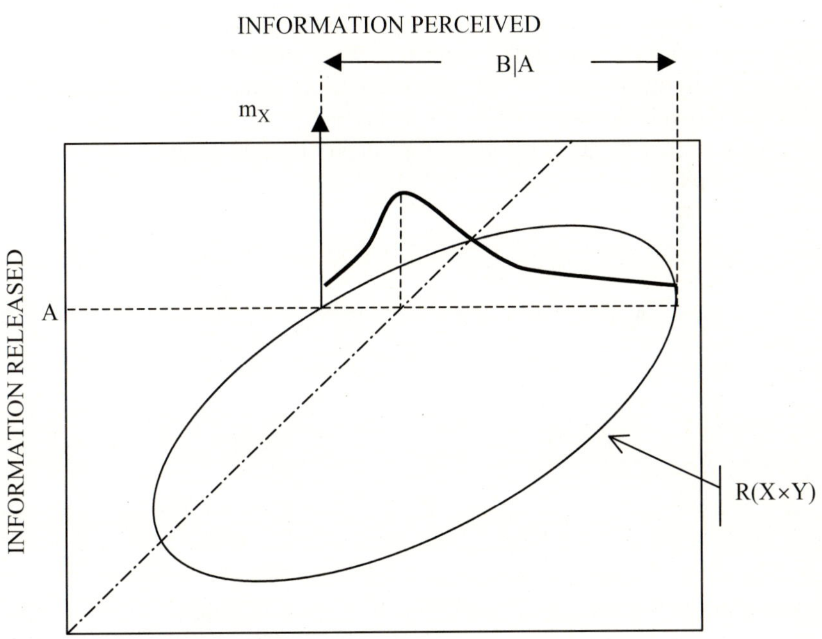

Fig. 3. Marginal basic probability assignment of information A

F be the set of all relations R induced by m on $X \times Y$;
F_X and F_Y be the sets of all focal elements induced by m_X and m_Y, respectively.

The Uncertainty related to the sets X and Y can be calculated as [17]:

$$U(X) = \sum_{A \in F_X} m_X(A) \cdot \log_2 |A| \tag{14}$$

$$U(Y) = \sum_{B \in F_Y} m_Y(B) \cdot \log_2 |B| \tag{15}$$

The relations (14) and (15) are called "simple uncertainties"; joint and conditional uncertainties can be also calculated as follows:

$$U(X,Y) = \sum_{A \times B \in F} m(A \times B) \cdot \log_2 |A \times B| \tag{16}$$

$$U(X|Y) = \sum_{A \times B \in F} m(A \times B) \cdot \log_2 \frac{|A \times B|}{|B|} \tag{17}$$

$$U(Y|X) = \sum_{A \times B \in F} m(A \times B) \cdot \log_2 \frac{|A \times B|}{|A|} \tag{18}$$

$U(X)$, $U(Y)$, $U(X, Y)$, $U(X|Y)$ and $U(Y|X)$ can be considered as some kind of measure associated with sets X and Y. Their meaning for the communication model is shown in Fig. 4.

$I(X,Y)$ is the so-called "information transmission" and represents, in this model, information somehow transmitted, while $U(Y|X)$ is residual uncertainty with respect to received information and $U(X|Y)$ is noise, unwanted information. The higher the value of information transmission is, the higher are quality and accuracy of informative system and the higher is the users' trust in provided information.

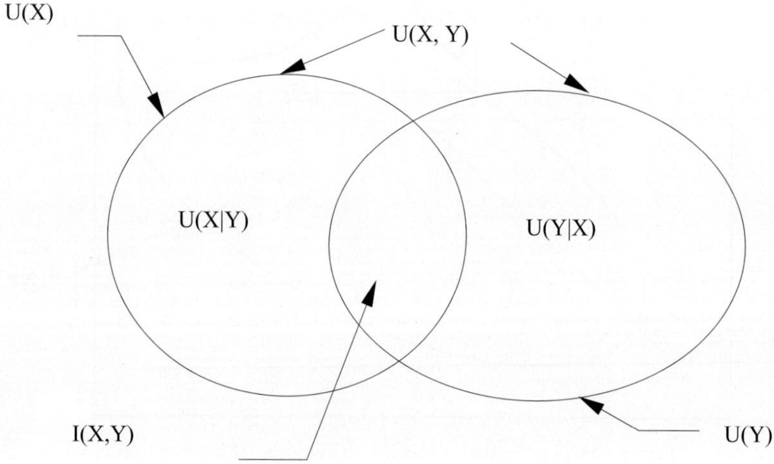

Fig. 4. Information transmission

If sets X and Y are disjoint, $I(X,Y) = \emptyset$; while if $X = Y, I(X,Y) = U(X)$, so that:

$$\emptyset \leq I(X,Y) \leq U(X) .$$

It is easy to see that:

$$I(X,Y) = U(X) + U(Y) - U(X,Y) \tag{19}$$

or

$$I(X,Y) = U(X) - U(X|Y) = U(Y) - U(Y|X) \tag{20}$$

Replacing (14) and (17) in (20), you obtain:

$$I(X,Y) = \sum_{A \in F_X} m_X(A) \cdot \log_2 |A| - \sum_{A \times B \in F} m(A \times B) \cdot \log_2 \frac{|A \times B|}{|B|} . \tag{21}$$

For any informative system with given noise characteristics, the maximum of information transmission, according to Shannon definitions, is its Capacity:

$$C = \max I(X,Y) \tag{22}$$

while the ratio $I(X,Y)/C$ is its Efficiency.

The problem of the system Capacity is a quite complex one, but an immediate consequence arises: increasing more and more the amount of released information would not be productive to improve knowledge of the transportation system, if the Capacity of informative system is reached.

5 Provision and Updating of Information

Urban traffic is becoming more and more congested in major city centers. Generally, countermeasures like expansion of existing road networks are infeasible, because of environmental impacts or physical structures of urban areas. Therefore, the research addresses to optimizing the use of existing roads. For this purpose, the introduction of new advanced information technologies seems at present one of the most promising tools.

Intelligent Transportation Systems (ITS) provide information both to users and to service providers to allow intelligent decisions. The strategies usually accepted for ITS are:

- Advanced Public Transportation Systems (APTS);
- Advanced Traveler Information Systems (ATIS);
- Advanced Traffic Management System (ATMS);
- Advanced Vehicle Control Systems (AVCS);
- Alternative Fuels (AF);
- Vehicle Safety Devices (SD) and
- Telecommuting (TC).

This paper evaluates the effects of information on private car drivers, so attention is confined to ATIS. An informative system, like ATIS, may provide information to users before they begin the trip (real-time pre-trip information) or while they are moving (real-time en-route information). Travelers combine this information with their own experience to obtain a prediction about the cost of each path and to choose the best one.

To incorporate information on the system conditions in the choice process, it is assumed that the user:

- has some experience about the attributes of the transportation system;
- uses information to update his experience;
- chooses an alternative according to his updated experience.

This kind of sequential model simulates updating of user knowledge through the following relationship [5, 13, 18]:

$$ET_{j,i}^{t+1} = \alpha_i \times I_j^{t+1} + (1 - \alpha_i) \times ET_{j,i}^{t} \tag{23}$$

where:

- $ET_{j,i}^{t+1}$ = cost perceived by i-th user for j-th alternative updated at time $t+1$;
- $ET_{j,i}^{t}$ = historical cost for i-th user and j-th alternative updated at time t;
- I_j^{t+1} = information on alternative j-th released by the system at time $t+1$;
- α_i = parameter representing the i-th user's trust in information released by the system, $\alpha_i \in [0,1]$.

The model (23) incorporates important aspects such as [18]:

- dynamic nature of information integration. The cost of an alternative at time $t+1$ is influenced by the historical cost (user's experience and memory) at time t;
- reliability of the informative system. The more reliable information is, the more important is the effect on the updated perception;
- non-linear relationship between information and updated perception. The parameter α itself is function of information, so that the perceived cost updated at time $t+1$ is a non-linear function of information;
- information quality. Additional vague information can lead to a more uncertain perception.

In this framework, it seems necessary to investigate more on the effect and the meaning of the parameter α. Experimental studies carried out in last years by different researchers [14, 23] have found different values, 0.2 to 0.7, for this parameter. Such range of variability could be justified assuming that α is affected by the level of uncertainty imbedded in information. Therefore we assume that:

- the user's trust in information decreases with the increasing of the uncertainty (negative elasticity);
- the need of additional information is proportional to the uncertainty.

On these hypotheses, the following relationship between α and the uncertainty level has been carried out:

$$(d\alpha/\alpha)/(dU/U) = -U \tag{24}$$

and hence:

$$\alpha = 1/\exp(U) \tag{25}$$

where U is the value of residual uncertainty about the perceived information, $U = U(Y.X)$.

Within the mathematical framework of the Evidence Theory, the conditional Uncertainty is expressed by (18); then, through the relations (23) and (25), the model simulates the updating of the costs of alternatives, according to the information released by the system.

Although the nature of the historical costs (experience) could be stochastic, in the updating processes such costs are elaborated by the human mind in terms of approximate values; therefore, also a possibilistic (fuzzy) choice model should be an appropriate framework to simulate the user's choice process [12].

6 Numerical Example

The Fig. 5 shows a link for the OD pair $i-j$; at the origin i there is an electronic device capable of displaying real-time traffic information, so-called Variable Message Signal (VMS). Assume that the users have the perception that the travel time for this OD pair lies in the range 12 to 22 minutes; or, in other words, they have the vague experience that the travel time related to the link $i-j$ is "approximately 17 min". In Fig. 6, the fuzzy set corresponding to this approximate value is depicted.

Through the VMS, an ATIS service can provide the drivers with the following real-time en-route information about the traffic conditions on the link:

(a) congestion;
(b) queue;
(c) road accident.

Assume additionally that drivers perceive those propositions like:

1. "travel time is 20 to 26 min";
2. "travel time is 22 to 26 min";
3. "travel time is 26 to 42 min".

The allocation of propositions is represented in Fig. 7, while Table 2 shows the joint basic probability assignments for focal elements.

Hence,

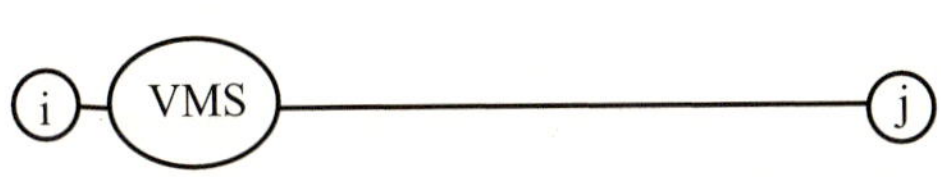

Fig. 5. Example link

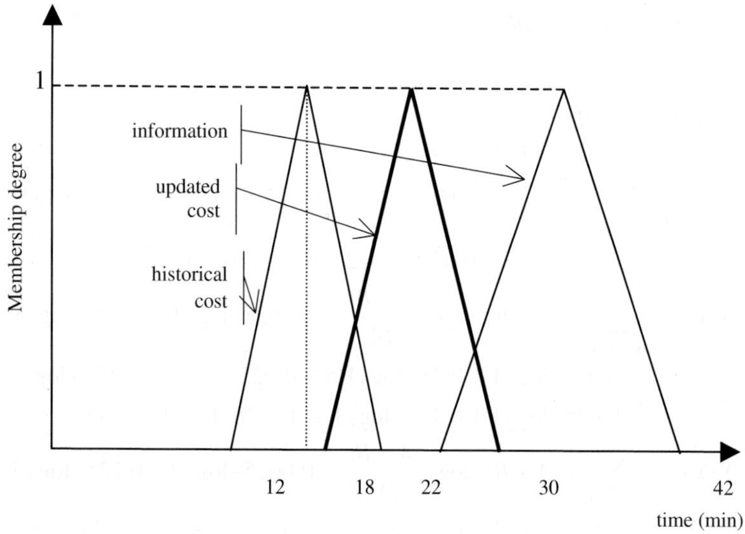

Fig. 8. Comparison between historical and updated costs

$$m(A_i) = r(x_i) - r(x_{i+1})$$

where $r(x_i)$ is the Possibility value for i-th element of a set ordered in decreasing order, for which $r(x_i) \geq r(x_{i+1})$.

Then, the Uncertainty is expressed by:

$$U = \sum [r(x_i) - r(x_{i+1})] \cdot \log_2 i \tag{26}$$

Calculating Uncertainty through the (26), for historical cost it results: $U = 1.98$; for updated cost: $U = 2.22$; for information: $U = 2.62$.

Applications of this methodology can be implemented for management of transportation networks: usually path choice models assume that users make choices comparing the costs of different alternatives; therefore, different additional costs could be provided through different VMS messages, so that it could be used as a kind of road pricing tool.

7 Conclusions

In this paper, the influence of Uncertainty in updating the knowledge of attributes of a transportation system, namely the expected travel time on a link, has been examined in detail considering the ATIS environment. The presented model points out the relevant role of the Evidence Theory in calculating the conditional Uncertainty and then users' trust level in information. In consequence, the importance of conditional Uncertainty in updating knowledge of the system results evident.

Through the Evidence Theory a modeling framework, which represents the uncertainties imbedded in the perception of travel attributes, has been developed. The model allows the quantitative calculation of the user's trust in information and, then, quantitative updating of expected travel time. In a wider framework, the outcomes of this paper can be used to carry out a path choice model that quantitatively takes into account information provision. Moreover, different additional costs can be calculated as perceived by users according to different messages provided by VMS, so that it could be used as a kind of road pricing tool.

References

1. J. L. Adler and V. J. Blue (1998) Toward the design of intelligent traveler information systems. Transportation Research 6-C157-172.
2. J. L. Adler W. W. Recker and M. G. A McNally (1992) Conflict Model and Interactive Simulator (FASTCARS) for Predicting Enroute Assessment and Adjustment Behavior in Response to Real-Time Traffic Condition Information. 71st TRB Annual Meeting.
3. T. Akiyama and H. Tsuboi (1996) Description of Route Choice Behaviour by Multi-Stage Fuzzy Reasoning. Proceedings of the Highways into The Next Century, Hong Kong 739–746.
4. W. R. Ashby (1972) Systems and Their Informational Measures. in: G. J. Klir ed., Trends in General Systems Theory (Wiley-Interscience, New York) 78–97.
5. M. Ben Akiva A. De Palma and I. Kaysi (1991) Dynamic models of peak period traffic congestion. Transportation Research 18-B(4/5), 339–355.
6. P. H. L. Bovy and E. Stern (1990) Route Choice: Wayfinding in Transportation Networks, Kluwers Academic Publishers, Boston.
7. E. Cascetta (1989) A stochastic process approach to the analysis of temporal dynamics in transportation networks. Transportation Research 23-B(1), 1–17.
8. E. Cascetta and G. E. Cantarella (1991) A day-to-day and within-day dynamic stochastic assignment model, Transportation Research 25-A(5), 277–291.
9. R. Conant (1981) Efficient Proofs of Identities in N-Dimensional Information Theory, Cybernetica 24 (3), 191–197.
10. C. Daganzo and Y. Sheffi (1977) On stochastic models of traffic assignment, Transportation Science, 253–274.
11. R. V. L. Hartley (1928) Transmission of Information. The Bell System Technical Journal **7**, 535–563.
12. V. Henn (1997) Fuzzy route choice model for traffic assignment, Paper presented at the 9th Mini-EURO Conference, Budva (Yugoslavia).
13. J. L. Horowitz (1984) The stability of stochastic equilibrium in a two-link transportation network, Transportation Research 18-B(1), 13–28.
14. Y. Iida T. Akiyama and T. Uchida (1992) Experimental analysis of dynamic route choice behavior. Transportation Research 26-B(1), 17–32.
15. R. Jayakrishnan H. Mahmassani and T. Y. Hugh (1994) An evaluation tool for Advanced Traffic Information and Management System in urban networks, Transportation Research C 2 (3) 129–147.
16. A. J. Khattak J. L. Schofer and F. S. Koppleman (1991) Commuters' Enroute Diversion and Return Decisions: IVHS Design Implications. Proc. of International Conference on Travel Behavior, Quebec City. Quebec, Canada.

17. G. J. Klir and T. A. Folger (1988) Fuzzy Sets, Uncertainty and Information, Prentice-Hall, Englewood Cliffs, New Jersey.
18. T. Lotan and H. Koutsopoulos (1993) Approximate reasoning models for route choice behavior in the presence of information, in: C. Daganzo ed., The 12th International Symposium on Transportation and Traffic Theory (ISTTT), Berkeley, CA, 71–88.
19. H. S. Mahmassani and D. O. Stephan (1988) Experimental investigation of route and departure time choice dynamics of urban commuters, Transportation Research Record 1203, 63–83.
20. J. Niittymäki, and M. Pursula (1997) Signal-Group Control Using Fuzzy Logic, Paper presented at the 9th Mini-EURO Conference, Budva (Yugoslavia).
21. V. Perincherry, and S. Kikuchi (1997) A planning model for large-scale infrastructure under uncertainty. Paper presented at the 9th Mini-EURO Conference, Budva (Yugoslavia).
22. G. Shafer (1976) A mathematical Theory of Evidence. Princeton University Press, Princeton, New Jersey.
23. K. Vaughn, M. Abdel-Aty, R. Kitamura, P. Jovanis, and H. Yang (1992) Experimental analysis and modeling of sequential route choice behavior under ATIS in a simplistic traffic network, Research Report UCD-ITS-RR-92-16, Institute of Transportation Studies, University of California, Davis CA.
24. M. Wenger, J. Spyridakis, M. D. Haselkorn, W. Barfield, and L Conquest, (1990) Motorist behavior and the design of motorist information systems, Transportation Research Record 1281, 159–167.

Part III

Evolutionary Algorithms I

Interactive Evolutionary Computation in Identification of Dynamical Systems

Janos Abonyi, Janos Madar, Lajos Nagy and Ferenc Szeifert

University of Veszprem, Department of Process Engineering, Veszprem, P.O. Box 158, H-8201, Hungary
abonyij@fmt.vein.hu

Summary. In practical system identification it is often desirable to simultaneously handle several objectives and constraints. In some cases, these objectives and constraints are often non-commensurable and the objective functions are explicitly/mathematically not available. In this paper, Interactive Evolutionary Computation (IEC) is used to effectively handle these identification problems. IEC is an optimization method that adopts evolutionary computation (EC) among system optimization based on subjective human evaluation. The proposed approach has been implemented in MATLAB (EAsy-IEC Toolbox) and applied to the identification of a pilot batch reactor. The results show that IEC is an efficient and comfortable method to incorporate a priori knowledge of the user into a user-guided optimization and identification problems. The developed EASy-IEC Toolbox can be downloaded from the website of the authors: http://www.fmt.vein.hu/softcomp/EAsy.

Key words: System identification, Interactive Evolutionary Computation

1 Introduction

The main objective of system identification is to identify a model with good prediction capabilities in the sense that it is able to accurately predict the system's response to a given class of excitations. Hence, a common identification objective is to minimize the mismatch between model prediction and observed data. However, it is often desirable to introduce additional objectives and constraints into the identification problem. There are several reasons for this [7]: In most practical cases the data sequence used for identification may be incomplete or uncertain, the model structure is not necessarily known and can be overparameterized, there may be some properties that we may want the model to have. Furthermore, there is often additional information available, and it makes sense to use this information to improve the accuracy and validity of the model [1]. Multi-objective system identification tools allow the incorporation of these objectives into the identification procedure. E.g., single constraints or penalties on the parameter space have been suggested, for example to ensure stability [1, 24], fulfillment of balance equations and steady-state data [23], and explicit belief about parameter values [10, 14]. These multi-objective identification problems

Janos Abonyi et al.: *Interactive Evolutionary Computation in Identification of Dynamical Systems*, Advances in Soft Computing **1**, 73–84 (2005)
www.springerlink.com

involve the simultaneous consideration of multiple performance criteria. These objectives are often non-commensurable and are frequently in conflict with one another [7]. Trade-offs exist between some objectives where advancement in one objective will cause deterioration in another. In most cases the cost function is defined prior to the optimization procedure; this requires in-depth information concerning the various trade-offs and valuation of each individual objective. However, in some cases, the objective functions are explicitly/mathematically not available and the identification problem can be decomposed only for large-scale problems.

This paper proposes the application of Interactive Evolutionary Computation (IEC) [21] to handle these identification problems, and its application to the identification of tendency models of batch reactors. In the proposed framework, human users evaluate the performances of the individuals (set of solutions) by selecting the best individuals. The human user does not (only) evaluate the mean square fitting error, or time constant of the model but he or she can simultaneously analyze these numerical values with the plotted dynamic and/or steady-state behavior of the model, e.g. step-responses, gains, etc. The proposed approach has been implemented in MATLAB (EAsy-IEC Toolbox) and applied to the identification of a pilot batch reactor.

The remaining part of the paper is organized as follow. In Sect. 2 the algorithm of Evolutionary Strategy is presented, which is tailored taking into account the specialities of IEC. Section 3 shows an application example when the proposed approach is applied to identification of a tendency model of a heating-cooling jacket. Finally, some conclusions are drawn in Sect. 4.

2 Description of the Algorithm

2.1 Introduction to Evolutionary Algorithms

Evolutionary Algorithms (EAs) [18] are stochastic optimization algorithms that mimic the process of natural selection. Evolutionary Algorithms handle several candidate solutions simultaneously. Thus, unlike conventional algorithms, EAs start form a pool of points (candidate solutions), usually referred to as individuals. Every individual represents a point in the search space. Every individual has a fitness value, which is calculated by the evaluation function. The fitness value determines the successfulness of the given individual appropriate to the objectives. The next generation of individuals is generated from the current population using genetic operators: selection, mutation and crossover. Likes in the natural world, the more successful individuals, which have bigger fitness values, have higher probability to transmit their genes into the new generation. Because the genetic operators are stochastic, EAs are stochastic optimization methods.

In the initialization step, EAs generate individuals randomly. In every evolutionary-loop, all of the individuals are evaluated and their fitness values are calculated based on an objective function. After that the selection operator selects randomly the best individuals to constitute the next generation. The crossover (recombination) operator generates two new individuals by taking two selected individuals and recombining them. The mutation operator changes randomly the individuals. Finally the old individuals are replaced by the new individuals.

Generally, EAs are automated algorithms using a given objective function to calculate the fitness values of the individuals and the selection operator performs automatically the selection based on these calculated fitness values (for example using roulette wheel selection). But sometimes instead of a given objective function, a human user can make the valuation of the individuals, which results in an interactive optimization.

Interactive Evolutionary Computation (IEC) is an optimization method that adopts evolutionary computation among system optimization based on subjective human evaluation [21]. In contrast to the automated evolutionary optimization algorithms, IEC cannot use many individuals and searching generations because of resulting human fatigue. Hence, there is a need to develop and tune an algorithm which effectively search with a few individuals within a few searching generations. For this purpose, this paper presents an algorithm based on Evolutionary Strategy (ES). In the following the IEC and the ES will be described and then the structure of the developed IEC-Tune Toolbox will be shown.

2.2 Interactive Evolutionary Computation

Most optimization techniques which work by improving a single solution step by step are not suited for interactive optimization approach. For example, in simple hill-climbing a user suggestion can either be accepted, which means the loss of results gained so far, or denied, which may lead the algorithm a wrong way. Evolutionary Algorithms and other population based optimization procedures are better suited for interactive optimization, because a suggestion can be put into the population to see whether it is worth further consideration or not. A good suggestion can be expected to survive and lead the optimization process towards a good solution, while a bad suggestion will be eliminated quite soon. This approach become well known as Interactive Evolutionary Computation (IEC) [19, 20].

Simply stated, IEC is a technique from the class of evolutionary algorithms (EAs), in which the fitness function is replaced by a human user. As in interactive evolution, the user selects one or more individual(s) which survive(s) and reproduce(s) (with variation) to constitute a new generation. Hence IEC uses two different spaces for its search: the human user evaluates the output of the target system according to the distance between the target goal and the system output in psychological space [20], on the other hand, the EA searches in the parameter space.

These techniques were applied already in computer graphics [9], animation, creating forms, textures, and motion [2]. Potential applications of interactive evolution include artificial life design, e.g., development of components of biological nature [6, 17] and engineering construction design [13].

Although, IEC has been applied in knowledge acquisition and data mining, according to our best knowledge, the applicability of this approach in system identification has not been studied yet. However, in recent years, the application of Evolutionary Computing to black-box and grey-box model identification has received considerable interest since the seminal paper by Kristinsson and Dumont [8]. Subsequent evolutionary system identification applications are applied to build dominant structural identification with local parametric tuning without the need of a differentiable performance index in the presence of noisy data. As the overview of Fleming [5] illustrates evolutionary techniques provide an excellent fitting performance and is capable of accommodating multiple objectives such as to examine the relationships between model complexity and fitting accuracy during the model building process. These approaches can be extended to the identification of gray-box models, to use of a priori knowledge [22].

2.3 Evolution Strategy

In contrast to automated evolutionary optimization algorithms, the IEC cannot use many individuals and searching generations because the human fatigue. Hence, there is a need to develop an algorithm which can effectively search with a few individuals within a few searching generations. For this purpose, this paper presents an algorithm based on Evolutionary Strategy. Evolution Strategy (ES) was developed by [15], with selection, mutation, and a population of size one. Schwefel [16] introduced recombination and populations with more than one individual, and provided a nice comparison of ES with more traditional optimization techniques. Evolutionary Strategy are typically applied to real-valued parameter optimization problems. The characteristic feature of ES lies in the self-adaptation of the standard deviation of the Gaussian distribution used in the mutation [3]. The basic idea is to add these adaptation parameters to the genotype and have them undergo evolution themselves. The main elements of the algorithm used in this paper are the following:

Representation

An ES-individual $\mathbf{a}_j = (\mathbf{x}_j, \sigma_j)$ consists of two components: the object variables $\mathbf{x}_j = [x_{j,1}, \ldots, x_{j,n}]$ and the strategy variables $\sigma_j = [\sigma_{j,1}, \ldots, \sigma_{j,n}]$. The strategy parameters allow the algorithm to adapt to the topology of the objective function by controlling the mutation step size.

Mutation

As small changes occur frequently but large ones only rarely in the nature, the mutation operator adds normal distributed random numbers $z_{j,i} \sim N(0, \sigma_{j,i})$ to the individuals:

$$x_{j,i} = x_{j,i} + z_{j,i}\,. \tag{1}$$

Before the object variables are changed, the standard deviations are mutated using a multiplicative normally distributed process:

$$\sigma_{j,i}^{(t)} = \sigma_{j,i}^{(t-1)} \exp(\tau' N(0,1) + \tau N_i(0,1)) , \tag{2}$$

with $\exp(\tau' N(0,1))$ as a global factor which allows an overall change of the mutability and $\exp(\tau N_i(0,1))$ allowing for individual changes of the mean step sizes $\sigma_{j,i}$. The τ' and τ parameters can be interpreted in the sense of global learning rates. Schwefel suggests to set them as [16]:

$$\tau' = \frac{1}{\sqrt{2n}}, \quad \tau = \frac{1}{\sqrt{2\sqrt{n}}} . \tag{3}$$

Recombination

Recombination in ESs can be either sexual, where only two parents are involved in the creation of an offspring, or global, where up to the whole population contributes to a new offspring. Sexual recombination of just two individuals is often called local while the contribution of all individuals is called global recombination. Traditional recombination operators are discrete recombination, intermediate recombination, and geometric recombination, all existing in a sexual and global form. When F and M denote two randomly selected individuals from the μ parent population, the following operators can be defined:

$$x_i' = \begin{cases} x_{F,i} & \text{no recombination} \\ x_{F,i} \text{ or } x_{M,i} & \text{discrete} \\ (x_{F,i} + x_{M,i})/2 & \text{intermediate} \\ \sum_{k=1}^{\mu} x_{K,i}/\mu & \text{global avarage} \end{cases} \tag{4}$$

$$\sigma_i' = \begin{cases} \sigma_{F,i} & \text{no recombination} \\ \sigma_{F,i} \text{ or } x_{M,i} & \text{discrete} \\ (\sigma_{F,i} + \sigma_{M,i})/2 & \text{intermediate} \\ \sqrt{(\sigma_{F,i}\sigma_{M,i})} & \text{geometric} \\ \sum_{k=1}^{\mu} \sigma_{K,i}/\mu & \text{global avarage} \end{cases} . \tag{5}$$

Throughout this work discrete recombination of the object variables and global intermediate combination of the strategy parameters were used.

Selection and Replacement

In the applied $(\mu + \lambda)$ Evolution Strategy, at a given generation (t), there are μ parents, and λ offspring are generated by recombination and mutation. Then, the $\mu + \lambda$ members of then next population are sorted according to their objective function values given by the user, and he or she selects the best μ to become the parents of the next generation.

2.4 IEC Integration to MATLAB

IEC systems seek to interface EA to human preference in order to create systems capable of evolving artifacts that require a human expertise that has not yet succumb to computation. Hence, the interfacing of human ability with machine computation requires resolving difficult issues [12]. Further, progress in the design of interactive evolutionary systems allows a glimpse into how very human abilities such as intuition, projection, and holistic perception interplay with the mechanics of machine computation. This section reports on one such interactive evolutionary system that seeks to combine human perception with the previously presented evolutionary strategy to evolve the parameters of MATLAB and Simulink models (see Fig. 1).

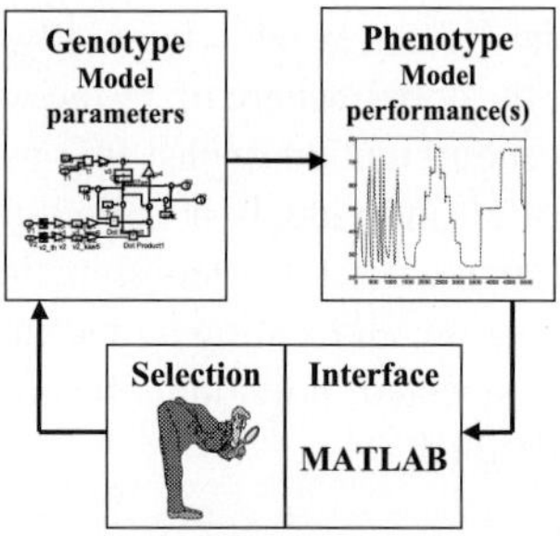

Fig. 1. Scheme of the Proposed IEC System Integrated to MATLAB

The EAsy-IEC Tune Toolbox is downloadable from the website of the the author: www.fmt.vein.hu/softcomp/EAsy, is designed to be applicable for different types of optimization problems (e.g., system identification, controller tuning, data mining).

The number of displayed individuals, $\mu + \lambda$ is set to 9, which can be displayed spatially. The number of searching generations is limited to 20 generations at the most, due to the fatigue of human operators. In the proposed framework human users evaluate the performances of the individuals (set of solutions). Not the genotypes and phenotypes are applied directly, but rather the output of the target system realized by the individuals. For example, the user does not (only) evaluate the mean square fitting error, or time constant of the model but he or she can simultaneously analyze these numerical values with the plotted dynamic and/or steady state behavior of the model, e.g. step-responses, gains, etc. An example of such evaluation figure is given in Fig. 4. Due to small number of individuals, the human evaluation was restricted to selecting the individuals which will survive and constitute the next generation.

3 Application Example

3.1 Process Description

In recent years our department focuses to the development of control strategies of multipurpose production plants. Hence, a prototype of pharmaceutical process

Fig. 2. The stirred tank reactor with heating-cooling jacket

systems was designed and installed in our laboratory. The central element of this process unit is a 50 liter stirred reactor with heating-cooling jacket (see Fig. 2).

This paper focuses to the identification of the tendency model of the heating-cooling jacket. Through the jacket the direct heating-cooling system allows cooling with chilled water or heating with steam in steam or hot water mode. In the hot water heating mode the water is circulated by a pump, the steam is introduced through a mixer, while the excess water is removed through an overflow.

3.2 Tendency Model of the Heating-Cooling Jacket

The jacket and the circulating liquid can be described using the common lumped parameter enthalpy or heat balances given in the chemical engineering literature. In the model, zero-volume distributors and mixers and are applied and the overflow as well as the feed of the steam and the fresh cooling water are taken into account. The obtained simplified first principle model can be regarded as the tendency model [4] of the most important phenomena and its scheme is given on Fig. 3.

$V1$ and $V2$ are the valve positions (0%–100%), TH is the temperature of the cooling water coming from the environment, TF is the equivalent steam temperature calculated from the boiling temperature, the latent heat and specific heat. The valve characteristics are given in form of second order polynomials ($VALVE1$ and $VALVE2$). The two first order transfer functions ($\frac{1}{T1s+1}$ and $\frac{1}{T2s+1}$) are obtained from the lumped parameter model of the jacket and the model of the thermometer at the jacket inlet. The $K1, K2$, and $K3$ are gains and the DT is the dead-time.

The identification problem is identifying these parameters of the model (gains, time constants, gains, dead-time and parameters of second order polynomials). For this purpose open-loop experiments were conducted on the whole operating range of the composite manipulated variable u (because of the split-range control: $V1$ and $V2$ are not opened at the same time, $u = 100\%$ means that $V1$ is opened fully, $u = 0\%$ means that $V2$ is opened fully).

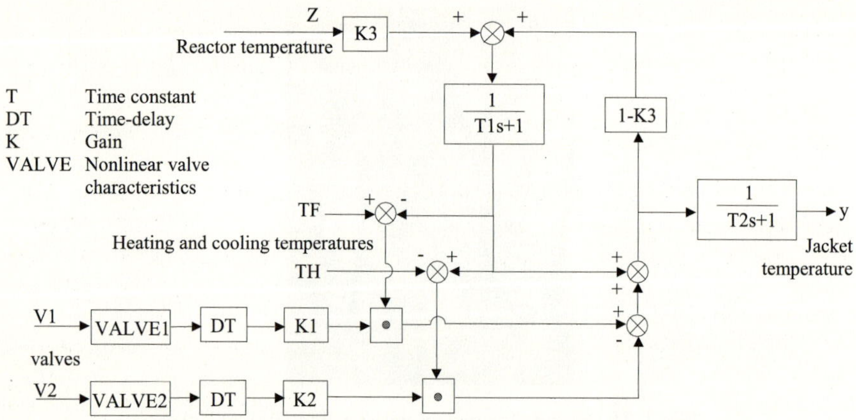

Fig. 3. Tendency model of the jacket of the stirred ractor

3.3 Results

Based on the collected input-output data, the identification problem can be formulated as an optimization problem and can be solved using common nonlinear optimization methods (e.g. SQP). To formulate an optimization problem we need a cost-function which is minimized under the optimization procedure. Usually for this purpose the mean square model error is used, $MSE = \sum_{k=1}^{n} (y(k) - \tilde{y}(k))^2/n$, where y is the observed output, $\tilde{y}$ is the predicted output of the model, n is the number of observation.

Beside the MSE, other objectives should be considered. E.g. the steady-state prediction error and the characteristics of the valves. But it is difficult to formulate these objectives and it is difficult to select the weights of the objectives in the cost-function. Hence it is problematical to introduce these additional objectives into the identification problem. The IEC handles this problem, because the human user can consider and balance these objectives.

For comparison, beside the proposed IEC optimization scheme, the direct minimization of the MSE prediction error is considered. For this purpose, Sequential Quadratic Programming (SQP) was used. The SQP method is a standard general purpose algorithm for solving NLP optimization problems. It is a gradient based method which calculates the search direction vector from the quadratic approximation of the objective function around the actual local point [11]. The SQP based identification method minimizes automatically the cost-function subject to the parameters of the model. The cost-function was evaluated by simulating the model and calculating the MSE.

In contrast to SQP based method, the user evaluates the model in the IEC based identification method. Firstly, the identified model was simulated and then the human user selected the best solutions based on the dynamic and the near steady-state behavior of the system, and the identified characteristic of the valves. Beside these

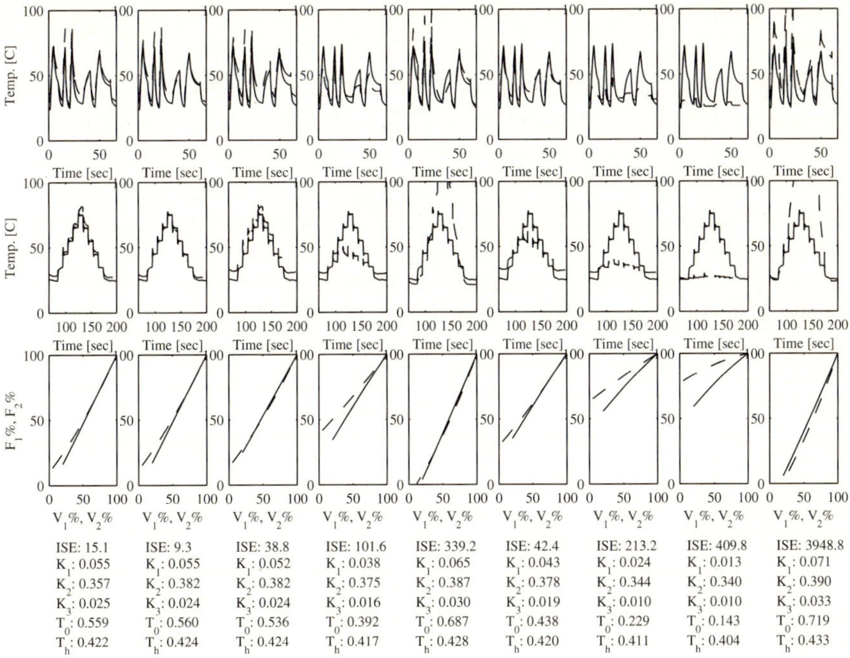

Fig. 4. MATLAB Evaluation figure in the IEC-Tune toolbox

plots, some numerical values were also given to the user to glimpse. Such an evaluation display is depicted in Fig. 4.

From an arbitrary initial value, the application of IEC resulted in 5.0 MSE prediction error after 14 generations. This corresponds to 146 evaluations of the model. If the same numbers of function calls are applied in SQP, the obtain model results in 26.0 MSE. The drawback of IEC is that is slower than the SQP method due to the time-requirement of the human evaluation. Hence while the IEC procedure consumed about 8.5 minutes, the SQP consumed only 2.5 minutes. (See the Table 1 and Figs. 5, 6.)

Table 1. Comparative results

Method	Evaulation Number	Time (sec)	MSE
IEC	146	508	5.0
SQP	149	150	26.0
SQP	506	501	1.5

Our experiences and these results show that the proposed IEC approach is extremely useful to find good initial conditions to SQP that is sensitive to stacking in local minima.

11. M. Papadrakakis N.D. Lagaros and G. Kokossalakis. Structural optimization using evolutionary algorithms. *Computers and Structures*, 80, 2002.
12. R.W. Noel and S. Acchione-Noel. Objet trouvé, holism, and morphogenesis in interactive evolution. CogSci, 2002, Fairfax Virginia, Aug., 2002.
13. M. Ohsaki and H. Takagi. Application of interactive evolutionary computation to optimal tuning of digital hearing aids. In *Int'l Conf. on Soft Computing (IIZUKA'98), Iizuka, Fukuoka, Japan*. World Scientic, 1998.
14. A. Pages, H. Pinguad, M. Meyer, and X. Joulia. From process simulation to general estimation. *Computers and Chemical Engineering*, 20:1413–1424, 1996.
15. I. Rechenberg. Case studies in evolutionary experimentation and computation. *Comput. Methods Appl. Mech. Engrg.*, 186:125–140, 2000.
16. H.P. Schwefel. *Numerical Optimization of Computer Models*. John Wiley and Sons, New York, 2nd edition, 1995.
17. J.R. Smith. Designing biomorphs with an interactive genetic algorithm. In *4th Int'l Conf. on Genetic Algorithms (ICGA'91), San Diego, CA, USA*, pp. 535–538. Morgan Kaufmann Publisher, 1991.
18. W.M. Spears, K.A. De Jong, T. Back, D.B. Fogel, and H. Garis. An overview of evolutionary computation. In *Proceedings of European Conference on Machine Learning*. 1993.
19. H. Takagi. Interactive evolutionary computation – cooperation of computational intelligence and human kansei. In *5th Int'l Conf. on Soft Computing (IIZUKA'98)*, pp. 41–50. World Scientic, Iizuka, Fukuoka, Japan, 1998.
20. H. Takagi. Interactive evolutionary computation: System optimization based on human subjective evaluation. In *IEEE Int'l Conf. on Intelligent Engineering Systems (INES'98)*, pp. 1–6. Vienna, Austria, 1998.
21. H. Takagi. Interactive evolutionary computation: Fusion of the capabilities of ec optimization and human evaluation. *IEEE*, 89(9), 2001.
22. K.C. Tan and Y. Li. Grey-box model identification vie evolutionary computing. *Control Engineering Practice*, 10:673–684, 2002.
23. M.L. Thompson and M.A. Kramer. Modeling chemical processes using prior knowledge and neural networks. *AIChE J.*, 40:1328–1340, 1994.
24. H.J.A.F. Tulleken. Grey-box modelling and identification using physical knowledge and bayesian techniques. *Automatica*, 29:285–308, 1993.

Replacement Strategies to Maintain Useful Diversity in Steady-State Genetic Algorithms

Manuel Lozano[1], Francisco Herrera[1] and José Ramón Cano[2]

[1] Dept. of Computer Science and Artificial Intelligence, University of Granada, 18071-Granada, Spain
lozano@decsai.ugr.es, herrera@decsai.ugr.es

[2] Dept. of Electronic Engineering, Computer Systems and Automatics, Escuela Superior de La Rábida, University of Huelva, 21819-Huelva, Spain
jose.cano@diesia.uhu.es

Summary. In this paper, we propose a replacement strategy for steady-state genetic algorithms that takes into account two features of the element to be included into the population: a measure of the contribution of diversity to the population and the fitness function. In particular, the proposal attempts to replace an element in the population with worse values for these two features. In this way, the diversity of the population is increased and the quality of its solutions is improved, simultaneously, maintaining high levels of useful diversity. Experimental results show that the use of the proposed replacement strategy allows significant performance to be achieved for problems with different difficulties, which regards to other replacement strategies presented in the literature.

1 Introduction

There are two primary factors in the search carried out by an genetic algorithm (GA): population diversity and selective pressure [30]. In order to have an effective search there must be a search criteria (the fitness function) and a selection pressure that gives individuals with higher fitness a higher chance of being selected for reproduction, mutation, and survival. Without selection pressure, the search process becomes random and promising regions of the search space would not be favored over non-promising regions. On the other hand, population diversity is crucial to a GA's ability to continue the fruitful exploration of the search space [16]. If the lack of population diversity takes place too early, a premature stagnation of the search is caused. Under these circumstances, the search is likely to be trapped in a region not containing the global optimum. This problem, called *premature convergence*, has long been recognized as a serious failure mode for GAs [6, 10].

Selective pressure and population diversity are inversely related [30]. Increasing selective pressure results in a faster loss of population diversity, while maintaining population diversity offsets the effect of increasing selective pressure. These two

Manuel Lozano et al.: *Replacement Strategies to Maintain Useful Diversity in Steady-State Genetic Algorithms*, Advances in Soft Computing **1**, 85–96 (2005)
www.springerlink.com

factors should be controlled in order to obtain their beneficial advantages simultaneously, allowing the most promising search space regions to be reached and refined. Under these circumstances, we will say that the population has achieved *useful diversity*, i.e., population diversity that in some way helps produce good solutions [18].

In *steady-state* GAs (SSGAs) usually only one or two offspring are produced in each generation. Parents are selected to produce offspring and then a decision is made as to which individuals in the population to select for deletion to make room for the new offspring. SSGAs are *overlapping* systems, since parents and offspring compete for survival. Different studies have shown that improved performance in SSGAs with regards to generational GAs (where entire population is replaced every generation by the offspring population) is due to their higher selection pressure and changes in the exploration/exploitation balance caused by using different parent selection and replacement strategies and is not due to the use of an overlapping model [4]. This justifies the study on SSGAs, and in particular, the design of effective replacement strategies having the aim of propitiating useful diversity.

In this paper, we propose a replacement strategy that takes into account two features of the element to be included into the population: a measure of the contribution of diversity to the population and the fitness function. It attempts to replace an element in the population with worse values for these two features. In this way, the diversity of the population is increased and the quality of its solutions is improved, simultaneously. The goal of this strategy is to protect those individuals that allow the highest levels of useful diversity to be maintained.

The paper is set up as follows. In Sect. 2, the SSGA model is described. In Sect. 3, different types of replacement strategies presented in the literature for promoting SSGA population diversity are explained. In Sect. 4, we propose the replacement strategy based on the contribution of diversity. In Sect. 5, different empirical experiments are carried out for studying the performance of this strategy. Finally, in Sect. 6, some conclusions are pointed out.

2 Steady-State Genetic Algorithms

The generational GA creates new offspring from the members of an old population using the genetic operators and places these individuals in a new population which becomes the old population when the whole new population is created. The SSGA is different to the generational model in that there is typically one single new member inserted into the new population at any one time. A replacement/deletion strategy defines which member of the population will be replaced by the new offspring. The basic algorithm step of SSGA is the following [30]:

1. Select two parents from the population.
2. Create an offspring using crossover and mutation.
3. Evaluate the offspring with the fitness function.
4. Select an individual in the population, which may be replaced by the offspring.
5. Decide if this individual will be replaced.

In step 4, one can chose the *replacement strategy* (e.g., replacement of the worst, the oldest, or a randomly chosen individual). In step 5, one can chose the *replacement condition* (e.g., replacement if the new individual is better, or unconditional replacement). A widely used combination is to replace the worst individual only if the new individual is better. In the paper, this strategy will be called *standard replacement strategy*. In [11], it was suggested that the deletion of the worst individuals induced a high selective pressure, even when the parents were selected randomly.

3 Favouring Population Diversity Throughout Replacement Strategies

There are different replacement strategies that attempt to preserve population diversity. Most of them are instances of the *crowding* methods [5]. They work as follows: new individuals are more likely to replace existing individuals in the parent population that are similar to themselves based on genotypic similarity. They have been used to locate and preserve multiple local optimum in multimodal problems.

An effective crowding method is the *restricted tournament selection* (RTS) [13]. RTS initially selects two element at random, A and B, from the population and perform crossover and mutation on these two elements resulting in a new element A'. Then, RTS scans w (window size) more members of the population and picks the individuals that most closely resemble A' from those w elements. A' then competes with this element, and if A' wins, it is allowed to enter the population.

Another type of crowding methods assume that the parents would be ones of the members of the population closest to the new elements. In this way, children compete with their parents to be included in the population, i.e., a *family competition* is held. These methods include *deterministic crowding* [17], *elitist recombination* [25], and *keep-best reproduction* [32].

- In deterministic crowding, each pair of parents, (P_i, P_j), undergoes crossover in combination with mutation to yield two offspring, (O_i, O_j), which compete against the parents for inclusion in the population through the following method of competition:

  If $[d(P_i,O_i)+d(P_j,O_j)] \leq [d(P_i,O_j)+d(P_j,O_i)]$ **then**
      If $f(O_i)$ is better than $f(P_i)$ **then** replace P_i with O_i.
      If $f(O_j)$ is better than $f(P_j)$ **then** replace P_j with O_j.
  Else
      If $f(O_i)$ is better than $f(P_j)$ **then** replace P_j with O_i.
      If $f(O_j)$ is better than $f(P_i)$ **then** replace P_i with O_j.

  where $f(\cdot)$ is the fitness function and $d(\cdot,\cdot)$ is a distance measure between two chromosomes.
- In elitist recombination, the best two of these four individuals (parents and offspring) go to the next generation.

- Finally, keep-best keeps the best parent and the best offspring in order to introduce good new genetic material into the population.

In addition, two replacement strategies for SSGAs have been designed for encouraging diversity in the population: the *first-in-first-out* (FIFO) strategy [9, 28] and the *conservative* strategy [29].

- In the FIFO strategy, the element to be replaced is the oldest in the population. In its elitist form, the replacement of the current best element is not allowed.
- The *conservative* strategy combines a FIFO replacement strategy with a modified deterministic binary tournament selection operator. In this case, a tournament is carried out between the oldest element in the population and another member picked at random. The worse is replaced. In this way, elitism is implicitly assured.

4 The Proposed Replacement Strategy

The population diversity versus selective pressure problem may be consider as a biobjective problem. The GA may be used for searching simultaneously the best chromosomes together with the ones that provide more diversity to the GA population, i.e., promoters of useful diversity.

In fact, there are some generational GA approaches that use this idea for avoiding premature convergence and improving GA behavior [15, 19, 23]. During the selection phase, they use the fitness function and some type of measures for evaluating the diversity introduced by the candidate chromosomes with the aim of favoring those individuals with both high fitness function values and high diversity contributions.

In this paper, we propose to solve the diversity versus selective pressure problem in the replacement phase of a SSGA. In particular, we present a replacement strategy based on the contribution of diversity of the offspring to the population where it will be included. The idea is to replace an element of the population with a worse fitness function value than the one of the offspring, and with a lower contribution of diversity than the one provided by the offspring. In this way, we deal with two underlying objectives simultaneously: to optimize the fitness function and to enhance population diversity, i.e., to promote useful diversity.

In Subsect. 4.1, we present measures of the contribution of diversity of an element to an SSGA population, and in Subsect. 4.2, we use these measures for defining a replacement strategy that tackles the biobjective diversity-convergence.

4.1 Contribution of Diversity

Next, we present two different types of measures of the contribution of diversity of a chromosome x to a population P ($x \notin P$), $ConDiv(x,P)$. They quantify the importance of a chromosome as promoter of diversity in the population.

- Measures based on distance. They are defined as the distance of x to a particular member of P. We have chosen two possibilities: 1) this member is the best one, and 2) it is the nearest neighbor of x in P.

  These two measures are the ones used in [23] and [3], respectively, and they will be denoted as *ConDiv-B* and *ConDiv-NN*, respectively.

  Under these definitions, a chromosome will provide a great contribution of diversity when it is very different from all the other chromosomes (*ConDiv-NN*) or from the best one (*ConDiv-B*).
- Measures based on population diversity. Now, we define the contribution of diversity as follows:

  $$ConDiv(x,P) = Div(P \cup \{x\}) - Div(P) ,$$

  where $Div(P)$ is a measure of the diversity in the population P. Measures of this type are to be found in [2, 18].

  In this case, the contribution of diversity of a chromosome will be high when its inclusion in the population allows a significant increasing of diversity to be achieved. In this paper, we employ the following diversity measure for a population with real-coded chromosomes, $C_j = (c_1^j, \cdots c_n^j), j = 1, \cdots N$, which is called *average interval usage*:

  $$AIU(P) = \frac{1}{n} \sum_{i=1}^{n} \frac{Max_i - Min_i}{b_i - a_i}$$

  where $Max_i = \max_{j=1,\dots N}\{c_i^j\}, Min_i = \min_{j=1,\dots N}\{c_i^j\}$, and a_i and b_i $(a_i < b_i)$ are the extremes of the action interval of the gene i $(i = 1, \cdots, n)$. The measure of the contribution of diversity derived from AIU will be called *ConDiv-AIU*.

4.2 The Proposed Replacement Strategy

Now, let us assume that an offspring o is returned from the recombination phase of an SSGA, then we propose to include o in the population by means of the following strategy:

1. Find the element z in P that fulfils:
   (i) $f(o)$ is better than $f(z)$, f being the fitness function.
   (ii) $\Delta_z > \Delta_w > 0$, for all w in P, where for any $t \in P$:

   $$\Delta_t = ConDiv(o, P - \{t\}) - ConDiv(t, P - \{t\}) .$$

2. If z does not exist, then apply the standard replacement strategy (Sect. 2).

First, this strategy attempts to find an element in the population with worse fitness than the offspring and whose contribution of diversity is the lowest with regards to the one provided by the offspring in a population where this element have been removed. This element will be replaced by the offspring. In this way, we introduce into the population an element with better fitness function and higher contribution of diversity, increasing the population diversity and improving the quality of its solutions.

If this individual does not exist, then the standard replacement strategy is applied. In this way, we ensure that the best elements found during the run are included in the population (elitist strategy).

5 Experiments

Minimization experiments on the test suite described in Sect. 5.1 were carried out in order to study the behavior of the replacement strategy presented in the previous section.

In Sect. 5.2, we propose different SSGA instances based on this strategy and compare its results with the ones of a standard SSGA. Finally, in Sect. 5.3, we compare the best SSGA of Sect. 5.2 with different SSGAs based on the replacement strategies described in Sect. 3.

5.1 Test Suite

The test suite that we have used for the experiments consists of six test functions and three real-world problems. The test functions are the following: *Sphere* model (f_{Sph}) [5, 22], *Generalized Rosenbrock's* function (f_{Ros}) [5], *Schwefel's Problem 1.2* (f_{Sch}) [22], *Generalized Rastringin's* function (f_{Ras}) [1, 26], *Griewangk's* function (f_{Gri}) [12], and *Expansion of* f_{10} (ef_{10}) [31]. The dimension of the search space is 10 for ef_{10} and 25 for the remaining test functions. The three real-world problems are the following: *Systems of Linear Equations* [8], *Frequency Modulation Sounds Parameter Identification Problem* [27], and *Polynomial Fitting Problem* [24]. For all the problems, the global optimum has a fitness value of zero.

5.2 SSGAs Based on the Proposed Replacement Strategy

We have implemented three instances of SSGA based on the replacement strategy proposed in Subsect. 4.2. They are called SGA-ConDiv-B, SGA-ConDiv-NN, and SGA-ConDiv-AIU, and use the *ConDiv-B*, *ConDiv-NN*, and *ConDiv-AIU* measures of the contribution of diversity defined in Subsect. 4.1, respectively. We compare these algorithms against a SSGA that uses the standard replacement strategy (see Sect. 2), which is called SGA-Standard.

All the algorithms use real coding [14] and apply the BGA mutation operator [20] and the BLX-α operator ($\alpha = 0.5$) [7]. The mutation probability is $\frac{1}{n}$ and the population size is 60 chromosomes. Parents are selected at random. They were executed 50 times, each one with a maximum of 100,000 evaluations. Table 1 shows the results obtained. The performance measures utilized are the following:

- **A** performance: average of the best fitness function found at the end of each run.
- **B** performance: best of the fitness values averaged as **A** performance. If the global optimum has been reached sometimes, this performance will represent the percentage of runs in which this happens.

Table 1. Results for SSGAs based on the proposed replacement strategy

Algorithm	f_{sph} A		f_{sph} B	f_{Ros} A		f_{Ros} B	f_{Sch} A		f_{Sch} B
SGA-ConDiv-AIU	1.2e−15	∼	2.9e−54	2.4e+01	∼	1.0e+01	3.4e−01	−	2.9e−02
SGA-ConDiv-NN	4.3e−50	∼	7.5e−52	2.0e+01	∼	1.9e+01	5.9e−02	−	6.9e−03
SGA-ConDiv-B	1.1e−32	∼	1.0e−41	2.7e+01	∼	1.6e+01	4.3e+00	∼	3.0e−01
SGA-Standard	2.2e−04		1.1e−55	2.3e+01		1.3e+01	2.9e+00		1.5e−01

Algorithm	f_{Ras} A		f_{Ras} B	f_{Gri} A		f_{Gri} B	ef_{10} A		ef_{10} B
SGA-ConDiv-AIU	6.8e−01	−	52.0 %	1.4e−03	−	62.0 %	6.8e−03	∼	6.2e−35
SGA-ConDiv-NN	4.4e−01	−	56.0 %	3.5e−04	−	86.0 %	4.8e−26	−	7.4e−27
SGA-ConDiv-B	1.1e+00	∼	30.0 %	9.6e−03	+	38.0 %	5.3e−03	∼	1.8e−27
SGA-Standard	1.2e+00		32.0 %	4.3e−03		4.0 %	6.3e−03		8.5e−37

Algorithm	P_{fms} A		P_{fms} B	P_{sle} A		P_{sle} B	P_{Cheb} A		P_{Cheb} B
SGA-ConDiv-AIU	7.3e+02	∼	1.3e+02	1.4e+02	∼	1.8e+01	1.5e+01	∼	6.5e−07
SGA-ConDiv-NN	2.9e+02	−	2.8e+00	3.1e+01	−	5.7e−01	1.5e+00	−	2.0e−26
SGA-ConDiv-B	2.9e+03	+	1.3e+02	3.1e+02	+	5.1e+01	1.3e+01	∼	2.2e−10
SGA-Standard	7.8e+02		1.2e+02	1.2e+02		1.2e+01	1.4e+01		3.2e−10

A two-sided *t*-test (H_o : means of the two groups are equal, H_a : means of the two group are not equal) at 0.05 level of significance was applied in order to ascertain if differences in the *A* performance for the standard SSGA are significant when compared against the ones for the SSGAs based on the proposed replacement strategy. The direction of any significant differences is denoted either by:

- A plus sign (+): the *A* performance of SGA-Standard is better than the one of the corresponding algorithm.
- A minus sign (−): the algorithm improves the *A* performance of SGA-Standard.
- An approximate sign (∼): non significant differences.

Taking into consideration these results, we would make the following comments:

- SGA-ConDiv-NN returns better *A* results than SGA-Standard (see t-test results) on all test functions except on the unimodal f_{sph} and f_{Ros}, where they are similar. Moreover, SGA-ConDiv-NN achieves a better *B* measure than SGA-Standard for the more complex functions: f_{Sch}, P_{Cheb}, P_{fms}, and P_{sle}.
  SGA-ConDiv-AIU provides better *A* results than SGA-Standard on the complex f_{Sch} and on the multimodal f_{Ras} and f_{Gri}. For the remaining function, their performance is similar.

These facts show that using the proposed replacement strategy, the results may be improved with regards to a standard SSGA. This means that its formulation (which has the aim of propitiate useful diversity) allows a profitable operation to be achieved. Furthermore, the measure of contribution of diversity *ConDiv-NN* arises as a very adequate measure for designing this strategy.

- On the other hand, SGA-ConDiv-B losses performance with regards to SGA-Standard on f_{Gri}, P_{sle}, and P_{fms}. Therefore, we may claim that the measure ConDiv-B is not suitable for building the proposed replacement strategy. The idea of favoring those elements in the population that are located far away from the current best individual is not a good diversification strategy, because it may obstruct the refinement of the best solutions, degrading the quality of the final solutions.

5.3 Comparison with Other Replacement Strategies

In this section, we compare the best algorithm based on the proposed replacement strategy, SGA-ConDiv-NN, against different SSGAs that use the replacement strategies described in Sect. 3. Table 2, shows these algorithms along with their main features.

Table 2. Algorithms based on other replacement strategies

Algorithm	Replacement Strategy
SGA-RTS	Restricted tournament selection ($\omega = 5$)
SGA-DC	Deterministic crowding
SGA-ER	Elitist recombination
SGA-KBR	Keep-best reproduction
SGA-FIFO	First-in-first-out
SGA-Conse	Conservative strategy

Table 3 has the results. A *t*-test was applied in order to ascertain if differences in the *A* performance for SGA-ConDiv-NN are significant when compared against the one for the other algorithms.

Table 4 was introduced in order to analyse the results. It shows the number of improvements, reductions, and non differences (according to the t-test results in Table 3) in the *A* performance for SGA-ConDiv-NN with regards to the one for the other algorithms.

The table shows that SGA-ConDiv-NN outperforms all the other algorithms, because it improves their results on a great number of functions and never achieves a worse *A* performance measure. The source of diversity for these algorithms is the same than for SGA-ConDiv-NN since they use the same mating and recombination strategies. The difference between them is the way in which diversity is retained by the replacement strategies. The promising results of the proposal reveal that it

Table 3. Results for the comparison

Algorithm	f_{sph} A	f_{sph} B	f_{Ros} A	f_{Ros} B	f_{Sch} A	f_{Sch} B
SGA-RTS	9.2e−17 +	5.7e−20	2.1e+01 ~	1.5e+01	1.6e+01 +	2.1e+00
SGA-DC	9.4e−32 +	6.5e−33	2.1e+01 +	2.0e+01	1.2e+01 +	3.8e+00
SGA-ER	8.0e−46 +	2.3e−47	2.1e+01 ~	2.0e+01	3.8e−01 +	7.6e−02
SGA-KBR	2.8e−12 ~	6.2e−14	2.3e+01 ~	1.5e+01	1.4e−01 +	1.9e−02
SGA-FIFO	1.9e−02 +	9.3e−12	2.5e+01 +	2.1e+00	5.5e+01 +	3.6e−01
SGA-Conse	7.0e−03 ~	2.2e−12	3.2e+01 +	4.5e+00	1.2e+02 +	7.8e−01
SGA-ConDiv-NN	4.3e−50	7.5e−52	2.0e+01	1.9e+01	5.9e−02	6.9e−03

Algorithm	f_{Ras} A	f_{Ras} B	f_{Gri} A	f_{Gri} B	ef_{10} A	ef_{10} B
SGA-RTS	6.8e−01 ~	3.3e−10	6.6e−03 +	1.1e−18	1.4e−02 +	1.3e−09
SGA-DC	2.2e+01 +	3.2e+00	4.3e−08 ~	90.0 %	9.6e−15 +	3.0e−15
SGA-ER	2.6e+00 +	8.0 %	8.9e−04 ~	84.0 %	4.7e−25 +	1.3e−25
SGA-KBR	5.4e−01 ~	1.9e−10	2.0e−03 ~	6.6e−11	8.8e−03 +	1.8e−03
SGA-FIFO	1.0e+00 +	1.2e−08	3.7e−02 +	8.7e−09	4.5e−01 +	4.0e−06
SGA-Conse	1.2e+00 +	2.8e−04	4.3e−02 +	1.5e−08	3.7e−01 +	3.1e−11
SGA-ConDiv-NN	4.4e−01	56.0 %	3.5e−04	86.0 %	4.8e−26	7.4e−27

Algorithm	P_{fms} A	P_{fms} B	P_{sle} A	P_{sle} B	P_{Cheb} A	P_{Cheb} B
SGA-RTS	5.7e+02 +	6.1e+01	7.0e+01 +	6.7e+00	4.8e+00 +	4.6e−15
SGA-DC	2.8e+02 ~	3.4e+01	2.4e+01 ~	2.5e+00	3.9e+00 +	2.4e−05
SGA-ER	3.2e+02 ~	2.1e+01	5.1e+01 +	1.3e+00	3.4e+00 ~	3.0e−17
SGA-KBR	3.3e+02 ~	5.5e+01	6.9e+01 +	3.0e+00	7.9e+00 +	1.1e−15
SGA-FIFO	1.6e+03 +	3.6e+02	2.2e+02 +	4.1e+01	1.7e+01 +	5.8e−07
SGA-Conse	1.5e+03 +	1.5e+02	2.5e+02 +	2.6e+01	1.8e+01 +	4.0e−04
SGA-ConDiv-NN	2.9e+02	2.8e+00	3.1e+01	5.7e−01	1.5e+00	2.0e−26

Table 4. SGA-ConDiv vs. the other algorithms

Algorithm	Num. Impr. (+)	Num. Red. (−)	Num. Non Diff. (~)
SGA-RTS	7	0	2
SGA-DC	6	0	3
SGA-ER	5	0	4
SGA-KBR	5	0	4
SGA-FIFO	9	0	0
SGA-Conse	8	0	1

conserves those elements (the more diverse and with the higher quality) that allow the search process to be driven towards better regions than by means of the other strategies. Thus, we conclude that it is an adequate replacement strategy for SSGAs.

6 Conclusions

This paper presented a replacement strategy for SSGAs that attempts to promote useful diversity by considering the merit of the offspring as solution and its contribution of diversity to the population. It attempts to replace an individual that is worse than the offspring with regards to these two factors, obtaining two benefits simultaneously: an increase of the population diversity and an improvement on the quality of its solutions. The principal conclusions derived from the results of the experiments carried out are the following:

1. The ConDiv-NN measure of the contribution of diversity is the best choice for designing the proposed strategy. However, other measures (such as ConDiv-AIU) allow the effectiveness of the strategy to be exhibited as well, with regards to a standard SSGA.
2. The proposal is a suitable replacement strategy for SSGAs. Its effects on SSGA performance are more decisive than the ones derived from the use of other replacement strategies presented in the GA literature.
3. These results confirm that the replacement strategy is a determinant issue to take into account for designing effective SSGAs.

Finally, we should point out that extensions of the proposed replacement strategy may be followed in four ways: 1) consider additional features of the chromosomes for deciding which individual should be replaced, such as their age, average distance to their k nearest neighbours, distance to a neighbouring subpopulation (in a distributed SSGA framework), etc., 2) design similar replacement strategies for working under other types of coding (such as binary-coding, order-based codings, codings used for the genetic programming, etc.), and 3) study the effects of the application of this strategy on nonstationary problems.

Acknowledgments

This research has been supported by the project TIC2002-04036-C05-01.

References

1. Bäck T (1992) Self-adaptation in genetic algorithms. In: F.J. Varela, P. Bourgine, (eds) Proc. of the First European Conference on Artificial Life. The MIT Press, Cambridge, MA, 263–271.
2. Bedau MA, Ronneburg F, Zwick M (1992) Dynamic of diversity in an evolving population. In: R. Männer, B. Manderick, (eds) Parallel Problem Solving from Nature 2. Elsevier Science Publichers, Amsterdam, 95–104.
3. Bonham CR, Parmee IC (1999) An investigation of exploration and exploitation within cluster oriented genetic algorithms (COGAs). In: Proc. of the Genetic and Evolutionary Computation Conference 1999. Morgan Kaufmann, San Francisco, California, 1491–1497.

4. De Jong KA, Sarma J (1993) Generation gaps revisited. In: L.D. Whitley (ed) Foundations of Genetic Algorithms 2. Morgan Kaufmann, San Mateo, CA, 19–28.
5. De Jong KA (1975) An Analysis of the Behavior of a Class of Genetic Adaptive Systems. Ph. dissertation, Univ. of Michigan, Ann Arbor, 1975.
6. Eshelman LJ, Schaffer JD (1991) Preventing premature convergence in genetic algorithms by preventing incest. In: R. Belew, L.B. Booker, (eds) Proc. of the Fourth Int. Conf. on Genetic Algorithms. Morgan Kaufmann, San Mateo, CA, 115–122.
7. Eshelman LJ, Schaffer JD (1993) Real-coded genetic algorithms and interval-schemata. In: L. Darrell Whitley (ed) Foundation of Genetic Algorithms 2. Morgan Kaufmann Publishers, San Mateo, 187–202.
8. Eshelman LJ, Mathias KE, Schaffer JD (1997) Convergence controlled variation. In: R. Belew, M. Vose, (eds) Foundations of Genetic Algorithms 4. Morgan Kaufmann, San Mateo, CA, 203–224.
9. Ghosh A, Tsutsui S, Tanaka H (1998) Function optimization in nonstationary environment using steady state genetic algorithms with aging of individuals. In: Proc. of the 1998 IEEE International Conference on Evolutionay Computation. IEEE Press, New York, 666–671.
10. Goldberg DE (1989) Genetic Algorithms in Search, Optimization, and Machine Learning. Addison-Wesley, Reading, MA.
11. Goldberg DE, Deb K (1991) A comparative analysis of selection schemes used in genetic algorithms. In: G.J.E. Rawlins (ed) Foundations of Genetic Algorithms. Morgan Kaufmann, San Mateo, CA, 69–93.
12. Griewangk AO (1981) Generalized descent of global optimization. Journal of Optimization Theory and Applications 34: 11–39.
13. Harik G (1995) Finding multimodal solutions using restricted tournament selection. In: Proc. 6th Int. Conf. Genetic Algorithms, L. J. Eshelman (Ed.), (Morgan Kaufmann, San Mateo, CA, 1995) 24–31.
14. Herrera F, Lozano M, Verdegay JL (1998) Tackling real-coded genetic algorithms: operators and tools for behavioural analysis. Artificial Intelligence Review 12(4): 265–319.
15. Lee C-Y (2003) Entropy-boltzmann selection in the genetic algorithms. IEEE Transactions on Systems, Man, and Cybernetics – Part B: Cybernetics 33(1): 138–149.
16. Li T-H, Lucasius CB, Kateman G (1992) Optimization of calibration data with the dynamic genetic algorithm. Analytica Chimica Acta 2768: 123–1342.
17. Mahfoud SW (1992), Crowding and preselection revised. In: R. Männer, B. Manderick, (eds) Parallel Problem Solving from Nature 2. (Elsevier, Amsterdam, 27–36.
18. Mahfoud SW (1995) Niching methods for genetic algorithms. IlliGAL report 95001, University of Illinois at Urbana-Champaign, IL, Illinois Genetic Algorithms Laboratory.
19. Mori N, Yoshida J, Tamaki H, Kita H, Nichikawa Y (1995) A thermodynamical selection rule for genetic algorithms. In: Proc. of the 2nd IEEE Conference on Evolutionary Computation (ICEC'95). IEEE Press, New York, 188–192.
20. Mühlenbein H, Schlierkamp-Voosen D (1993) Predictive models for the breeder genetic algorithm I. continuous parameter optimization. Evolutionary Computation 1: 25–49.
21. Schlierkamp-Voosen D, Mühlenbein H (1994) Strategy adaptation by competing subpopulations. In: Y. Davidor, H.-P. Schwefel, R. Männer (eds) Parallel Problem Solving from Nature 3. Springer-Verlag, Berlin, 199–208.
22. Schwefel H-P (1981) Numerical optimization of computer models. Wiley, Chichester.
23. Shimodaira H (1999) A diversity control oriented genetic algorithm (DCGA): development and experimental results. In: Proc. of the Genetic and Evolutionary Computation Conference 1999. Morgan Kaufmann, San Francisco, California, 603–611.

24. Storn R, Price K (1995) Differential evolution – a simple and efficient adaptive scheme for global optimization over continuous spaces. Technical Report TR-95-012, International Computer Science Institute, Berkeley, CA.
25. Thierens D (1997) Selection schemes, elitist recombination, and selection intensity. In: T. Bäck, (ed) Proc. of the Seventh Int. Conf. on Genetic Algorithms. Morgan Kaufmann Publishers, San Mateo, 152–159.
26. Törn A, Antanas Ž (1989) Global optimization. Lecture Notes in Computer Science 350, Springer, Berlin.
27. Tsutsui S, Fujimoto Y (1993) Forking genetic algorithm with blocking and shrinking modes. In: S. Forrest (ed) Proc. of the Fifth Int. Conf. on Genetic Algorithms. Morgan Kaufmmann, San Mateo, CA, 206–213.
28. Vavak F, Fogarty TC (1996) Comparison of steady state and generational genetic algorithms for use in nonstationary environments. In: Proc. Of the 1996 IEEE International Conference on Evolutionary Computation. IEEE Press, New York, 192–195.
29. Vavak F (1997) Genetic algorithm based self-adaptive techniques for direct load balancing in nonstationary environments. PhD thesis, University of the West of England.
30. Whitley D. (1989), The GENITOR algorithm and selection pressure: why rank-based allocation of reproductive trials is best. In: J. David Schaffer (ed) Proc. of the Third Int. Conf. on Genetic Algorithms. Morgan Kaufmann Publishers, San Mateo, 116–121.
31. Whitley D, Beveridge R, Graves C, Mathias K (1995) Test driving three 1995 genetic algorithms: new test functions and geometric matching. Journal of Heuristics 1: 77–104.
32. Wiese K, Goodwin SD (1999) Convergence characteristics of keep-best reproduction. In: Selected Areas in Cryptography, 312–318.

Hierarchical Modelling of Evolutionary Computation

Christian Veenhuis, Mario Köppen, and Katrin Franke

Fraunhofer Institute for Production Systems and Design Technology, Department Pattern Recognition, Pascalstr. 8-9, 10587 Berlin, Germany
{veenhuis|koeppen|franke}@ipk.fhg.de

Summary. Nowadays a lot of libraries and software systems for Evolutionary Computation (EC) are in existance. Unfortunately, an EC created by one of those can rarely be exchanged between them. In recent works different Soft Computing methods are modelled by using XML notation to overcome this problem. But to be able to model ECs in, e.g., XML notation, a concept is needed which allows the hierarchical description of ECs. This paper proposes a hierarchically modelling concept for Evolutionary Computation which allows the description of ECs as EC-hierarchy on an abstract level. A created EC-hierarchy can be mapped on every notation which is capable of representing tree-like structures like XML or hierarchy-oriented programming languages. Such a hierarchical model could be used for exchanging ECs between software systems, libraries as well as researchers.

Key words: EC-Hierarchy, Hierarchical Model, EC-Model, Evolutionary Computation, Algorithm Design

1 Introduction

The term *Evolutionary Computation* (EC) subsumes all computational models that are inspired by natural evolution like Genetic Algorithms (GA), Genetic Programming (GP), Evolutionary Strategies (ES) and so on. Each of them represents a huge variety of algorithms, which differ in their configuration and used operators.

If one wants to adopt an implemented or newly published algorithm, one has to reprogram or translate the algorithm in the worst case (especially, if one wants to deploy it on another platform or in another programming language). Furthermore, created and defined ECs can rarely be exchanged between software systems and EC libraries.

In recent works Soft Computing methods are modelled by using XML, as shown in [4], to overcome the mentioned problems. XML [5] is a platform-independent and text-based notation for describing data or structures hierarchically and is in general considered as one of the major future achievements for distributing and exchanging documents and data via internet. For instance, EAML [3] is used to model Evolutionary Algorithms (EAs) hierarchically with XML notation. Every EA is built up by a

Christian Veenhuis et al.: *Hierarchical Modelling of Evolutionary Computation*, Advances in Soft Computing **1**, 97–111 (2005)
www.springerlink.com

hierarchical order of XML elements which represent the different genetic operators and methods. This way, a semantic model is constructed: the kind of order and the nesting of the elements represent the specific EA.

To be able to model or develop ECs in, e.g., XML notation, a concept is needed which allows the hierarchical description of ECs. This paper proposes a hierarchically modelling concept for ECs which allows the description of ECs as EC-hierarchy on an abstract level.

Generally speaking, a hierarchical model specifies elements of a domain to describe an instance of this domain by arranging the appropriate components in a tree-like manner, whereby the nested order possesses semantics like *contains*, *is made of*, *kind of* and *next step(s)* as depicted in Fig. 1.

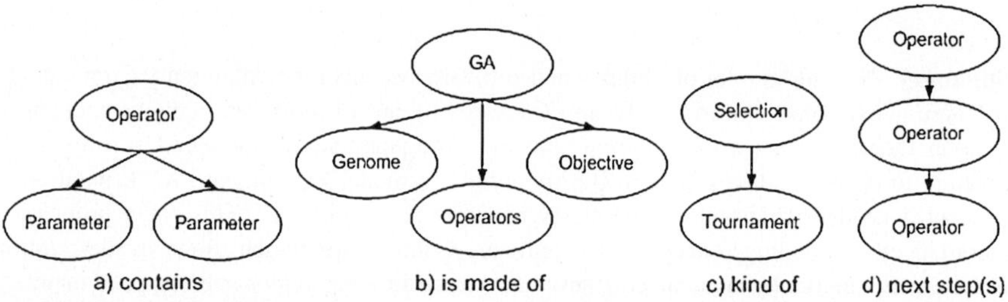

Fig. 1. Nested order possesses semantics like (**a**) contains, (**b**) is made of, (**c**) kind of and (**d**) next step(s)

A complete EC can be described by arranging its components hierarchically forming this way an EC-hierarchy. The flow of individuals is specified by an operator-hierarchy describing this way a major part of the EC. Each genetic operator is used to create a certain number of offspring. The produced offspring can be created by single operators or being processed by further nested genetic operators. Such mechanisms are used, e.g., in Simple GAs, where, after one-point crossover, the individuals are additionally mutated.

The proposed hierarchically modelling concept allows the description of ECs on an abstract level. Thereby, only the algorithm of an EC is described. The operators and methods are considered as atomic units, i.e., their internal procedure and structure aren't described. A created EC-hierarchy can be mapped on every notation which is capable of representing tree-like structures like XML or even hierarchy-oriented programming languages like Lisp. Such a mapped EC-hierarchy can be used as a model of the EC for exchange and interpretation by software systems.

This paper is organized as follows. Section 2 derives the hierarchical model and notation. The usage together with an example is shown in Sect. 3. In Sect. 4 some conclusions are drawn with an outlook to future works.

2 Hierarchical Model

2.1 Representation

Let REP be the representation of genomes realized as an ordered rooted m-ary tree with directed edges, i.e., a tree with exactly one root node and every internal node can have up to m children, whereby these children are ordered from left to right. Equation (1) defines REP as tree with a set V of N_{nodes} nodes (represented by unique indices) and a set E of directed edges. Here, an edge is a relation defining an ordered tuple $\langle a,b \rangle$ where node a directs to node b. All edges directing to the children of a given node x are a set of edges $E_{node}(x)$ with a maximum cardinality of m. The node x is a leaf if $E_{node}(x) = \emptyset$. If $0 \langle \| E_{node}(x) \| \leq m$ then the node x is an internal node of the tree. A node x can have a value which can be obtained with $VAL_{node}(x)$.

$$
\begin{aligned}
REP &= (V,E) \\
V &= \{1,\ldots,N_{nodes}\}, \quad E \subset V \times V = \{\langle a,b \rangle | \forall a,b \in V\} \\
E_{node}(x) &= \{\langle a,b \rangle | a = x\} \subseteq E \\
VAL_{node}(x) &: V \to \mathbb{R} \ , \ VAL_{node}(x) := \text{value of node x}
\end{aligned} \tag{1}
$$

In Fig. 2 a sample tree as REP is shown. The defined representation is the general basis of all possible representations used for an EC. Every further representation for a specific EC has to be derived from REP by defining a structure onto the tree as depicted in (2).

$$
\begin{aligned}
\mathcal{S}(REP) &= \langle N_{dec} \ , \ DEFS = \{DEF(1),\ldots,DEF(n)\}\rangle \\
DEF(i) &= \langle ID(i), VAL(i)_{\min}, VAL(i)_{\max}, ARITY(i)_{\min}, ARITY(i)_{\max}\rangle \\
VAL(i) &= \{v \mid v \in [VAL(i)_{\min}, VAL(i)_{\max}] \ \wedge \ v \cdot 10^{N_{dec}} \in Z\} \\
VAL_{node}(i) &\in VAL(i)
\end{aligned} \tag{2}
$$

The structure of a representation $\mathcal{S}(REP)$ is a pair consisting of the number of allowed decimal places N_{dec} for the values of the nodes ($VAL_{node}(x)$) and a set $DEFS$

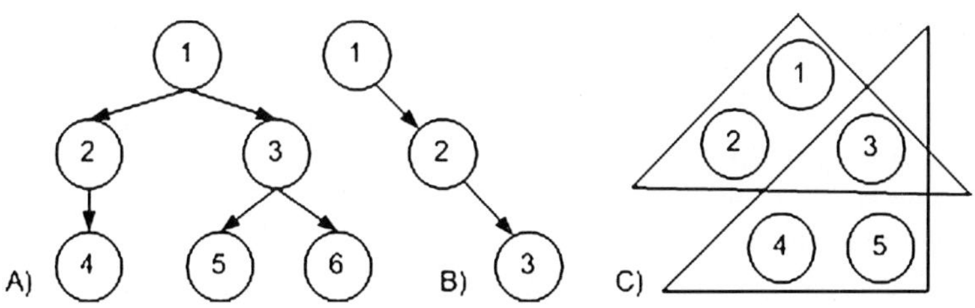

Fig. 2. (**A**) A sample tree with $V = \{1,2,3,4,5,6\}$ and $E = \{\langle 1,2 \rangle, \langle 1,3 \rangle, \langle 2,4 \rangle, \langle 3,5 \rangle, \langle 3,6 \rangle\}$. E.g., the internal node 3 has the children $E_{node}(3) = \{\langle 3,5 \rangle, \langle 3,6 \rangle\}$. (**B**) A sample tree representing a list with $V = \{1,2,3\}$ and $E = \{\langle 1,2 \rangle, \langle 2,3 \rangle\}$. The node 3 is a leaf, because it has no children ($E_{node}(3) = \emptyset$). (**C**) A sample hypergraph $H = (X, \varepsilon)$ with $X = \{1,2,3,4,5\}$ and $\varepsilon = (E_i)_{i \in \{0,1\}} = \{\{1,2,3\},\{3,4,5\}\}$

containing any number of DEF tuples. A DEF tuple contains an ID and two intervals: one for the valid range of symbols (values) for this definition and one for the possible number of children nodes (the arity) for which DEF represents the parental node. A DEF tuple can be understood as a kind of a node. To avoid illegability by using to much indices the notation $DEF(i)$ is used instead of DEF_i. With $ID(i)$, e.g., a name for this definition can be given as string, symbol or whatever. The value of a node x determined by $VAL_{node}(x)$ must be an element of the $VAL(x)$ set as depicted in (2) (by considering the maximum number of decimal places N_{dec}).

To describe a structure of representation $\mathcal{S}(REP)$ hierarchically, the following notation can be used:

```
representation decs =...
 def id =... valmin =... valmax =... aritymin =... aritymax =...
 def id =... valmin =... valmax =... aritymin =... aritymax =...
```

It's introduced with the term for this EC component (`representation`) which represents $\mathcal{S}(REP)$ followed by the parameter decs which defines N_{dec}. After this, with an appropriate indentation, the $DEF(i)$ tuples follow (each a line). Each parameter of the $DEF(i)$ tuples is given with the following meaning: id $= ID(i)$, valmin $= VAL(i)_{\min}$, valmax $= VAL(i)_{\max}$, aritymin $= ARITY(i)_{\min}$ and aritymax $= ARITY(i)_{\max}$.

To describe a tree for the operations multiplication and minus-sign, e.g., for a Genetic Programming method, the following structure could be defined:

$$\mathcal{S}(REP) = \langle 0, \{\langle \text{“}*\text{”}, 1, 1, 2, 2\rangle, \langle \text{“}-\text{”}, 2, 2, 1, 1\rangle, \langle \text{“}X\text{”}, 3, 3, 0, 0\rangle\}\rangle \tag{3}$$

The operation with the ID "*" possesses exactly two children nodes, because it's a binary operation. The sign (as in -8) possesses only one child node. And the terminal node with ID "X" possesses no children nodes. Each $DEF(i)$ tuple has its unique $VAL(i)$ set ($VAL(i)_{\min} = VAL(i)_{\max}$) as usual in Genetic Programming. For GP approaches all $VAL(i)$ sets should be disjoint, because in this case $VAL(i)$ represents a single symbol. Therefore, N_{dec} is set to 0.

In hierarchical notation this would look like:

```
representation  decs = 0
 def id = ‘‘*"  valmin = 1 valmax = 1  aritymin = 2 aritymax = 2
 def id = ‘‘-"  valmin = 2 valmax = 2  aritymin = 1 aritymax = 1
 def id = ‘‘X"  valmin = 3 valmax = 3  aritymin = 0 aritymax = 0
```

There are three lines for the $DEF(i)$ tuples ($\| DEFS \| = 3$). Each of them defines a tuple with all its attributes.

To describe a list (or vector) derived from REP, as depicted in Fig. 2, e.g., for a Genetic Algorithm approach, the following structure could be defined:

$$\begin{aligned} \mathcal{S}(REP)_{list} &= \langle \infty, \{\langle INTERNAL, -\infty, +\infty, 1, 1\rangle, \langle LEAF, -\infty, +\infty, 0, 0\rangle\}\rangle \\ \mathcal{S}(REP)_{bit-string} &= \langle 0, \{\langle INTERNAL, 0, 1, 1, 1\rangle, \langle LEAF, 0, 1, 0, 0\rangle\}\rangle \end{aligned} \tag{4}$$

This shows that the ranges defined with $VAL(i)$ intervals/sets are allowed to overlap. In the proposed modelling concept a list is understood as special case of a tree:

all internal nodes posses exactly one child node up to the leaf node. In $\mathcal{S}(REP)_{list}$ two types of nodes are defined: INTERNAL and LEAF and each of them allow every value of $\mathbb{R}$ as value for the node x (i.e., $VAL_{node}(x) \in (-\infty, +\infty)$). $\mathcal{S}(REP)_{bit-string}$ defines a bit-string as list with a range for $VAL_{node}(x)$ of $VAL(x) = [0,1]$. But bit-strings uses only the symbols 0 and 1 and not every possible value between them. Therefore, N_{dec} is set to 0 which leads to $VAL(x) = \{0,1\}$.

In hierarchical notation $\mathcal{S}(REP)_{list}$ would look like:

```
representation  decs = inf
 def id = INTERNAL  valmin=-inf valmax=+inf  aritymin=1 aritymax=1
 def id = LEAF      valmin=-inf valmax=+inf  aritymin=0 aritymax=0
```

The $\mathcal{S}(REP)_{bit-string}$ can be derived from $\mathcal{S}(REP)_{list}$ by only changing the $VAL(i)$ set and N_{dec}:

```
representation  decs = 0
  def id = INTERNAL  valmin=0 valmax=1  aritymin=1 aritymax=1
  def id = LEAF      valmin=0 valmax=1  aritymin=0 aritymax=0
```

If all representations are derived from REP by describing $\mathcal{S}(REP)$, each genetic operator and further method working on REP representations needs to be defined and implemented only one time. Most EC libraries implement their operators multiple times for each used and fix-implemented data-structure. This is not very efficient. By using the introduced concept one needs to define the operators only one time and can use them on every data-structure derived from REP.

2.2 Individuals

The main objects considered in ECs are the individuals. An individual has the structure as depicted in (5).

$$IND_x = \langle REP_x, EVAL_x, FIT_x \rangle \tag{5}$$

Each individual possesses the data structure represented by a REP tree. Additionally, it has an evaluation and fitness value. The objective function of the EC computes the evaluation value. The fitness value represents the feasibility of this individual to solve a given task and is computed by the fitness function. Up to now, individuals aren't part of the structure of an EC. Therefore, no hierarchical notation is needed.

2.3 Genetic Operators

A genetic operator O can be understood as a transformation of instances of REP representations of individuals. Primarily, three types of genetic operators are used: unary, binary and n-ary operators. An n-ary operator O_{n-ary} working on n individuals can be defined as shown in (6). It takes n individuals IND as well as the defined structure of the representation $\mathcal{S}(REP)$ and returns n new (modified) individuals valid according to $\mathcal{S}(REP)$. Unary operators (e.g., for mutation) as well as binary operators (e.g., for crossover) are special cases of n-ary operators with $n = 1$ ($O_{unary} = O_1 : \mathcal{S}(REP) \times \{IND\} \rightarrow \{IND\}$) and $n = 2$ ($O_{binary} = O_2 : \mathcal{S}(REP) \times \{IND\} \times \{IND\} \rightarrow \{IND\} \times \{IND\}$), respectively.

$$O = O_{n-ary} = O_n : \mathcal{S}(REP) \times \Pi_{i=1}^{n}\{IND\}_i \rightarrow \Pi_{i=1}^{n}\{IND\}_i \tag{6}$$

$\mathcal{S}(REP)$ is needed for operators which want to change values of nodes or the structure of the tree (e.g., mutation operators). These modifications must be done according to the $VAL(x)$ and ARITY(x) intervals to maintain the validity of REP representations.

All used genetic operators O of an EC have to be modelled within the proposed hierarchical model. For this, an operator-hierarchy is introduced which contains all operators as lists and sequences. The hierarchical order defines the dependencies between the operators. Furthermore, the individual flow is described which yields to the new recombined individuals for the next generation. In (7) the operator-hierarchy $\mathcal{OH}$ is defined as finite sequence of operator nodes ON. This sequence represents the top-level of the operator-hierarchy. Each operator node ON consists of the actual genetic operator (O), its name (ON_{name}), a rate (ON_{rate}), its probability (ON_{prob}), a set of parameters (ON_{param}) and an ordered sequence of other subordinated operator nodes (ON_{succ}) building this way the hierarchy.

$$\begin{aligned}
\mathcal{OH} &= (ON_i)_{i=1}^{k} \\
ON &= (O, ON_{name}, ON_{rate}, ON_{prob}, ON_{param} = \{PAR_i\}, ON_{succ} = \langle ON_i \rangle_{i=1}^{s}) \\
ON_{rate} &\in [0;1]\,,\; ON_{prob} \in [0;1] \\
PAR &= (PAR_{name}, PAR_{value})
\end{aligned} \tag{7}$$

In the hierarchical model, an operator-hierarchy can be defined as follows:

```
operators
  operator name=... rate=... prob=...
    param name=... value=...
    operator name=... rate=... prob=...
  operator name=... rate=... prob=...
```

$\mathcal{OH}$ is represented by the term (`operators`) which can contain any number of `operator` nodes (ON). All subordinated operators are defined by using an appropriate indentation. The resulting individuals of one `operator` node are used as input sub-population for the operators in sequence ON_{succ}. As input for all `operator` nodes directly nested under `operators` the whole current population is used. The proportioning of the input individuals to the `operator` nodes on the same level is ruled according to the rates defined with ON_{rate}.

2.4 Initializer

Initializing functions ("initializer") are defined as shown in (8). They get n parameters as well as a defined structure of representation and return valid trees according to $\mathcal{S}(REP)$. In this sense they are "tree generators".

$$\mathcal{I}: \{PAR\}^n \times \mathcal{S}(REP) \rightarrow REP \tag{8}$$

Luke and Panait showed in a comparative study [1] that the choice of the used initializing operator has no influence on the achievable best fitness. The differences of the operators are, e.g., the size of the generated trees or the distribution of the nodes. Therefore, a complex hierarchy of initializing methods is not neccessary. In the hierarchical model, initializer can be defined as follows:

```
initializer
  method name=...
    param  name=mindepth  value=...
    param  name=maxdepth  value=...
```

An initializing method possesses a name, and optionally a minimum (`mindepth`) and a maximum depth (`maxdepth`). The name means the type of initializing operator (e.g., growth, full or ramped-half-and-half). The minimum and maximum depth define an interval of valid numbers of level the generated trees may have. A typical initializer for GP tasks is:

```
initializer
  method  name=ramped-half-and-half
    param  name=mindepth  value=2
    param  name=maxdepth  value=6
```

If you set `mindepth` = `maxdepth` you define trees of the same length. This is neccessary if you use a $\mathcal{S}(REP)_{list}$ or $\mathcal{S}(REP)_{bit-string}$ description and you want to create fixed length lists. For this, you would need the full method which creates full trees. The following initializer creates lists with a fixed length of 10:

```
initializer
  method  name=full
    param  name=mindepth  value=10
    param  name=maxdepth  value=10
```

2.5 Evaluator

Each new initialized and modified individual has to be evaluated to determine its feasibility for a given problem. The evaluator function $\mathcal{E}$ gets an individual and returns its evaluation value as shown in (9). In the literature this evaluator function is often called the *objective* function.

$$\mathcal{E}: IND \rightarrow \mathbb{R} \tag{9}$$

Because the objective function is unique to each given problem no standard operators can be defined in a reasonable way. Although no operators can be given, the evaluator can be described at least by using comments:

```
evaluator
  comments on the used approach (compute the hamming
                                            distance etc.)
```

2.6 Fitness Functions

After evaluating the individuals, their actual fitness have to be computed. This fitness value is used by all other processes like selection, sorting, replacement etc. as the final value of feasibility of an individual. Equation (10) shows the definition of fitness functions. They take n parameters as well as the actual individual to be processed.

$$\mathcal{F} : \{PAR\}^n \times \{IND\} \rightarrow \mathbb{R} \tag{10}$$

In the hierarchical model, fitness functions can be defined as follows:

```
fitness
  method  name=...
    param  name=...  value=...
```

They only consist of the name of the used method as well as optional parameters. The normalizing fitness could simply look like:

```
fitness
  method  name=normalized
```

2.7 Population

In EC terms a population is a set of λ individuals and can be defined as shown in (11).

$$\mathcal{P} = (IND_1, \ldots, IND_\lambda) \tag{11}$$

Although standard ECs only use an unstructured population (each individual can mate with each other), in recent works structures are defined onto populations (each individual can mate only with its neighbors).

Sprave proposed in [2] to use hypergraphs for modelling population structures. In (1) *REP* is defined as tree. There, an edge is an ordered pair $\langle a, b \rangle$ defining a directed connection between vertices. A graph is the general concept of a tree: each edge is usually defined as (unordered) pair (a, b). But each edge possesses exactly two vertices. Instead of a pair, a hypergraph uses a set for each edge. This way, a single edge can "group" together any number (> 0) of vertices. The pair $H = (X, \varepsilon)$ is a hypergraph as depicted in Fig. 2. X is a set of vertices and each vertex is the index number of an individual. $\varepsilon = (E_i)_{i \in I}$ is a family of subsets, whereby each subset E_i is an edge.

In the hierarchical model, populations can be defined as follows:

```
population size = ...
  type or description of structure
```

Sprave's modelling concept can be used directly in the proposed hierarchical model. Each edge E_i can be understood as a group of individuals. A graph can be represented by using an incidence matrix whose lines are the vertices and columns are the edges. If a vertex (individual) is contained in an edge (group), a 1, otherwise a

0 is set to the position in the matrix, as shown below. Incidence matrices can be used as description of population structures. Each line is represented by the index number of an individual. Thus, only the edges need to be defined. But only those entries need to be defined which has a 1 in the matrix (whereby the appropriate index numbers of the individuals are used):

```
population size = 5
  group
    1 2 3
  group
    3 4 5
```

$$\begin{array}{c} \\ 1 \\ 2 \\ 3 \\ 4 \\ 5 \end{array} \begin{array}{c} \begin{array}{cc} E_0 & E_1 \end{array} \\ \begin{pmatrix} 1 & 0 \\ 1 & 0 \\ 1 & 1 \\ 0 & 1 \\ 0 & 1 \end{pmatrix} \end{array} = \text{incidence matrix}$$

This example is shown in Fig. 2 and uses five individuals and defines two groups (edges): $E_0 = \{1,2,3\}$ and $E_1 = \{3,4,5\}$. For instance, the neighbors of individual 2 are 1 and 3 ($2 \in E_0$). But the neighbors of 3 are all defined individuals because $3 \in E_0 \wedge 3 \in E_1$.

Of course, implementations of this concept should pre-define the frequently used structures, like the following for unstructured (panmictic) populations:

```
population size = 1000
  panmictic
```

$\Leftrightarrow$

```
population size = 1000
  group
    1 2 ... 999 1000
```

A panmictic population structure possesses only one edge (group) containing all vertices (individuals): $H = (X, \varepsilon), X = \{1, \ldots, \lambda\}, \varepsilon = (X)$.

2.8 Selection

The selection function as depicted in (12) picks out an individual from a population according to its fitness. It takes n parameters as well as the population to be processed. The first individual for mating is selected out of the global population. All further individuals (e.g. the second one for crossover operators) have to be selected locally relative to the first individual according to the defined population structure. For this, only a (sub-)population, consisting of the neighboring individuals, is given to the selection function.

$$\mathcal{S} : \{PAR\}^n \times \mathcal{P} \rightarrow IND \tag{12}$$

In the hierarchical model, selection functions can be defined as follows:

```
selection
  method  name=...
    param  name=...  value=...
```

The well-known tournament selection with, e.g., four competitors can be described as follows:

```
selection
  method  name=tournament
    param  name=number  value=4
```

2.9 Replacement Scheme

After creating offspring individuals, they have to be inserted into the population of the next generation. For this, replacement schemes as defined in (13) are used. A scheme gets n parameters and the current population $\mathcal{P}^{(t)}$ as well as the produced offspring and returns the new population $\mathcal{P}^{(t+1)}$ as next generation.

$$\mathcal{R} : \{PAR\}^n \times \mathcal{P}^{(t)} \times \mathcal{P}_{offspring} \rightarrow \mathcal{P}^{(t+1)} \tag{13}$$

In the hierarchical model, replacement schemes can be defined as follows:

```
replacement local=true|false
  method name=...
    param name=... value=...
```

They need an attribute which determines, whether the replacement shall be done relative to the whole population (local="false") or to the selected parents of the offspring (local="true"). For instance, the *Generational Replacement* is thought to replace the whole population:

```
replacement local=false
  method name=delete-N-last
    param name=N value=100%
```

The *Steady-State Replacement* works only on the sub-population of selected parental individuals:

```
replacement local=true
  method name=delete-N-last
    param name=N value=2
```

Replacement schemes can be modelled by determining the set of individuals (whole population, parental sub-population) and the method of replacing a certain number of individuals. Please notice that both replacement schemes could be modelled with the same *Delete-N-Last Replacement* method as known from literature.

2.10 Evolutionary Computation Model

In the previous sections all components of an EC are defined. Based on these, an EC can be finally defined as 9-tuple as shown in (14).

$$\mathcal{EC} = (PARAM = \{PAR_i\}, \mathcal{S}(REP), \mathcal{P}, \mathcal{S}, \mathcal{OH}, \mathcal{R}, \mathcal{I}, \mathcal{E}, \mathcal{F}) \tag{14}$$

An EC needs some global parameters (*PARAM*), the structure of the representation ($\mathcal{S}(REP)$), a population ($\mathcal{P}$), a selection method ($\mathcal{S}$), an operator-hierarchy ($\mathcal{OH}$), a replacement scheme ($\mathcal{R}$), an initializing method ($\mathcal{I}$), an objective function ($\mathcal{E}$) and a fitness function ($\mathcal{F}$).

This leads to the basic skeleton of the hierarchical model:

```
EC
  param name=... value=...

  representation decs=...

  population size=...

  selection

  operators

  replacement local=...

  initializer

  evaluator

  fitness
```

The hierarchical notation of each component is nested into the root called `EC`. The order of components is unimportant. An example for this hierarchical model is given in the next section.

3 What's the Use of It?

The hierarchical modelling allows a more abstract and programming-language-independent description of ECs. This is neccessary if one wants to exchange ECs between different software systems or libraries. But for a better understanding a simple example is designed in the following.

Assumed that an EC is needed which is capable of evolving a sequence of ten bits (bit-string), like 1010101010, to solve a given problem. For this, the well-known Simple GA (SGA) shall be used. A suitable EC-hierarchy can be defined as follows:

```
EC
  param name=number_of_elitists  value=1
  param name=number_of_offspring value=100

  representation decs=0
    def id = INTERNAL  valmin=0 valmax=1  aritymin=1 aritymax=1
    def id = LEAF      valmin=0 valmax=1  aritymin=0 aritymax=0
  population size=100
    panmictic
  selection
```

```
    method  name=roulettewheel
  operators
    operator name=onepointcrossover rate=100 prob=0.9
      param name=sameCuttingPoint value=true
      operator name=pointmutation prob=0.01
  replacement local=false
    method name=delete-N-last
      param name=N value=100%
  initializer
    method  name=full
      param name=mindepth value=10
      param name=maxdepth value=10
  evaluator
    ...comments on your used objective function...
  fitness
    method  name=identity
```

As representation the structure of a bit-string is defined. The range for $VAL_{node}(x)$ is set to $VAL(x) = \{0,1\}$. In SGA each individual can mate with each other, i.e., the population is panmictic. The proportional selection (roulettewheel) is used in SGA. The offspring is generated by applying the one-point crossover followed by a point mutation on the results of the crossover operator. This is described by nesting the genetic operators appropriately. Additionally, the crossover operator uses the same cutting point for each individual maintaining this way the same length of the bit-strings (lists). The Generational Replacement is used in SGA. The initial population is initialized by creating individuals of the same length (=10). This hierarchical model describes the needed EC.

To use this concept with a given EC library, one has several possibilities. For instance, an interpreter can be implemented which uses the functionality of a library to realize the EC-hierarchy. It's imaginable to implement a software generator which generates source-code which uses the functionality of a library to realize the EC-hierarchy. One could implement a procedure which simply configures a library or software system (if possible) to support the EC-hierarchy functionality.

Another way to use it in practice is to write it down, e.g., as XML document. An XML document consists of so-called elements written in the following notation:

```
< NameOfElement
  Attribute[i].name = "Attribute[i].value"
>
  content (...further nested elements...)
</ NameOfElement>
```

Each element is introduced by a start-tag (*<NameOfElement>*) and finished with an end-tag (*</NameOfElement>*). Enclosed in these both tags the element-content is written. Depending of the element-type an attribute-list containing further parameters can be specified. Attributes are able to carry default-values and haven't to be specified explicitly. Some elements might also contain further nested elements building this way an element-hierarchy. But if an element contains no further nested ones, the following shortcut-notation can be used:

```
< NameOfElement
  Attribute[i].name = "Attribute[i].value"
/>
```

The EC-hierarchy can be mapped directly to an element-hierarchy:

```
< EC>
  < param name="number_of_elitists"  value="1" />
  < param name="number_of_offspring" value="100" />

  < representation decs="0">
    <def id="INTERNAL"  valmin="0" valmax="1"
                  aritymin="1" aritymax="1" />
    <def id="LEAF"      valmin="0" valmax="1"
                  aritymin="0" aritymax="0" />
  </ representation>
  < population size="100"> <panmictic /> </ population>
  < selection> <method  name="roulettewheel" /> </ selection>
  < operators>
    <operator name="onepointcrossover" rate="100"
                                  prob="0.9" >
      <param name="sameCuttingPoint" value="true" />
      <operator name="pointmutation" prob="0.01" />
    </operator>
  </ operators>
  < replacement local="false">
    <method name="delete-N-last">
      <param name="N" value="100%" />
    </method>
  </ replacement>
  < initializer>
    <method  name="full">
      <param name="mindepth" value="10" />
      <param name="maxdepth" value="10" />
    </method>
  </ initializer>
  < evaluator> ...comments on your used objective
                        function... </ evaluator>
  < fitness> <method  name="identity" /> </ fitness>
</ EC>
```

An EC described this way can be used in various ways. One can give this XML document directly to an interpreter which realizes the appropriate functionality. But

XML documents can also be translated into runnable source-code [3] covering the modelled functionality or into latex or HTML files realizing this way an automated documentation.

To translate XML documents into source-code one can develop or use appropriate software generators or style-sheets created to produce source-code. These style-sheets can be used directly with so-called XSL(T) processors which are programs able to translate them. But XSL(T) processors are also used within WEB servers as extensions. Thus, one can imagine that it's possible to setup WEB servers able to translate specific EC models written in XML. Furthermore, one can imagine to perform the modelled functionality of an EC model server-sided and to return the results.

4 Conclusions and Future Work

This paper introduced a hierarchical way of specifying ECs. All representations are based on a tree. Different genome structures (e.g., lists) can be realized by defining a structure onto the tree. The individuals can be grouped together as neighbors by defining a structure onto the population. Genetic operators build an operator-hierarchy to realize the flow of individuals. All components of an EC can be arranged in a tree-like manner forming this way the EC-hierarchy.

Such an EC-hierarchy can be used to describe and transfer ECs on an abstract level. XML can be used as one possible notation for representing EC-hierarchies (whereby every notation capable of describing hierarchical structures can be used). A hierarchical model can be used for exchanging ECs between software systems, libraries as well as researchers. They can even be interpreted or translated by software generators.

The proposed concept can only be a first step towards hierarchical modelling of ECs. A lot of advanced mechanisms aren't considered yet. E.g., it should be extended for multiple objective optimization tasks. For this, an IND_x should contain not only a single $EVAL_x$, but a finite sequence of $EVAL_x$ values. Furthermore, multiple populations should be integrated for parallel ECs. As the individuals of one population, the structure of a set of populations themself could be described by using hypergraphs, too. Finally, the definition of grammars for the representation should be possible, too.

References

1. Sean Luke and Liviu Panait, *A Survey and Comparison of Tree Generation Algorithms*, Proceedings of the Genetic and Evolutionary Computation Conference, 2001
2. Joachim Sprave, *A Unified Model of Non-Panmictic Population Structures in Evolutionary Algorithms*, Proceedings of the Congress of Evolutionary Computation, 1999
3. C. Veenhuis, K. Franke, M. Köppen, *A Semantic Model for Evolutionary Computation*, Proc. 6th International Conference on Soft Computing, Iizuka, Japan, 2000

4. C. Veenhuis, M. Köppen, *XML based Modelling of Soft Computing Methods*, Proc. 7th Online World Conference on Soft Computing in Industrial Applications, On the Internet (World Wide Web), 2002
5. W3C (World Wide Web Consortium), *Extensible Markup Language (XML) 1.0 (Second Edition)*, URL: http://www.w3.org/TR/2000/REC-xml-20001006, April 14, 2003

An ALife-Inspired Evolutionary Algorithm for Dynamic Multiobjective Optimization Problems

P. Amato[1] and M. Farina[2]

[1] Soft Computing, Si-Optics and Post-Silicon Technologies Corporate R&D, STMicroelectronics, Via Remo De Feo 1, 80022 Arzano (NA), Italy
paolo.amato@st.com

[2] Soft Computing, Si-Optics and Post-Silicon Technologies Corporate R&D, STMicroelectronics, Via C. Olivetti, 2, 20041, Agrate(MI), Italy,
marco.farina@st.com

Summary. Several important applications require a time-dependent (on-line) in which either the objective function or the problem parameters or both vary with time. Several studies are available in the literature about the use of genetic algorithms for time dependent fitness landscape in single-objective optimization problems. But when dynamic multi-objective optimization is concerned, very few studies can be found. Taking inspiration from Artificial Life (ALife), a strategy is proposed ensuring the approximation of Pareto-optimal set and front in case of unpredictable parameters changes. It is essentially an ALife-inspired evolutionary algorithm for variable fitness landscape search. We describe the algorithm and test it on some test cases.

1 Introduction

Several important applications require a time-dependent (on-line) multiobjective optimization in which either the objective function or the problem parameters or both vary with time. In handling such problems, there exist not many algorithms and certainly there is a lack of test problems to adequately test a dynamic multi-objective evolutionary algorithm.

In this paper we refer to (eventually on-line) of time-varying systems, where (i) the optimal controller is time-dependent (because the system's properties are time-dependent), and (ii) several objectives have to be optimized at the same time. In [6] the authors and K. Deb gave a full formulation of the resulting multiobjective dynamic nonlinear optimization problem, and they formulated some continuous and discrete test problems where the time dependent Pareto-optimal solutions are known analytically.

Optimal design of controllers is a classical field of application for evolutionary computation and. Once closed loop stability is assured, several additional criteria for performances improvement can be considered such as maximum overshooting minimization, settling time minimization and rise time minimization, in order to design

P. Amato and M. Farina: *An ALife-Inspired Evolutionary Algorithm for Dynamic Multiobjective Optimization Problems*, Advances in Soft Computing **1**, 113–125 (2005)
www.springerlink.com

stable and powerful controllers. Several examples of such optimization procedure are available in literature in case of static design problems, that is when the optimization is to be performed off-line and when the model of the system (the plant or the device) is not time dependent. Two early examples can be found in [7] where some controllers (among which an $\mathcal{H}_2/\mathcal{H}_\infty$ one) are optimized with an EMO algorithm. Another classical application of EMO for static controllers optimization consider fuzzy rule set optimization for fuzzy controllers, some examples can be found in [8, 9].

When considering dynamic single-objective optimization problems, several studies are available in the literature [5, 10, 11, 12] about the use of genetic algorithms for. Major modifications in the operators are required for a prompt reaction to time dependent changing. Moreover, several non-GA strategies for dynamic optimization procedure for single objective problems are also proposed in the literature. But when is concerned, very few studies are available in literature [13, 14, 15].

In [16], the authors introduced artificial-life inspired algorithm for dynamic single-objective optimization problems. may be defined as a lower bound for AI following the idea that "the dumbest smart thing you can do is stay alive". This funny motto has deep meaning when ALife is considered for computational purposes [17, 18]. If life and interactions among individuals in a changing environment is itself a type of intelligence, it may be exploited for developing searching algorithms. While classical evolutionary algorithms (GA and ES) consider Darwinian evolution as a type of intelligence to be exploited [19], the proposed method uses life of individuals in a population as a basic form of intelligence and exploits this for search in a dynamic environment.

In this paper, we make an attempt to extend this approach to dynamic multi objective test cases.

2 Problem Setting and Test Cases Description

A dynamic non-linear multiobjective problem can be defined in the following way

Definition 1. *Let t be the time variable,* $\mathbf{V}$ *and* $\mathbf{W}$ *be n-dimensional and M-dimensional continuous or discrete vector spaces,* $\mathbf{g}$ *and* $\mathbf{h}$ *be two functions defining inequalities and equalities constraints and* $\mathbf{f}$ *be a function from* $\mathbf{V} \times t$ *to* $\mathbf{W}$. *A* dynamic non-linear multi-criteria (minimum) optimization problem *with M objectives is defined as:*

$$\begin{cases} \min\limits_{\mathbf{v}\in\mathbf{V}} \mathbf{f} = \{f_1(\mathbf{v},t),\ldots,f_M(\mathbf{v},t)\} \\ \text{s. t. } \mathbf{g}(\mathbf{v},t) \leq 0,\, \mathbf{h}(\mathbf{v},t) = 0\,. \end{cases}$$

In problem 1 some variables are available for optimization (v) and some other (the time t) are imposed parameters being independent from optimization variables; both objective functions and constraints are parameter-dependent. A more general definition of the problem can be found in [15].

Definition 2. *We call* at time t $(\mathcal{S}_P(t))$ *and* at time t $(\mathcal{F}_P(t))$ *the set of Pareto-optimal solutions at time t in design domain and objective domain, respectively.*

Unlike in the single-objective optimization problems, here we are dealing with two different search spaces: decision variable space and objective space. Therefore, the following are the four possible ways a problem can dynamically change. **Type I**: The Pareto-optimal set (optimal decision variables) $\mathcal{S}_P$ changes, whereas the pareto-optimal front (optimal objective values) $\mathcal{F}_P$ does not change. **Type II**: Both $\mathcal{S}_P$ and $\mathcal{F}_P$ change. **Type III**: $\mathcal{S}_P$ does not change, whereas $\mathcal{F}_P$ changes. **Type IV**: Both $\mathcal{S}_P$ and $\mathcal{F}_P$ do not change, although the problem can dynamically change.

A straightforward extension of ZDT and DTLZ test problems developed earlier [20, 21] for two and higher objectives can be considered in order to insert time dependance factors into multiobjective optimization test cases [15]. As it is well known, ZDT and DTLZ problems provide different difficulties which may be encountered when considering real-life multiobjective optimization problems: non-concavity, discontinuity, deceptiveness, presence of local fronts, etc.

When solving dynamically changed problems, such difficulties may transform themselves from one of the above features to another with random sudden jumps or with a gradual change. A generic test problem for such a dynamic situation is presented in the following equation:

$$\min_{\mathbf{X}}(f_1(\mathbf{x}), f_2(\mathbf{x})) = (f1(\mathbf{x}_I), g(\mathbf{x}_{II}) \cdot h(\mathbf{x}_{III}, f_1, g)) \tag{1}$$

where $\mathbf{x}_I$,$\mathbf{x}_{II}$ and $\mathbf{x}_{III}$ are subsets of design variables set $\mathbf{x}$. In the above test problem, there are three functions f_1, g, and h. In the original paper, the following functions were suggested:

$$f_1(\mathbf{x}_I) = x_1,\ g(\mathbf{x}_{II}) = \sum_{x_i \in \mathbf{X}_{II}} x_i^2,\ h(f_1, g) = 1 - \left(\frac{f_1}{g}\right)^2. \tag{2}$$

Each of them can change dynamically or in combination. In dynamic multi-objective test cases the functions f_1, g and h are re-defined in terms of three new time dependent functions F, G and H.

In this paper we consider only one test case the FDA1 (see [22]):

Definition 3 (FDA1). *Type I, convex POFs*

$$\begin{cases} f1(\mathbf{x}_I) = x_1, \\ g(\mathbf{x}_{II}) = 1 + \sum_{x_i \in \mathbf{X}_{II}} (x_i - G(t))^2, \\ h(f_1, g) = 1 - \sqrt{\frac{f_1}{g}}, \\ G(t) = \sin(0.5\pi t), \quad t = \frac{1}{n_t} \lfloor \frac{\tau}{\tau_T} \rfloor, \\ \mathbf{x}_I = (x_1) \in [0, 1], \quad \mathbf{x}_{II} = (x_2, \ldots, x_n) \in [-1, 1]\,. \end{cases} \tag{3}$$

Here, τ is the generation counter, τ_T is the number of generation for which t remains fixed, and n_t is the number of distinct steps in t. The suggested number of variables

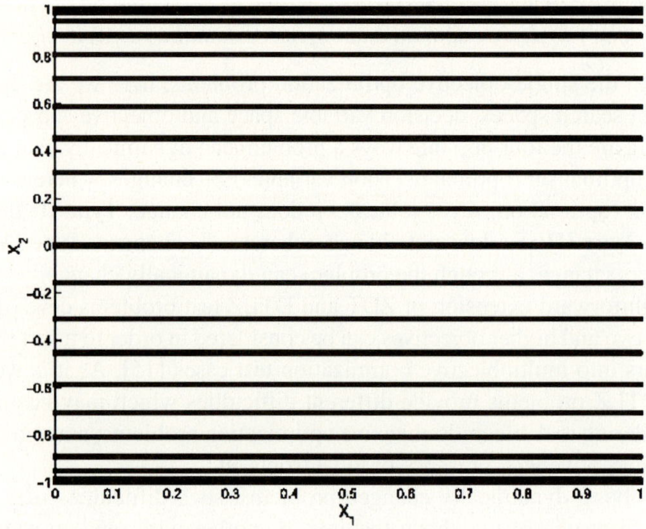

Fig. 1. $\mathcal{S}_{\mathrm{P}}(t)$ for FDA1, first two decision variables, 24 time steps

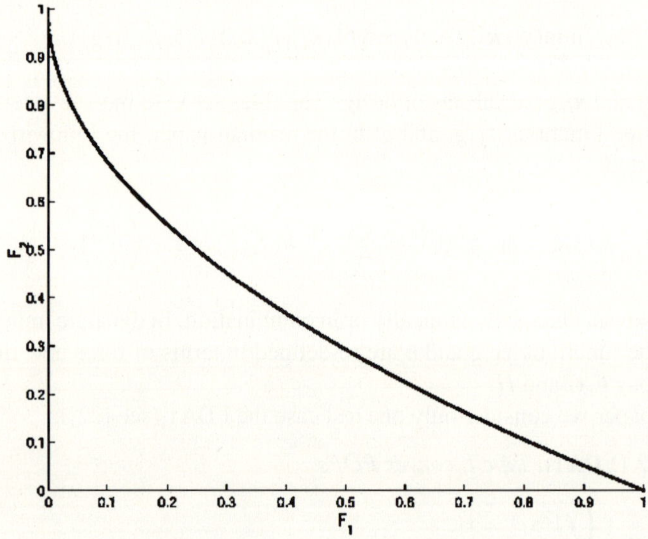

Fig. 2. $\mathcal{F}_{\mathrm{P}}(t)$ for FDA1, 24 time steps

is $n = 20, \tau_T = 5$, and $n_T = 10$. In this problem the Pareto optimal front does not change, while the optimal set (in search space) suddenly change over time every τ_T iterations (as shown in Fig. 2 and Fig. 1 respectively). The task of a dynamic MOEA would be to find the same Pareto-optimal front $f_2 = 1 - \sqrt{f_1}$ every time there is a change in t.

3 Outline of the ALife-Inspired Algorithm for Dynamic Multiobjective Optimization Problems

GAs are based on the simulation of nature evolution and the exploitation of Darwinian natural selection operators; they consider coded strings as genotypes of individuals. For this reason they may be defined a low level evolution imitation, where artificial operators are considered imitating natural operators on genes. On the contrary, the proposed algorithm is based on a population level evolution: coded strings are considered as individuals interacting in a population, and artificial operators imitate interactions between individuals (like meeting, fight and reproduction). In this approach there is no a priori selection; each individual has the same probability of meeting another individual. In some cases, they will procreate two sons, which are added to the population without eliminating the parent. In other cases, they will fight and the stronger (i.e., the one that dominates in Pareto sense the other) will kill the other one. Moreover, individuals which do not encounter anybody else can reproduce in asexual way; hence a new individual (a mutation of his parent) is inserted in the population. As a consequence of all these operators, the population size is variable.

The aim of the algorithm we propose is not to definitively converge, but to be able to "sense" the changing of the Pareto optimal set or front and then automatically follow it.

The general behaviour of the algorithm is depicted in Fig. 3. Each time an individual is considered, he can meet or not another individual according to a probability

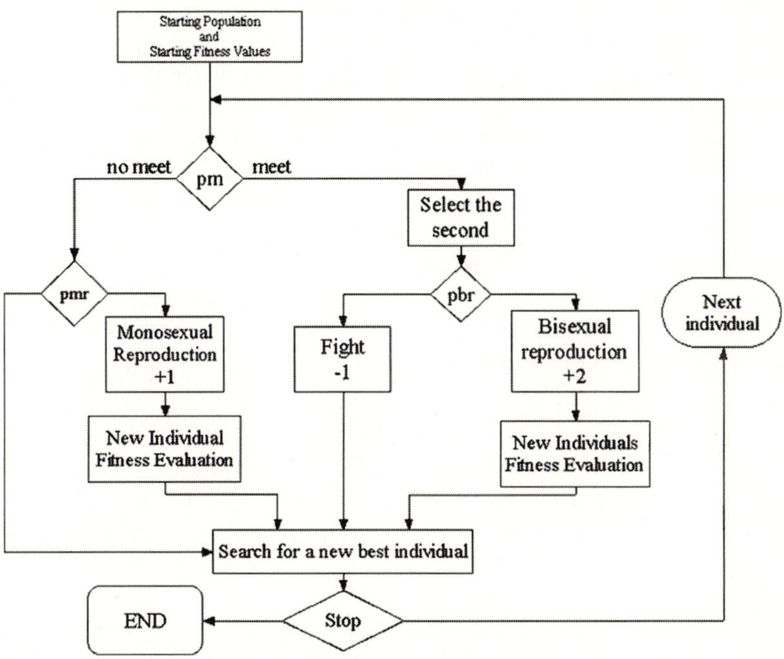

Fig. 3. Principle flowchart of the algorithm

p_m. If he meets someone else, either reproduction or competition can occur. Otherwise can take place or nothing happens. The *meeting probability* p_m is defined in terms of the actual size N_i of the population at iteration i and the maximum size $N_{\max}$ (fixed a priori) in the following way:

$$p_{\mathrm{m}} = \frac{N_i}{N_{\max}} \tag{4}$$

Thus for each individual a value r in $[0,1]$ is randomly chosen. If $r > p_{\mathrm{m}}$ meeting occurs, no meeting otherwise. In this way, when the maximum individual number is approached the meeting probability is very high and viceversa. Consequences of this will be clearer later on.

When the meeting probability is satisfied a new individual is randomly selected for meeting with the current individual. When two individuals meet *either bisexual reproduction* or *fight* can occur; the probability for bisexual reproduction p_{br} is the following:

$$p_{\mathrm{br}} = 1 - p_{\mathrm{m}} \tag{5}$$

Two new individuals are then added to the population, (for further details on bisexual reproduction see the dedicated paragraph below). If bisexual reproduction does not occur, fight is performed between the two selected individuals; the better kills the other one. This operator thus reduces the population size by one.

If meeting does not occur either asexual reproduction or nothing happens. Asexual reproduction, which is performed with probability p_{ar} (equal to p_{br}), adds a new individual to the population. More details on the operators can be found in [16].

In the actual implementation of the algorithm the bisexual reproductions is the standard random crossover operation. Thus the fight operator is the only one that involves the evaluation of the objective functions. When fight occurs, all the objective functions f_j are evaluated for both the individuals. The dominating individual in Pareto sense[1] survives, while the dominated one dies and is eliminated from the population. If nobody dominates the other, the algorithm eliminates the individual with a greater number of individual in a given neighborhood. This happens in order to preserve diversity among individuals.

The probabilistic routing strategy leads to the population size N behavior shown in Fig. 4. As can be seen it oscillates around a probabilistic computable value satisfying the following logistic formula:

$$N_{i+1} = N_i \left(1 - 2\frac{N_i}{N_{\max}}\right) + \frac{1}{N_{\max}}; \tag{6}$$

where i is the iteration index. Two limit behaviors correspond to $N \sim N_{\max} \Rightarrow p_{\mathrm{m}} \sim 1$ and $N \sim 0 \Rightarrow p_{\mathrm{m}} \sim 0$. In the former case meeting and fight always occurs and population size reduces. In the latter meeting never occurs and asexual reproduction

[1] Let $\mathbf{v}_1, \mathbf{v}$ be two candidate solutions (individuals). Then $\mathbf{v}_1$ is said to *dominate* $\mathbf{v}_2$ in the Pareto sense if and only if the following conditions hold: (i) $f_i(\mathbf{v}_1) \leq f_i(\mathbf{v}_2)$ for all $i \in \{1,2,\ldots,M\}$, (ii) $f_j(\mathbf{v}_1) < f_j(\mathbf{v}_2)$ for at least one $j \in \{1,2,\ldots,M\}$.

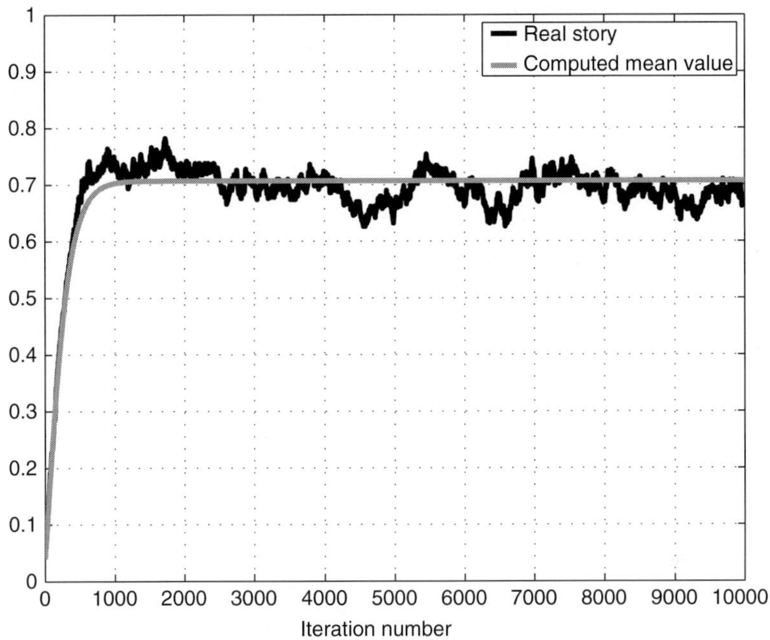

Fig. 4. Population size story compared with the computed analytical mean value

always adds one individual; consequently population size increase by one at each iteration.

The proposed strategy is supposed to run for an indeterminate time following system changing, without definitely converging towards a final optimum unless a static system is considered. For test problems a fictitious maximum iteration or generation number is imposed but it only has an obvious practical meaning. The algorithm is extremely flexible because probability threshold values are updated at each iteration.

4 Application of the ALife-Inspired Algorithm to Test Case FDA1

For the sake of clarity, we consider a problem with a two dimensional input space (X_1, X_2) and two objective functions (f_1, f_2). In this way we can easily plot the problem in both domains. Moreover we consider only two changes of the Pareto optimal set.

Since the population size is not fixed and the algorithm proceeds individual by individual, it is not properly correct to speak of "epochs". However we will use this term to signify the application of the algorithm a number of times (iterations) equal to the average size of the population.

The starting population is uniformly distributed in the search space (Fig. 5), while the analytical optimal set is the straight line $x_2 = 0$ (dotted line in Fig. 6). After 45 epochs (Figs. 7, 8), there is a good approximation of the Pareto-optimal set and front.

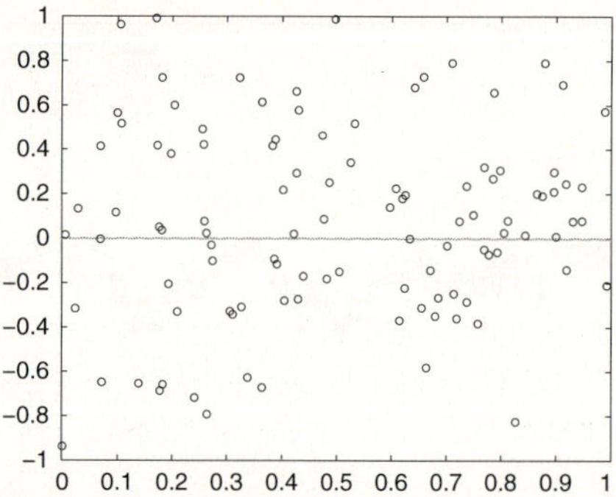

Fig. 5. Starting population in search space

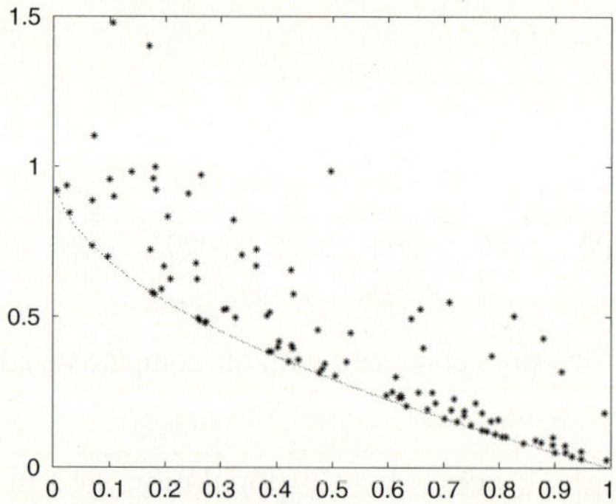

Fig. 6. Starting Population in objective space

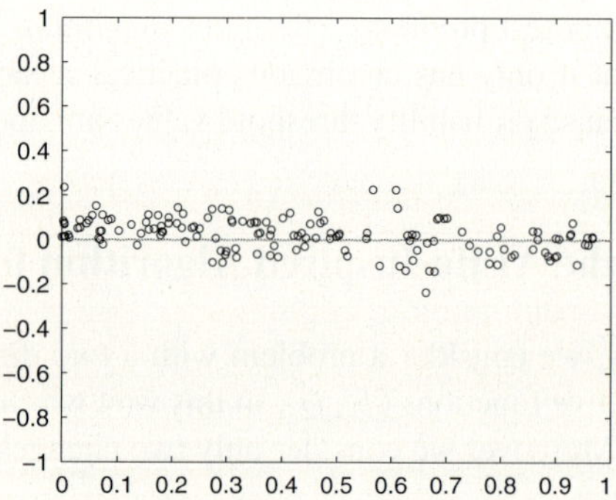

Fig. 7. Population in search space at epoch 45

As it is well known a good approximation of Pareto-optimal front (and set) requires the solution to be (i) close to the exact front (and set), and (ii) as distributed as possible on the front (and set); this two requirements are usually clashing in multiobjective evolutionary algorithms. Due to the absence of selection pressure (there is no fitness

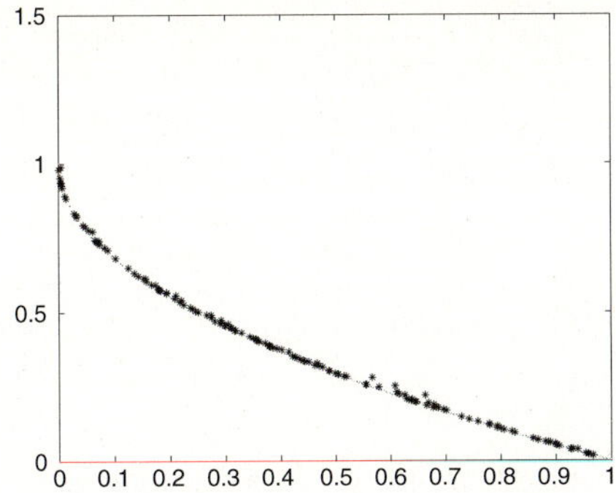

Fig. 8. Population in objective space at epoch 45

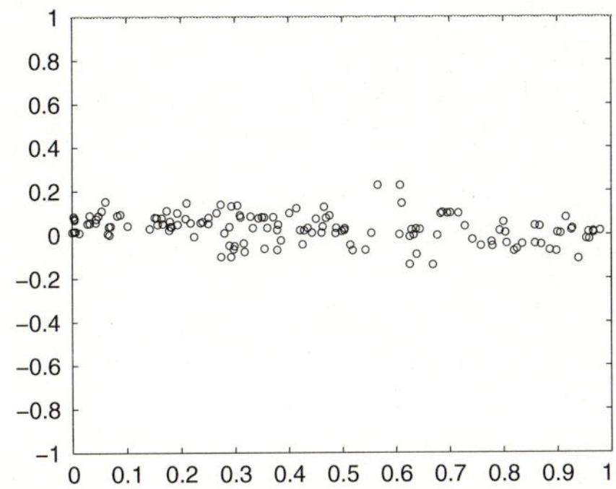

Fig. 9. Epoch 46: Sudden change of the optimal front in search space; now it becomes the straight (*dotted*) line $x_2 = 1$

based selection in the proposed algorithm) the population covers the entire front and set. Moreover the absence of fitness based selection is one of the main differences between the proposed algorithm and the evolution based multiobjective optimization algorithms. During the 46th epoch there is a sudden change of the Pareto-optimal set; now it is the straight line $x_2 = 1$. Consequently there is a big approximation error both in search and objective space (Figs. 9, 10). Finally after 100 epochs the population is again a good approximation of Pareto-optimal set and front (Figs. 11, 12). As evident the absence of fitness based selection is a drawback when speed of reaction to time-dependent changes is concerned.

The main drawback of this algorithm is that, in general, it converges slowly (in term of number of epochs). Moreover the converge velocity strongly depends on the distribution of the population in search space. In fact after the sudden change (epoch 46), the convergence to the new optimal set is much more slower than the previous one. Slowness of ALife algorithm is a known problem. For single-objective optimization problem there is a speed-up by introducing crossover operators that privilege the better parent (see [16]). However this kind of operators cannot be easily

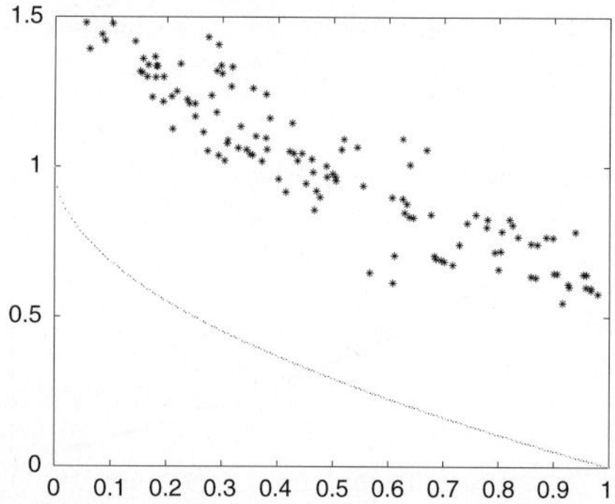

Fig. 10. Population in objective space at epoch 46

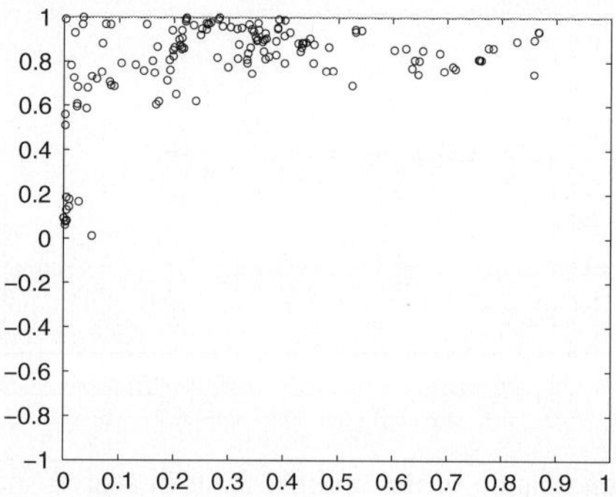

Fig. 11. Population in search space at epoch 100

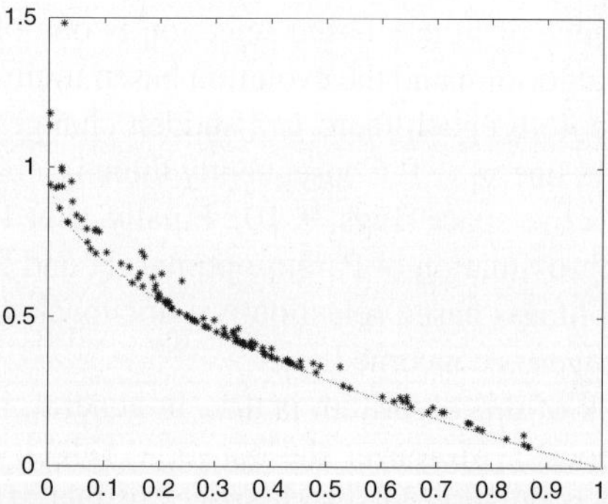

Fig. 12. Population in objective space at epoch 100

introduced in the Multiobjective algorithm, because we have to keep the diversity among the individuals (both in search and objective space).

On the other side, this algorithm has two main advantages. The first is that it is able to automatically follow the changes a dynamic Multiobjective optimization problem, without any external help.

The second one is that the evaluation of the objective function (being usually a computationally expensive task or requiring a measure on the system) is needed only when the meeting probability is satisfied and not at each iteration, as it is required for fitness based selection in an evolution based algorithm.

5 Conclusion

In this paper we introduced an ALife-inspired evolutionary algorithm for dynamical multiobjective optimization problems. Although not flawless, this algorithm is simple to implement and able to detect the change of objectives or constraints of the problem, and then to follow the Pareto-optimal set and fronts.

Our work is only at a preliminary state. Further developments may be concerned with the following considerations.

On one side, the convergence to the optimal front and set could be fastened (in terms of number of iteration) by exploiting the information about Pareto optimality. For example the worst individuals in Pareto sense could be automatically eliminated (without having to fight). Or, on the other side, the Pareto optimal individuals could gain some advantages (re-introducing in this way a kind of selection pressure).

On the other side, the number of objective function evaluation could be decreased by changing the probability route leading to a fight (the only operator that requires objective evaluation). The corresponding increasing of population size (fight is also the only operator that decrease population size) may be balanced by introducing, for example, a kind of spontaneous decay (as in ant systems) – i.e. the automatic elimination of the individuals that lived for more than a given iteration threshold.

References

1. M. A. Lee and H. Esbensen. Fuzzy/Multiobjective Genetic Systems for Intelligent Systems Design Tools and Components. In Witold Pedrycz, editor, *Fuzzy Evolutionary Computation*, pp. 57–80. Kluwer Academic Publishers, Boston, Massachusetts, 1997.
2. P.M. Reed and B.S. Minsker. Discovery & Negotiation using Multiobjective Genetic Algorithms: A Case Study in Groundwater Monitoring Design. In *Proceedings of Hydroinformatics 2002*, Cardiff, UK, 2002.
3. Kalyanmoy Deb and Tushar Goel. A Hybrid Multi-Objective Evolutionary Approach to Engineering Shape Design. In Eckart Zitzler, Kalyanmoy Deb, Lothar Thiele, Carlos A. Coello Coello, and David Corne, editors, *First International Conference on Evolutionary Multi-Criterion Optimization*, pp. 385–399. Springer-Verlag. Lecture Notes in Computer Science No. 1993, 2001.

4. K.C. Tan, K. Sengupta, T.H. Lee, and R. Sthikannan. Autonomous Registration of Disparate Spatial Data via an Evolutionary Algorithm Toolbox. In *Congress on Evolutionary Computation (CEC'2002)*, volume 1, pp. 31–36, Piscataway, New Jersey, May 2002. IEEE Service Center.
5. F. Vavak, K. A. Jukes, and T. C. Fogarty. Performance of a genetic algorithm with variable local search range relative to frequency of the environmental changes. *Genetic Programming 1998: Proceedings of the Third Annual Conference*, 1998.
6. M. Farina, P. Amato, and K. Deb. Dynamic multi-objective optimization problems: Test cases, approximations and applications. *IEEE Transactions on Evolutionary Computation*, 8(5):425–442, 2004.
7. Carlos Manuel Mira de Fonseca. *Multiobjective Genetic Algorithms with Applications to Control Engineering Problems*. PhD thesis, Department of Automatic Control and Systems Engineering, University of Sheffield, Sheffield, UK, September 1995.
8. Jessica M. Anderson, Tessa M. Sayers, and M. G. H. Bell. Optimization of a Fuzzy Logic Traffic Signal Controller by a Multiobjective Genetic Algorithm. In *Proceedings of the Ninth International Conference on Road Transport Information and Control*, pp. 186–190, London, April 1998. IEE.
9. Anna L. Blumel, Evan J. Hughes, and Brian A. White. Fuzzy Autopilot Design using a Multiobjective Evolutionary Algorithm. In *2000 Congress on Evolutionary Computation*, volume 1, pages 54–61, Piscataway, New Jersey, July 2000. IEEE Service Center.
10. Christopher Ronnewinkel, Claus O. Wilke, and Thomas Martinetz. Genetic algorithms in time-dependent environments. In L. Kallel, B. Naudts, and A. Rogers, editors, *Theoretical Aspects of Evolutionary Computing*, pp. 263–288, Berlin, 2000. Springer.
11. J. Branke. Evolutionary approaches to dynamic optimization problems – A survey. *Juergen Branke and Thomas Baeck editors: Evolutionary Algorithms for Dynamic Optimization Problems*, 13:134–137, 1999.
12. J.J. Grefenstette. Evolvability in dynamic fitness landscapes: A genetic algorithm approach. *Proc. Congress on Evolutionary Computation (CEC99) Washington DC IEEE press*, pp. 2031–2038, 1999.
13. Zafer Bingul, Ali Sekmen, and Saleh Zein-Sabatto. Adaptive Genetic Algorithms Applied to Dynamic Multi-Objective Problems. In Cihan H. Dagli, Anna L. Buczak, Joydeep Ghosh, Mark Embrechts, Okan Ersoy, and Stephen Kercel, editors, *Proceedings of the Artificial Neural Networks in Engineering Conference (ANNIE'2000)*, pp. 273–278, New York, 2000. ASME Press.
14. Kazuo Yamasaki. Dynamic Pareto Optimum GA against the changing environments. In *2001 Genetic and Evolutionary Computation Conference. Workshop Program*, pp. 47–50, San Francisco, California, July 2001.
15. M. Farina, K. Deb, and P. Amato. Dynamic multiobjective optimization problems: Test cases, approximation and applications. *To be published in the Proceedings of EMO'2003*, 2003.
16. P. Amato, M. Farina, G. Palma, and D. Porto. An alife-inspired evolutionary algorithm for adaptive control of time-varying systems. In *Proceedings of the EUROGEN2001 Conference, Athens, Greece, September 19-21, 2001*, pp. 227–222. International Center for Numerical Methods in Engineering (CIMNE), Barcelona, Spain, March 2002.
17. C.G. Langton. *Artificial life: an overview*. MIT Press, 1995.
18. C. Adami. *Introduction to Artificial life*. Springher-Verlag, 1998.
19. M. Mitchell and S. Forrest. Genetic algorithms and artificial life. *Santa Fe Institute Working Paper 93-11-072*. (to appear in Artificial Life).
20. Kalyanmoy Deb. Multi-Objective Genetic Algorithms: Problem Difficulties and Construction of Test Problems. *Evolutionary Computation*, 7(3):205–230, Fall 1999.

21. Kalyanmoy Deb, Lothar Thiele, Marco Laumanns, and Eckart Zitzler. Scalable Test Problems for Evolutionary Multi-Objective Optimization. Technical Report 112, Computer Engineering and Networks Laboratory (TIK), Swiss Federal Institute of Technology (ETH), Zurich, Switzerland, 2001.
22. Marco Farina, Alessandro Bramanti, and Paolo Di Barba. A GRS Method for Pareto-Optimal Front Identification in Electromagnetic Synthesis. *IEE Proceedings–Science, Measurement and Technology*, 2002. (In Press).

Part IV

Evolutionary Algorithms II

Optimum Tests Selection for Analog Circuits with the Use of Genetic Algorithm

Jerzy Rutkowski, Lukasz Zielinski, and Bartlomiej Puchalski

Faculty of Automatic Control, Electronics and Computer Science, Silesian University of Technology, ul. Akademicka 16, 41-100 Gliwice, Poland
jr@boss.iele.polsl.gliwice.pl, lukasz.zielinski@comverse.com, bartlomiej.puchalski@centertel.pl

Summary. This paper deals with the problem of optimum test program development in analog circuit testing process. Pre-production testing and production testing are taken into account by the presented algorithm. Genetic algorithms are used as a new method for optimum test selection. This approach enhances quality and speeds up finding suboptimal solutions by finding more then one good result in each genetic algorithm cycle. Results for a hypothetical example are given to clarify and discuss the method and they seem to be very promising.

1 Introduction

Testing of analog circuits is a very complex process. Different types of faults may occur. Generally they are grouped into two main categories: catastrophic faults (shorts, opens and other topological changes in a circuit) and parametric faults (variations of parameters due to imperfect parametric control of manufacturing process) [1, 2]. Such a variety of possible failure modes requires dedicated testing techniques. They can be divided into three categories: simulation before test (SBT), simulation after test (SAT) and functional testing (FT). Tests belonging to SBT and SAT are usually simple tests targeted to detect most probable failures at the earliest possible stage. Following [1], SBT is especially effective in detection of catastrophic faults while SAT is suitable for diagnosis of parametric faults. Functional testing checks whether a given circuit meets the design requirements. It could be possible to detect all faults by performing only functional testing. However, especially in case of mature products, it is better to perform simple tests targeted to detect most probable faults, before functional testing is performed. The whole process is terminated as soon as any failure is detected. Thanks to this, testing time is minimized. On the other hand in the initial phase of manufacturing process the information about possible failures of the circuit must be maximized. Minimization of necessary testing time is less important.

For the first time this problem was investigated in [1, 3]. An algorithm for optimum test selection was presented. A new criterion for test selection, utilizing

Jerzy Rutkowski et al.: *Optimum Tests Selection for Analog Circuits with the Use of Genetic Algorithm*, Advances in Soft Computing **1**, 129–135 (2005)
www.springerlink.com

information channel concept was proposed. Both, the number of detected faults, and time effort of each test were considered. Results presented in [3] will be referenced to.

In this paper a new approach to the problem of optimum test selection will be described. Application of genetic algorithm is considered. It has already been proven that application of these evolutionary techniques is very useful in case of analog circuits diagnosis [4]. In Sect. 2 a brief description of new method using genetic algorithms is given. Section 3 contains computational examples. Different stages of testing process are considered. Conclusions are given in Sect. 4.

2 Method Description

Finding optimal test set problem has been solved using genetic algorithm, which are state-space search techniques modelled on natural evolutionary mechanisms. Genetic algorithm [5, 6] is a compromise between random and informed search method. Variables are mapped to chromosomes, which consist strings of genes and genetic operators create new generations, such as: selection, crossover and mutation. From the whole population only some best-fitted chromosomes are taken into account in next generation. To ensure this, a selection operator is introduced. There are many known algorithms of selection. One of the most popular is stochastic universal sampling and roulette wheel selection. Crossover is an operator that combines randomly chosen portions of two chromosomes. Mutation randomly changes the values in each gene in a chromosome. In more advanced genetic algorithms inversion operator is used (randomly changing position of gene in one chromosome). The whole process is executed until a termination condition is satisfied. In this paper, the termination condition is satisfied when a given number of generations is reached. A typical flow of genetic algorithm used for finding optimal test set is presented below:

1. Randomly create a population P of n chromosomes. Evaluate their fitness values and store the best chromosome. Create an empty subpopulation Q
2. Select two chromosomes ch_i and ch_j from the population P with probability:

$$P_{chrom} = \frac{f_i}{\sum f_k} \tag{1}$$

   where f_i is the fitness value of chromosome ch_i
3. Crossover ch_i and ch_j to generate two new chromosomes $ch_i^{'}$ and $ch_j^{'}$
4. Mutate $ch_i^{'}$ and $ch_j^{'}$ with a certain probability (from the range 0–0,1)
5. Evaluate the fitness values of $ch_i^{'}$ and $ch_j^{'}$ and add them in Q. If Q contains less then m chromosomes, go to step 2
6. Replace m chromosomes in P with the ones in Q and empty Q Update the best chromosome and increment the generation counter. If the generation counter reached a pre-specified number, terminate the process and return the best chromosome. Go to step 2, otherwise

This algorithm prevents premature convergence of population and reaches an optimal solution with smaller number of fitness evaluations.

In each generation the state of algorithm is called the population, consisting of many strings (chromosomes). Each string represents a candidate solution to the optimization problem. In this paper, bit-string representation was chosen. The ith bit in a string is equal 1 if the ith test is included in the set. For instance, the chromosome [100010010] informs that tests 1, 5 and 8 are taken into account. The length of each chromosome is equal to number of all available tests.

It is assumed that test database is available, i.e. list of all realizable tests, their time efforts and information gained are known. Acquisition of such database is not the subject of this research. To properly evaluate best chromosomes among population, a fitness function has to be determined. Such function should take into account all criteria of optimization, such as: number of selected tests, their time efforts and information gained. A proper determination of a fitness function is the algorithm crucial point. Function described by (2) has been proposed.

$$F_{fit} = a \cdot \frac{N_{rt}}{N_{at}} + b \cdot \frac{N_{ff}}{N_{af}} + c \cdot \frac{T_{rt}}{T_{at}} \tag{2}$$

where:

N_{rt}–number of reduced tests (not included in the set)
N_{at}–number of all tests
N_{ff}–number of detected faults
N_{af}–number of all faults
T_{rt}–total time effort of reduced tests
T_{at}–total time effort of all tests

Fitness function consists of two parts: information factor which is represented by weighs a and b, and time factor which is represented by weigh c. By selecting appropriate weighs we can decide which part is prevailing in function. The weighs must fulfill the following equation:

$$a + b + c = 1 \tag{3}$$

In pre-production testing information about possible failure modes should be predominant. On the other hand, in case of production testing, time reduction should be the predominant factor. According to the defined fitness function, value of the best chromosome should be very close to 1.

3 Computational Examples

The method operation will be illustrated by a hypothetical example, originally considered in [3]. There are $N_{at} = 15$ tests available and $N_{af} = 22$ faults selected. For each test, all 23 simulated CUT states (22 faults: $x_1, \ldots, x_2$+"healthy": x_0) have been grouped into ambiguity sets, numbered 0,1, ... Assignment of states to ambiguity

Table 1. Hypothetical tests results and time efforts

Circuit State	Test Results T_1	T_2	T_3	T_4	T_5	T_6	T_7	T_8	T_9	T_{10}	T_{11}	T_{12}	T_{13}	T_{14}	T_{15}
$x_0(nom)$	0	0	0	0	2	0	2	0	0	5	0	0	9	0	5
x_1	1	1	0	0	1	0	1	6	5	1	0	1	0	9	0
x_2	1	2	5	0	2	0	3	6	3	2	0	2	0	9	0
x_3	1	3	3	0	3	0	4	2	2	2	0	1	0	9	0
x_4	1	4	5	0	0	0	2	3	1	2	0	1	3	1	2
x_5	1	5	1	0	0	0	3	5	8	3	0	2	2	2	1
x_6	1	6	2	0	5	0	0	2	9	5	3	1	1	3	4
x_7	1	3	3	0	2	0	1	3	7	4	2	2	4	3	3
x_8	1	3	4	0	3	0	3	1	5	6	3	1	6	3	3
x_9	1	4	5	0	6	0	3	2	6	4	2	2	6	4	3
x_{10}	1	6	6	5	1	0	2	0	4	3	1	3	9	5	3
x_{11}	1	6	7	1	3	0	1	0	3	0	1	3	9	6	3
x_{12}	1	6	8	2	4	0	3	6	2	0	1	3	5	7	3
x_{13}	1	2	0	3	0	0	0	5	5	7	1	3	8	8	5
x_{14}	1	2	0	4	0	4	2	6	7	6	1	2	7	9	5
x_{15}	1	2	2	5	1	2	3	3	8	5	2	2	1	9	5
x_{16}	1	1	7	6	3	1	4	3	9	4	2	1	10	9	5
x_{17}	1	2	3	7	5	2	2	6	3	8	4	2	10	9	5
x_{18}	1	4	0	8	2	1	3	4	0	3	3	4	10	5	5
x_{19}	1	4	0	0	7	2	4	1	8	4	2	3	10	4	4
x_{20}	0	0	1	0	2	3	2	6	6	4	1	3	10	3	4
x_{21}	0	0	1	0	7	4	1	2	2	4	2	3	10	4	4
x_{22}	0	0	1	0	3	5	2	6	3	3	2	2	10	9	4
Time effort c_i	1.2	1	3	2.7	1.1	2.4	1.6	7.3	3.2	6.3	3.2	1.8	2.9	4	3.7

sets has been depicted in Table 1, together with the normalized time efforts of tests. Origin of these data is immaterial for the intended optimization of test program. A genetic algorithm is applied to find an optimum set of tests. The assumed constants of the algorithm are as follows:

- number of individuals (chromosomes) in population: $n = 30$
- number of generations: 500
- mutation probability: 0,05
- crossover: single point crossover with probability 0,7
- selection: Stochastic Universal Sampling

Recombination and competitive selection pressure leads each succeeding generation to have individuals of better fitness. Stochastic universal sampling shows better results then wheel selection.

At first, the following values of weighs are assumed: $a = 0.5$, $b = 0.5, c = 0$. Circuit state Test results For such weights, information about failure modes is maximized, while time effort is not taken into account. This corresponds to pre-production

Table 2. Optimum test set found by genetic algorithm (only information factor is considered)

	Test Number														
Solution	1	2	3	4	5	6	7	8	9	10	11	12	13	14	15
1	0	0	0	0	0	0	0	0	1	0	1	1	0	0	0
2	0	0	0	0	0	0	0	0	1	0	1	0	0	1	0
3	0	0	1	0	0	0	0	0	1	1	0	0	0	0	0
4	0	0	0	0	0	0	0	1	0	1	0	1	0	0	0
5	0	0	0	0	0	0	0	0	1	1	0	0	1	0	0
6	0	0	0	0	0	0	0	1	1	1	0	0	0	0	0
7	0	1	0	0	0	0	0	1	0	1	0	0	0	0	0
8	0	0	0	0	1	0	0	0	1	0	0	1	0	0	0
9	0	0	1	0	1	0	0	0	0	1	0	0	0	0	0
10	0	0	0	0	0	0	0	0	1	1	0	1	0	0	0

testing. The obtained results are presented in Table 2. Fitness value of all solutions equals 0.900 As can be seen, several equivalent solutions are found. Number of tests, enabling detection of all faults, can be reduced from 15 to 3. Solution obtained by means of information channel method [3] is contained among solutions of Table 2. It is a solution of the 5th row, with T_{13} performed at first, as this test gives the maximum mutual information. It is worth to mention, that application of genetic algorithm gives many alternative solutions, while information channel based method only one. Solution of the GA optimization gives a set of the most desirable tests, under conditions defined by the fitness function weights. This optimization does not take into account test order of performance, also very important feature of a test program. Having a list of alternative solutions, a design engineer may intuitively select the best set of tests and establish the order of performance. Of course, the less costly is the test the sooner it should be performed. Test T_2 is the least costly and should be performed at first, if solution of the 7th row is selected. However, two other tests of this solution are the most costly and therefore it is difficult to say whether this solution is better or worse than any other (in Table 2).

Next, significance of time effort has been increased, however, it still has smaller value than information factor. The following values of weighs are assumed: $a = 0.45$, $b = 0.45$, $c = 0.1$. The obtained results are presented in Table 3. Two equivalent solutions have been found, with the fitness of 0.899. Both selected sets of tests enable

Table 3. Optimum test set found by genetic algorithm (time effort factor is introduced)

	Test Number														
Solution	1	2	3	4	5	6	7	8	9	10	11	12	13	14	15
1	0	1	0	0	0	1	1	0	0	0	0	0	0	0	0
2	0	1	0	0	1	0	0	0	0	0	0	0	1	0	0

detection of all faults, same as sets of Table 2, however, time effort of these sets is less than time effort of sets contained in Table 2.

Finally, significance of time effort has been assumed greater than information about possible failures. Values of weighs are: $a = 0.25$, $b = 0.25$, $c = 0.5$. Only one solution has been found, with the fitness of 0.899. The obtained result is presented in Table 4. This time, a number of tests was reduced to 2. This solution offers minimum testing time. However, in this case it is not possible to detect all the faults. As can be noticed, there is a tradeoff between the reduction of testing time and maximization of failure information. The obtained solution is suitable at early stages of production.

Table 4. Optimum test set found by genetic algorithm (time effort factor is prevailing)

	Test Number														
Solution	1	2	3	4	5	6	7	8	9	10	11	12	13	14	15
1	0	1	0	0	0	0	0	0	1	0	0	0	0	0	0

4 Conclusions

Selection of optimum test set is very important in optimization of analog circuit testing process. Optimum test selection is an NP-hard problem. It can be solved using different heuristic methods. It has been shown that application of genetic algorithms enables to solve the problem. Moreover, using this approach, several alternative solutions can be obtained, in contrary to other methods of optimization. The presented test selection strategy takes into consideration information about failure modes and time effort of evaluated tests. Adequate manipulation of the fitness function weights enables finding of solution(s) of the desired fault detection level and time effort.

Some other hypothetical examples have been tested and the obtained results compared with the results obtained by the information channel method [3]. In all cases the results were comparable. To check whether the obtained solution is the optimum one, performance of an exhaustive search is necessary. Unfortunately, such search is enormously time consuming, even for a moderate size example, like the one presented in this paper. Total number of solutions to be checked by an exhaustive search is:

$$N = \sum_{i=1}^{N_{at}} C_i^{N_{at}} \tag{4}$$

References

1. Milor L (1998) A tutorial introduction to research on analog and mixed-signal circuit testing. IEEE Trans. Circuits and Systems, Vol. 45. 1389–1407.

2. Bandler J W, Salama A (1985) Fault diagnosis of analog circuits. Proceedings of the IEEE, Vol. 73. 1279–1325.
3. Rutkowski J, Puchalski B (2002) Optimum test selection for analog circuits with the use of information channel concept. Proceedings of International Conference on Signals and Electronic Systems. Wroclaw–Swieradow Zdroj 325–330.
4. Golonek T, Rutkowski J (2002) Use of Evolution Strategies to Analog Testing with Time Domain Stimuli. Proceedings of International Conference on Signals and Electronic Systems. Wroclaw–Swieradow Zdroj 307–312.
5. Goldberg D E (1989) Genetic Algorithms in Search, Optimization and Machine Learning. Addison–Wesley Publishing Company.
6. Muhlenbein H (1989) Parallel genetic algorithms, population genetics and combinatorial optimization. Proc. 3rd ICGA. 416–422.

Unsupervised Pixel Clustering in Multispectral Images by Genetic Programming

I. De Falco[1], A. Della Cioppa[2], and E. Tarantino[1]

[1] ICAR, National Research Council of Italy, Via P. Castellino 111, 80131 Naples, Italy
{ivanoe.defalco,ernesto.tarantino}@na.icar.cnr.it
[2] Associato INFM, INFM, Dept. of Computer Science and Electrical Engineering, University of Salerno, Via Ponte don Melillo 1, 84084 Fisciano (SA), Italy
adellacioppa@unisa.it

Summary. In this paper an innovative approach to Spectral Pattern Recognition for multispectral images based on Genetic Programming is introduced. The problem is faced in terms of unsupervised pixel classification. Given an image consisting in B bands, the goal is to find the optimal number of clusters and the positions of their centres in the B-dimensional hyperspace, which allow the best possible description of the image. The pixels are then assigned to the clusters according to "minimum distance to means" principle. Furthermore the system is endowed with mechanisms able to avoid that cluster centres may be too close one another, which would favour an excessive increase in their number. As a result a good-quality clustered image is achieved. The output consists of the image divided into clusters, the proposed number of clusters, the centre coordinates and the spectral signature for any such cluster and solution fitness value. The results are compared against those achieved by another system, MultiSpec, which performs supervised classification, yet it is endowed with some features typical of an unsupervised classification system.

Key words: multispectral images, pixel clustering, Genetic Programming

1 Introduction

Remote Sensing [1, 2, 3] consists in deriving information about Earth's land and water areas from images taken at a distance, either from satellites or from airplanes. It relies on measurement of electro-magnetic energy reflected or emitted from the objects of interest at the surface of the Earth. Fields of application include disaster assessment, urban trends monitoring, pollution detection, land use development, water management, erosion assessment, weather forecast, climate changes studies, forest inventarization, and others [2, 3]. Given an energy source which illuminates target area, Remote Sensing can be accomplished by means of the following steps: record the reflected energy by means of a sensor, transmit recorded information to a receiving station, process data into a digital image and, finally, interpret and analyze

I. De Falco et al.: *Unsupervised Pixel Clustering in Multispectral Images by Genetic Programming*, Advances in Soft Computing **1**, 137–149 (2005)
www.springerlink.com

this image. This can be made at different wavelengths, resulting in a multispectral digital image.

As concerns the analysis step, a crucial task is feature classification. Classification procedures can be broadly subdivided into *supervised* and *unsupervised* classification. The objective is to assign all pixels in the image to particular classes or themes (e.g. water, coniferous forest, deciduous forest, corn, wheat, etc.). The resulting classified image is comprised of a mosaic of pixels, each of which belongs to a particular theme, and is essentially a thematic "map" of the original image.

A human analyst attempting to classify features in an image uses the elements of visual interpretation to identify homogeneous groups of pixels which represent various features or land cover classes of interest. On the contrary, digital image classification uses the spectral information represented by the digital numbers in one or more spectral bands, and attempts to classify each individual pixel based on this spectral information. This type of classification is termed *spectral pattern recognition.*

Recently researchers have started to apply evolutionary techniques to fulfill several image understanding tasks. As regards spectral pattern recognition, however, at present only little research is reported about supervised pixel classification, and we know only one paper dealing with unsupervised approach [4].

This paper introduces an innovative approach based on Genetic Programming (GP) [5] to unsupervised pixel classification for multispectral images. Unlike other unsupervised classification approaches, ours may automatically determine the most suitable number of clusters.

The paper is organized as follows. Section 2 describes the problems of spectral pattern recognition and of unsupervised pixel classification. Section 3 briefly reports on the use of Evolutionary Algorithms in image understanding and particularly in spectral pattern recognition. Section 4 focuses on our GP-based approach. In order to assess its feasibility, Sect. 5 reports preliminary results on the application of our tool to a publicly available LANDSAT multispectral image with 7 bands and $169 \cdot 169$ pixels. It represents an agricultural area containing different kinds of crops. The resulting image is compared against that achieved by MultiSpec [6] clustering system. The conclusions describe both the positive features and the current limitations of our method. Finally, future works are outlined.

2 Spectral Pattern Recognition

The aim of spectral pattern recognition is twofold. The first goal is the division of all the multispectral image pixels into clusters, based on statistical features of the pixels themselves. This also results in the identification of the optimal number of clusters. The second goal is the association of any found cluster with the material it represents. This can be accomplished based on the fact that the amount of solar radiation which is reflected, absorbed or transmitted by any given material varies with wavelength. This property of matter makes it possible to identify different substances or classes and separate them by their spectral signatures (spectral curves) identifying uniquely any given material. Huge catalogs exist which contain thousands of spectral signatures, and experts can tell, based on them and on their own experience, which material corresponds to a given spectral curve.

When talking about classes, we need to distinguish between information classes and spectral classes. Information classes are those categories of interest that the analyst is actually trying to identify in the imagery, such as different kinds of crops, different forest types or tree species, different geologic units or rock types, etc. Spectral classes are groups of pixels that are uniform (or near-similar) with respect to their brightness values in the different spectral channels of the data. The objective is to match the spectral classes in the data to the information classes of interest. Rarely is there a simple one-to-one match between these two types of classes. Rather, unique spectral classes may appear which do not necessarily correspond to any information class of particular use or interest to the analyst. Alternatively, a broad information class (e.g. forest) may contain a number of spectral sub-classes with unique spectral variations. Using the forest example, spectral sub-classes may be due to variations in age, species and density, or perhaps as a result of shadowing or variations in scene illumination. It is the analyst's job to estimate the utility of the different spectral classes and their correspondence to useful information classes.

Common classification procedures can be broken down into two broad subdivisions based on the method used: supervised and unsupervised classification. Basic step in classification is the choice of a decision rule. This can be either parametric or non-parametric. The former is based on known statistical properties, like mean vector and covariances; examples of deriving classification techniques are maximum likelihood, Bayes, Ward [1]. The latter, instead, relies on (non)linear functions or mathematical/geometrical subdivision of the feature space, and results in classification techniques known as minimum distance to means, nearest neighbor, parallelepiped box, and so on [1].

In supervised classification, the task of determining the right number of clusters is left to analysts. In fact, the analyst identifies in the imagery homogeneous representative samples of the different surface cover types (information classes) of interest. These samples are referred to as training areas. The selection of appropriate training areas is based on the analyst's familiarity with the geographical area and his/her knowledge of the actual surface cover types present in the image. Thus, the analyst is "supervising" the categorization of a set of specific classes. The numerical information in all spectral bands for the pixels comprising these areas are used to "train" the system to recognize spectrally similar areas for each class. A special algorithm (of which there are several variations) is used to determine the numerical "signatures" for each training class. Once the algorithm has determined the signatures for each class, each pixel in the image is compared to these signatures and labeled as the class it most closely "resembles" digitally. Thus, in a supervised classification we are first identifying the information classes which are then used to determine the spectral classes which represent them. The main drawback of such a method is that determining a good ground truth may be a very difficult task, since analysts must be sure that any pixel belonging to it is fully representative of the material related to the cluster, and does not contain any feature typical of another material. This may take place because if pixel resolution is not sufficiently high, different elements may be contained in a same pixel. For example, in a pixel representing a $3 \cdot 3$ meters area, we might have a road, a car and tree branches together. This gives origin to the so-called pixel unmixing problem [3].

Unsupervised classification in essence reverses the supervised classification process. Spectral classes are grouped first, based solely on the numerical information in the data, and are then matched by the analyst to information classes when possible. As regards the clusterization phase, classical iterative programs, called *clustering algorithms*, are typically used to determine the natural (statistical) groupings or structures in the data. Firstly, the analyst must specify how many groups or clusters are to be looked for in the data. In addition to specifying the desired number of classes, the analyst may also specify parameters related to the separation distance among the clusters and the variation within each cluster. The final result of this iterative clustering process may result in some clusters that the analyst will want to subsequently combine, or clusters that should be broken down further; each of these situations requires a further application of the clustering algorithm. Once clustering has terminated, matching between the found clusters and the materials they represent must be carried out. This constitutes a very critical task for unsupervised classification, and is accomplished thanks to spectral signatures achieved and to analyst's experience.

3 Evolutionary Algorithms in Spectral Pattern Recognition

There exist many techniques for exploiting the spectral content of multispectral imagery. We can recall here at least Tassel Cap, Atmospherically Resistant Vegetation Index, Normalized Difference Vegetation Index, Principal Component Analysis [1]. Unfortunately these techniques face difficulties when the number of bands increases. It is beyond the scope of this paper to explain into details why it is, suffice it to say here that many of these methods are based on band rationing, and that, given B bands, there exist $B \cdot (B-1)$ different possible ratios. As the number of bands increases the number of possible combinations becomes rapidly unmanageable. So, approaches based on innovative methods, which can help to automize as much as possible the classification task, are welcome.

Recently, researchers have started to take into account evolutionary approaches. In the following we briefly describe some recent developments. Please note that all of the papers are recent indeed, and the oldest one dates back to 1998, and since then interest in the technique has been increasing.

Viana and Malpica [7] use Genetic Algorithms (GAs) to project a high dimensional space (hyperspectral space) to one with few dimensions in unsupervised classification. Firstly, since the experience shows that bands that are close in the spectrum have redundant information, groups of adjacent bands are taken and a GA is applied in order to obtain the best representative feature for each group, in the sense of maximizing the separability among clusters. Then the GA is applied again, but this time context information is included in the process.

Yu et al. [8] use a feature selection technique based on GAs to reduce feature space dimensionality and to select features on 224-band Remote Sensing data generated from the NASA/JPL Airborne Visible/InfraRed Imaging Spectrometer (AVIRIS). GAs are combined with Fuzzy Nearest Neighbours Classifiers.

Hung et al. [4] perform unsupervised learning for multispectral image pixel classification by means of a hybrid technique based on GAs and on Differential Competitive Learning performed by an Artificial Neural Network.

Benson et al. [9] examine the evolution of automatic target detection algorithms and their application to the detection of shipping in spaceborne Synthetic Aperture Radar (SAR) imagery. They apply GP which turns out superior to other techniques used in the field.

Stanhope and Daida [10] use GP for both the generation of rules for the target/clutter supervised classification on a set of infrared military vehicles images obtained with SAR, and for the identification of tanks in a set of SAR images. To these aims, previously defined feature sets are generated on the various images, and GP is used to select relevant features and methods of analyzing these features.

Brumby et al. [11] apply a GP algorithm to image feature extraction in Remote Sensing. They aim at finding open waters amidst vegetation. They claim theirs are just preliminary results, and are investigating the GP algorithm parameter space, and the relative importance of crossover and mutation. Their work is based on supervised learning, and on a truth plane.

Fonlupt [12] applies GP to the ocean color problem. This consists in evaluating ocean components concentration (phytoplankton, sediment and yellow substance) from sunlight reflectance values at selected wavelengths in the visible band. He performs supervised learning. Two different sets of experiments are carried out, related to open ocean and coastal waters respectively. GP results to outperform traditional polynomial fits.

Howard and Roberts [13] use a staged supervised GP strategy to automate the task of visual inspection of images aimed to detect objects of interest. They evolve a ship detector for SAR images of the English Channel and a recognizer of motorized vehicles in infrared imagery.

Rauss et al. [14] describe an initial use of GP as a discovery engine that performs supervised classification from 28-band spectral imagery, aiming at discovering the most useful bands for a specific classification task. Their system finds out, for example, that for grass only 18 out of the 28 bands are helpful.

4 Our GP Approach

In the present paper we introduce an innovative approach to unsupervised spectral pattern recognition based on GP. An important problem when performing unsupervised learning is that we do not know a priori which the right number of clusters is. Differently from the classical methods, in our approach such a number is found by the system rather than being set by the human analyst. Therefore, the aim of our automatic system for unsupervised pixel classification is to find both the most suitable number of clusters for the image and the coordinates of any cluster centre. Our method leaves cluster matching task to experts, nonetheless the system allows to provide them with spectral signatures for all clusters it finds in the image, thus easing their job.

We have decided to make use of GP because it allows to easily evolve individuals with different numbers of clusters. Also a GA would be able to perform the same task, but it would be more awkward to manage the existence of individuals with different numbers of clusters (thus, with different genotype lengths). Actually, to effectively face the problem at hand our approach is based on an adaptation of the canonical GP scheme as explained in the following.

Each individual in the population is a tree: its root contains information about the number of clusters, and as many pointers to cluster information nodes as there are clusters. Any given cluster information node has below it exactly B nodes, B being the number of frequency bands in the multispectral image. Any such band node contains an integer number. More specifically, the set of node types making up trees is very simple: it consists of the three types *Number_of_clusters*, *Cluster* and *Band*. The root node can only be a *Number_of_clusters*, yielding an integer value n_c in the range $[n_{c_{min}}, n_{c_{max}}]$; this value expresses the number of clusters for the pixel classification represented by the tree. Under a *Number_of_clusters* node only *Cluster* nodes are allowed, and their number is exactly n_c. Any *Cluster* node has exactly B *Band* nodes under it, each of them being a terminal node with a constant integer value and representing one of the coordinates of the cluster centre. This means that any tree has just three levels, and that the structure is strongly constrained. Figure 1 shows the example of a tree. It is the GP's task to find out the most suitable number of clusters and the most adequate coordinates in the hyperspace for any such cluster centre. The data related to the multispectral image composed by $M \cdot N$ pixels in B bands is stored in an $M \cdot N \cdot B$ matrix P, often referred to in literature as "data cube". Each element $P(i, j, k)$ of the cube is an integer value in the range $[0, 255]$ expressing reflectance value of pixel (i, j) at k-th wavelength. Given a tree representing a number of n_c clusters, any pixel is assigned to one and only one among the clusters, namely to the

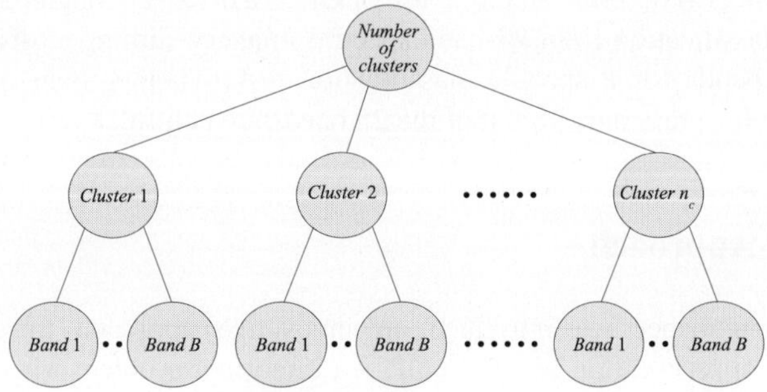

Fig. 1. Example of a tree

cluster z such that the euclidean distance in B dimensions between the pixel and the centre c_z of the cluster z is minimal. In formulae:

$$assign\ \ (i,j) \rightarrow z\ \ if$$

$$d((i,j),c_z) = \min_{z=1}^{n_c} \sqrt{\sum_{k=1}^{B} (P(i,j,k) - c_z(k))^2}$$

where $c_z(k)$ represents the centre of cluster z at k-th wavelength. Let us denote this minimal value with $d_{min}(i,j)$. The fitness ϕ of a tree is then given by the sum above all pixels of pixel distance from the centre of the cluster it is assigned to:

$$\phi = \sum_{i=1}^{M} \sum_{j=1}^{N} d_{min}(i,j)$$

and the problem becomes a minimization task.

For any pixel the system records the number corresponding to the cluster it has been assigned to. By doing so, an output image can be drawn which assigns the same color to all pixels belonging to the same cluster.

Our program is based on freeware *lil–gp* Genetic Programming Software version 1.02 [15], yet strongly modified to best serve our purposes. Thus, it is based on a set of parameters like number of individuals in the population, number of generations, on several selection methods and on operators of crossover, mutation and reproduction, each with an application probability (breed rate). The software also contains parameters concerning internal and external crossover probabilities, and internal and external mutations as well. By means of them crossover and mutation points can be chosen more frequently among either internal nodes or leaves. Other *lil–gp* parameters like initial tree depth range and maximum allowable tree depth are useless in our case, since all legal trees must have a depth of exactly three levels. To be sure that any new tree has three levels, a transition table has been used for mutations, allowing only legal trees to be generated. A further feature of our approach is that we, differently from other methods, do not make use of data pretreatment techniques like Principal Component Analysis (PCA) or others which are useful to reduce the number of bands to be taken into account. This is for two reasons. The first reason is that our approach aims to manage images with a high number of bands, and in those situations PCA, based on $B \cdot (B-1)$ comparisons between bands, becomes computationally expensive in terms of time. The second reason is that we aim to investigate GP ability to understand on its own if there are some frequencies which are more discriminating and, consequently, to find on its own the most suited division of the hyperspace related to the image.

The program gives as output five files containing the output image, the number of clusters and the centre position for each of them, the number of pixels assigned to any cluster, the spectral signature for each cluster and the evolution of the genetic system. Moreover, experiments performed in a preliminary study [16] have shown that our system, when left totally free to evolve, tends to create a number of clusters which becomes higher and higher as the number of generations increases. This may imply that some clusters actually represent a same material, so a postprocessing phase might require a grouping for those clusters. To get rid of this problem we have endowed the system with a further parameter, *min_dist*, which represents the minimum distance which can exist between the centres of any two clusters in the B-dimensional hyperspace. This prevents that too close clusters may be created, which might represent a same material. For low *min_dist* values we obtain the above described situation, whereas for high values we achieve a reduced number of clusters.

5 Experimental Results

We have downloaded from Purdue University [17] internet site a publicly available LANDSAT multispectral image with $169 \cdot 169$ pixels and 7 spectral bands. The image is reported in Fig. 2. The former three bands are in the visible and represent respectively the blue ($[0.45 - 0.52]\,\mu$m), green ($[0.52 - 0.60]\,\mu$m) and red ($[0.63 - 0.69]\,\mu$m) components. Next three bands are in the infrared and represent the near ($[0.76 - 0.90]\,\mu$m), medium ($[1.55 - 1.75]\,\mu$m) and far ($[2.08 - 2.35]\,\mu$m) components. Finally there are the data for the thermal band ($[10.4 - 12.5]\,\mu$m). It is an image related to Agriculture, and represents a set of fields cultivated with different kinds of crop. The fields are crossed by some roads.

As regards the GP parameters, we have chosen a population size of 1,000 and a maximum number of generations of 500. The breed rates are 0.4 for crossover, 0.5

Fig. 2. The original image

for mutation and 0.1 for reproduction. Selection chosen is tournament with size of ten. Internal crossover value is 0.8, while external crossover value is 0.2, meaning crossover being more probable on *Cluster* nodes rather than on *Band* ones. Internal and external mutation probabilities are set to 0.2 and 0.8, respectively. The values for $n_{c_{\min}}$ and $n_{c_{\max}}$ have been set to 2 and 20, respectively. A total number of eight runs has been carried out up to now by using different values for the *min_dist* parameter. A typical interval for this parameter can be $[0.00 - 0.15]$. The time needed for one experiment is about two days on a Sun 20 workstation.

Figure 3 contains the output image we have obtained at the end of program execution for the best run (in terms of lower final fitness) we have performed at present.

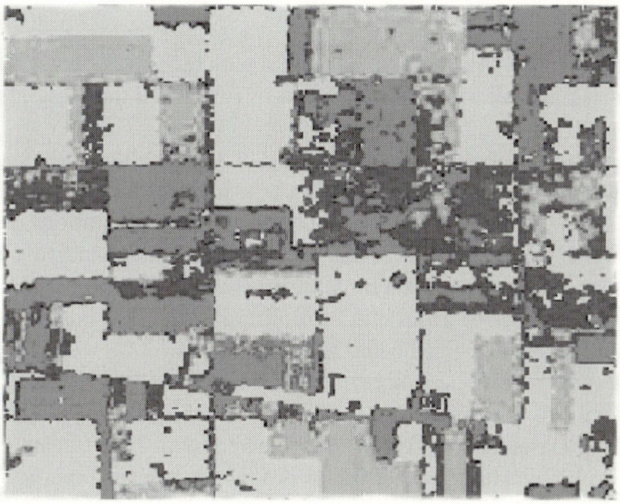

Fig. 3. Output image for the best run

The output image contains 4 different clusters. This run makes reference to a *min_dist* value equal to 0.07. The system has proved capable of reconstructing the image with very good quality, though only experts can say, based on spectral signatures and on existing catalogs, what any cluster (color in the output image) represents. If we pay attention to original image contents and to the output image, fields which are assigned to a same cluster seem to be actually similar in contents. Also roads seem quite clear. An example of multispectral signatures achieved in the best run, and related to cluster number 4, is shown in Fig. 4 (left). From the evolutionary point of view, Fig. 4 (right) shows the evolution of the best fitness value during the best run performed. As it can be noted, initial generation starts with a best value of about 14,000. The decrease in fitness values shows a first remarkable quasi-linear phase, until about generation 50, when a value of about 6,000 is reached. Then fitness improvement continues more slowly until generation 320, where a tree with a fitness of about 5,500 is reached. Since then a new quasi-linear phase (lasting

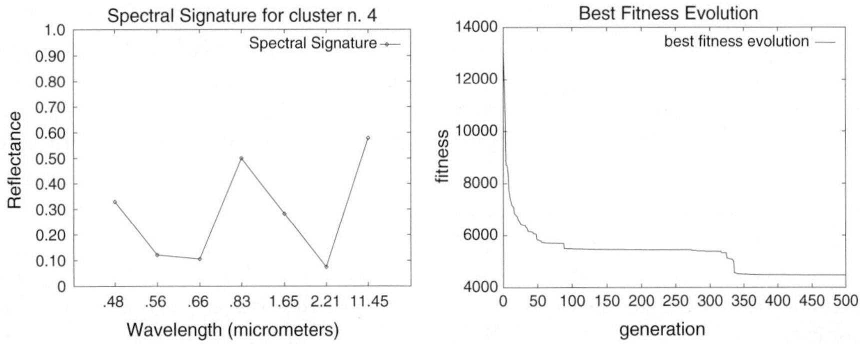

Fig. 4. Spectral signature achieved for cluster 4 (*left*) and best fitness during best run (*right*)

about 30 generations) starts in which a solution with fitness of about 4,600 is reached. Finally decrease gets slower again until end of run, when the best value reached is about 4,500.

It is interesting to report here that during first generations the system provides us with trees having only few clusters (the best individual in initial generation has just two clusters). As the number of generation increases, solutions consisting of a higher and higher number of clusters are found (three, four, then five and so on, until a number of six is found). At that point, nonetheless, the mechanism limiting cluster number not only allows to avoid cluster proliferation, it also makes this number to reduce to four, yet resulting in a better fitness value.

With the aim to evaluate more deeply the quality of the obtained solution, we have taken into account the clustering system MultiSpec. It is basicly a tool for supervised classification, yet it is endowed with some features typical of unsupervised classification: in fact it can propose to create some new clusters not hypothesized by the user, to divide a cluster into two or more parts and to group some clusters. The execution of this system, by exploiting also its nonsupervised features, allows to obtain as a result the image reported in Fig. 5. As it can be seen, the resulting image is very similar to ours, and the fields are grouped in four clusters in about the same way as our system does. It is worth noting that MultiSpec has created a fifth cluster, represented in black in the figure, which represents the roads. The similarity of our results to those achieved by a widely utilised and highly reliable system allows us to hypothesize that our approach is valid.

6 Conclusions and Future Work

Aim of this paper is to test the feasibility of a Genetic–Programming based approach to unsupervised spectral pattern recognition for multispectral images.

The experiments reported here have been conducted by using a publicly available LANDSAT multispectral image with $169 \cdot 169$ pixels and 7 spectral bands. Achieved results seem positive and encouraging. The output image is of good quality, and

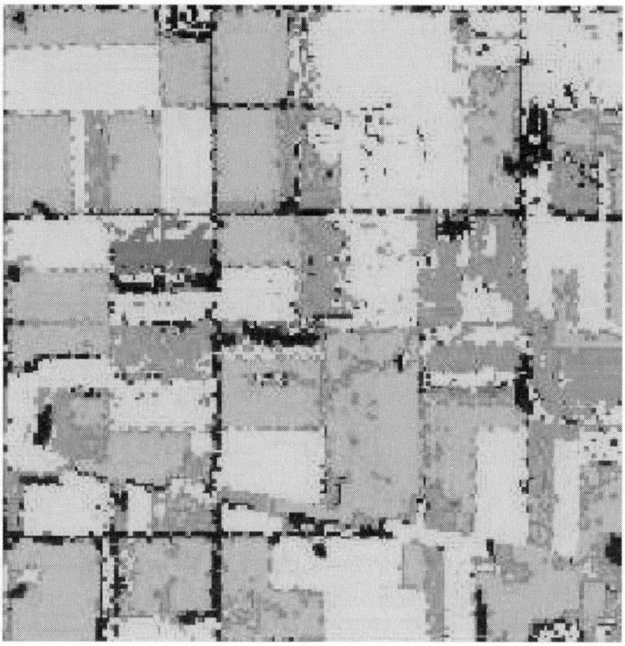

Fig. 5. Output image achieved by MultiSpec system

material related to any obtained cluster can be determined by means of the yielded spectral signature. Nonetheless we are well aware that these experiments are just preliminary, and are part of an undergoing effort. To obtain a better version of our system and to achieve better results, firstly we need to optimize the tool from the evolutionary point of view. This means to search for the most suitable parameter set. To accomplish this goal, many more runs are needed with the same and with other images.

This leads us to a current drawback of our system: the execution time for one run is still high. Therefore, we will do our best in order to reduce execution time. Firstly, code shall be optimized to reduce computing time. Secondly, a parameter set suitable for as many images as possible shall be found. Thirdly, we shall make use of parallel versions of Genetic Programming on Multiple Instruction Multiple Data (MIMD) parallel machines.

From the application point of view, after this tuning phase with publicly available images, we aim to apply our system to hyperspectral (about 90 bands, $512 \cdot 512$ pixels) forestry images coming from European Space Agency (ESA).

Acknowledgments

Dr. A. Della Cioppa gratefully acknowledges National Institute for the Physics of Matter (INFM, Italy) for supporting his research under the network "Theoretical and Computational Physics, Cybernetics".

References

1. Buiten H. J., Clevers J. P. G. W. (1993) Land Observation by Remote Sensing, Theory and Applications, Gordon and Breach Science Publishers.
2. Campbell J. B. (1987) Introduction to Remote Sensing, The Guilford Press, New York.
3. Lillesand T. M., Kiefer R. K. (1994) Remote Sensing and Image Interpretation, John Wiley and Sons Inc., New York.
4. Hung C. C., Coleman T. L., Scheunders P. (1998) Using Genetic Differential Competitive Learning for Unsupervised Training in Multispectral Image Classification Systems. In: IEEE Intl. Conf. on Systems, Man and Cybernetics, San Diego, CA, October 11–14, IEEE Press, 4482–4485.
5. Koza J. R. (1992) Genetic Programming – On the Programming of Computers by means of Natural Selection, MIT Press.
6. Landgrebe D., Biehl L. (2001) An Introduction to MultiSpec, School of Electrical and Computer Engineering, Purdue University, USA. Available at: http://www.ece.purdue.edu/∼biehl/MultiSpec/Intro5_01.pdf.
7. Viana R., Malpica A. J. (1999) Genetic Algorithm for Accomplishing Feature Extraction of Hyperspectral Data using Texture Information. In: Proceedings of SPIE: Image and Signal Processing for Remote Sensing V, Vol. 3871, 367–372.
8. Yu S., De Backer S., Scheunders P. (2000) Genetic Feature Selection Combined with Composite Fuzzy Nearest Neighbor Classifiers for High-Dimensional Remote Sensing Data. In: IEEE International Conference on Systems, Man and Cybernetics, Nashville, TN, IEEE Press, 1912–1916.
9. Benson K., Booth D., Cubillo J., Reeves C. (2000) Automatic Detection of Ships in Spaceborne SAR Imagery. In: Genetic and Evolutionary Computation Conference (GECCO), Las Vegas, Nevada, July 8–12, Morgan Kaufmann, 767.
10. Stanhope S. A., Daida J. M. (1998) Genetic Programming for Automatic Target Classification and Recognition in Synthetic Aperture Radar Imagery. In: Evolutionary Programming VII: Procs. of the Seventh Annual Conf. on Evolutionary Programming, Lecture Notes in Computer Science, n. 1147, Springer, 693–702.
11. Brumby S. P., Theiler J., Perkins S. J., Harvey N., Szymanski J. J., Bloch J. J., Mitchell M. (1999) Investigation of Image Feature Extraction by a Genetic Algorithm. In: Proceedings of SPIE: Applications and Science of Neural Networks, Fuzzy Systems and Evolutionary Computation II, Vol. 3812, 24–31.
12. Fonlupt C. (2001) Solving the Ocean Color Problem using a Genetic Programming Approach. Applied Soft Computing, 1:63–72.
13. Howard D., Roberts S. C. (1999) A Staged Genetic Programming Strategy for Image Analysis. In: Genetic and Evolutionary Computation Conference (GECCO), Orlando, Florida, Morgan Kaufmann, 1047–1052.
14. Rauss P. J., Daida J. M., Chaudhary S. (2000) Classification of Spectral Imagery using Genetic Programming. In: Genetic and Evolutionary Computation Conference (GECCO), Las Vegas, Nevada, July 8–12, Morgan Kaufmann, 726–733.

15. Zongker D, Punch W. (1995) “lilgp version 1.02”. Lansing, Michigan State University, GA Research and Applications Group. Available at: http://isl.cps.msu.edu/GA/software/lil-gp.
16. De Falco I., Della Cioppa A., Tarantino E. (2002) Unsupervised Spectral Pattern Recognition for Multispectral Images by means of a Genetic programming Approach. In: IEEE Congress on Evolutionary Computation (IEEE–CEC2002), IEEE Press, 231–236.
17. Multispec Clustering System. Available at: http://dinamo.ecn.purdue.edu/~biehl/MultiSpec.

A Genetic Programming System for Time Series Prediction and Its Application to El Niño Forecast

I. De Falco[1], A. Della Cioppa[2], and E. Tarantino[1]

[1] Institute of High Performance Computing and Networking – CNR, Via P. Castellino 111, 80131 Naples, Italy
ivanoe.defalco@na.icar.cnr.it

[2] Dept. of Computer Science and Electrical Engineering, University of Salerno, Via Ponte don Melillo 1, 84084 Fisciano (SA), Italy
adellacioppa@unisa.it

Summary. In this paper a system based on Genetic Programming for forecasting nonlinear time series is outlined. Our system is endowed with two features. Firstly, at any given time t, it performs a τ-steps ahead prediction (i.e. it forecasts the value at time $t+\tau$) based on the set of input values for the n time steps preceding t. Secondly, the system automatically finds among the past n input variables the most useful ones to estimate future values. The effectiveness of our approach is evaluated on El Niño 3.4 time series on the basis of a 12-month-ahead forecast.

Key words: Genetic Programming, time series forecasting

1 Introduction

The prediction of spatio-temporal patterns is substantial to a variety of applications in fields ranging from engineering to economics. Different methods have been investigated over the years [1]. The first approaches were devoted to linear models for which the theory is known and consequently many algorithms for model building are available. The most used linear regression methods have been the autoregressive (AR) and the autoregressive moving average (ARMA) models [2, 3]. An example of more complex regression method is the multivariate adaptive regression splines [4]. These stochastic methods are fast but of limited applicability. In fact, time series produced by systems with nonlinear dynamical behavior have proven notoriously resistant to prediction by conventional techniques.

This is the reason why nonlinear methods have become widely utilized. Among these Artificial Neural Networks (ANNs) represent an attractive approach for time series prediction problems [5, 6, 7, 8]. In fact, in addition to their ability to representing complex nonlinear functions, many different types of ANNs, such as Multi-Layer Perceptron and Radial Basis Function networks, can effectively

I. De Falco et al.: *A Genetic Programming System for Time Series Prediction and Its Application to El Niño Forecast*, Advances in Soft Computing **1**, 151–162 (2005)
www.springerlink.com

construct approximations for unknown functions by learning from examples [9]. In time series prediction with ANNs the main problems are the network structure design and the effectiveness of the learning phase. These limitations have been mitigated exploiting the training adaptability of evolutionary computations [10, 11, 12]. Though these approaches have revealed to be more powerful than stochastic methods, they present drawbacks in the neural model design and in the capability of providing explicit forecasting models.

As an alternative, the flexibility of evolutionary search methods has been exploited. In the last years the Genetic Programming (GP) framework [13] has been applied in various fields like engineering, medicine, economics and so forth with excellent results [10, 14, 15, 16, 17, 18, 19, 20]. The objective is to exploit GP's flexible tree structure for building a time series prediction model. The motivation of this choice is that GP has many advantages which allow to mitigate some of the limitations typical of neural models. These advantages can be summarized as follows: (i) generation of explicit model representations amenable to easy human comprehension, (ii) automatic discovering of the model structure from the given data, (iii) adaptive evolutionary search that allows to escape trapping in suboptimal, unsatisfactory local solutions, (iv) absence of specific knowledge. Moreover, GP is more attractive than traditional Genetic Algorithms for problems that require the construction of explicit models. In fact, with the variable length program tree representation, GP evolves program contents and structures at the same time.

Our GP-based system is able to modelise and forecast time series. In particular, the modeling and the forecasting of El Niño is carried on. Its prediction is of great importance as the global influence of this phenomenon on climate determines strategic planning in areas such as agriculture, management of water resources and reserves of grain and fuel oil before all in the more directly involved countries.

Our system, at any given time t, performs a τ-steps ahead prediction (i.e. it forecasts the value at time $t+\tau$) based on the set of input values for the n time steps preceding t. Furthermore, it automatically finds among the past n input variables the most useful ones to estimate future values.

It is known that a τ-steps ahead prediction can be performed in two different ways. A first approach consists in carrying out prevision recursively, by evaluating the forecast at $t+1$, then using it to forecast $t+2$, and so on, until $t+\tau$ is obtained as a function of these intermediate values. Another approach, instead, consists in directly forecasting the value at $t+\tau$ by using the values known at time t, i.e. $t-1, t-2$ and so on. Our system is based on this latter approach.

The paper is organized as follows. In Sect. 2 our GP-based forecasting system is outlined together with implementation details. In Sect. 3 the description of El Niño time series and the related previous studies are reported. In Sect. 4 the experimental results achieved are shown and discussed. Section 5 contains final comments and prospects of future work.

2 The Genetic Programming System

The aim is the implementation of a genetic system able to automatically provide the modeling of a nonlinear time series and its prediction.

Given a fitness function, the forecasting problem becomes the search of the model which best describes the essential characteristics of the time series. It is evident that an exhaustive search by enumerating all the possible descriptions is computationally impracticable. Hence we appeal to GP which is a powerful and flexible search method inspired by natural selection. It does not guarantee to find the global optimum, nonetheless it usually allows to retrieve a suboptimal solution in a reasonable computation time.

The evolving population is constituted by "programs" representing the potential forecasting models in the form of trees with variable depth. Each model is composed by elementary functions and numerical operators. Moreover, at each time t the prediction function $\tilde{x}(t+\tau)$ for time $t+\tau$ is simply searched as a function of the n past states of $x(t)$ in the n time steps preceding t [22]; this means that:

$$\tilde{x}(t+\tau) = f(x(t-1), \quad x(t-2), \ldots, x(t-n))$$

where n and τ are model parameters to be set.

A population of these candidate models is maintained and gradually improved by constructing new fitter ones until a model of sufficient precision is found or other stopping criteria are satisfied.

To construct the prediction model, data is partitioned into four sets: the processing, the training, the validation and the prediction sets. The processing set contains a number of n time series known elements, the training set contains the values to be approximated during the learning phase, while the validation set is used to evaluate the generalization ability of the found model, and real "blind" forecasting will be performed on the prediction set.

2.1 Encoding

The individuals are composite functions encoded as tree structures for which limits on the tree depth can be specified. The tree nodes are either functions or terminals. The function set consists of elementary functions and numerical operators as reported in Table 1. This set can be easily extended.

The terminal set has a fixed number of elements: the set of n time series past values in input and an Ephemeral Random Constant (ERC). ERC is a special terminal with a defined value. When a terminal ERC is generated a constant value is associated to that terminal. A range of variation $[ERC_{\min}, ERC_{\max}]$ can be specified for this value, depending on the problem at hand. The terminal set is reported in Table 2.

2.2 Genetic Operators

The new elements in the population are generated by means of three operators: *crossover*, *reproduction* and *mutation*:

Table 1. Set of numerical operators and elementary functions with the related arity

Symbol	Arity	Description
+	2	Addition
−	2	Subtraction
*	2	Multiplication
/	2	Protected division (returns 1 if the denominator is 0)
sin	1	Sine
cos	1	Cosine
exp	1	Exponential
rlog	1	Protected logarithm (rlog(0) is 0)

Table 2. Set of terminal symbols

Symbol	Arity	Description
x_1	0	input value of time series at time $(t-1)$
		...
x_i	0	input value of time series at time $(t-i)$
		...
x_n	0	input value of time series at time $(t-n)$
R	0	ERC in the range $[ERC_{\min}, ERC_{\max}]$

- **Crossover.** Two parent individuals are selected and a subtree is picked on each one. Then crossover swaps the nodes and their relative subtrees from one parent to the other. This operator must ensure the respect of the depth limits. If a condition is violated the too-deep offspring is simply replaced by one of the parents. There are other parameters that specify the frequency with which functions or terminals are selected as crossover points.
- **Reproduction.** The reproduction operator simply chooses an individual in the current population and copies it without changes into the new population.
- **Mutation.** The mutation operator can be applied to either a function node or a terminal node. A node in the tree is randomly selected. If the chosen node is a terminal it is simply replaced by another terminal. If it is a function and *point mutation* is to be performed, it is replaced by a new function with the same arity. If, instead, *tree mutation* is to be carried out, a new function node (not necessarily with the same arity) is chosen, and the original node together with its relative subtree is substituted by a new randomly generated subtree. A depth ramp is used to set bounds on size when generating the replacement subtree. Obviously it is to check that this replacement does not violate the depth limit. If this happens mutation just reproduces the original tree into the new generation. Further parameters specify the probability with which functions or terminals are selected as mutation points.

2.3 Fitness Function

To evaluate the accuracy of a forecasting model, researchers make normally reference to statistical indices. One of the most used is the Mean Squared Error (*MSE*). In formula this can be devised as follows:

$$MSE = \frac{\sum_{i=1}^{l}(x(i) - \tilde{x}(i))^2}{l} \tag{1}$$

where l is the number of the discrete points in the set under examination.

For our GP tool we have chosen as fitness function for any proposed model the value of MSE computed on the training set (MSE_t). With this choice the problem becomes a minimization task, with the global optimum equal to 0.

It is important to note here that when we perform an evolutionary search for a model, we hope to find a model with good generalization capability, i.e. it can predict unseen values with same quality as values on which training is being carried out.

Usually an arbitrarily good prediction performance can be achieved on training data, yet it often happens that the system specializes itself on those data, and may poorly predict other data of the series not shown (the so-called *overfitting* problem).

To overcome this drawback, we have decided to divide the set of data as mentioned earlier and to follow the approach proposed in [23]. Thus, we evaluate the model on the training set, and we monitor its performance over the validation set. It would be highly desirable to find by evolution models showing an error on the validation set lower than or equal to (or even slightly higher than) that on the training set. Therefore at the end of each generation we take into account its best individual and evaluate the errors on the training set (MSE_t), on the validation set (MSE_v) and on the set composed by points in previous two sets (MSE_0). This latter error is a weighted average of MSE_t and MSE_v. Intuitively, overfitting takes place when MSE_v increases and is higher than MSE_t. Whenever a proposed model shows errors such that $MSE_t \cong MSE_0$, this means that $MSE_t \cong MSE_v$, and we can save it as the current best model. Since it usually happens that MSE_t gets lower and lower as long as evolution takes place, every time we save a new model this will show better quality on validation set than the previous one. At the end of the evolution the lastly saved model will be used to perform forecasting.

2.4 The Genetic Time Series Algorithm

Our Genetic Time Series algorithm (GTS) is based on a freeware *lil-gp* Genetic Programming software [21]. The software has been adapted to our purposes. Once defined the fitness function and the operators, GTS operates as follows:

1. load the time series values;
2. generate at random an initial population of composite functions representing potential forecasting models;
3. evaluate the models by using the fitness;
4. at each generation

- select the functions to undergo the mechanism of reproduction;
- apply the genetic operators *crossover*, *reproduction* and *mutation* to produce new composite functions;
- insert these offspring in the new population;
- evaluate the models by means of fitness;
- evaluate MSE_t, MSE_v and MSE_0 for the best individual and, if is the case, save it as best-so-far model;

5. repeat step 4 until the specified maximum number of generations has been reached.

Even though the individuals in the initial population are randomly created by selecting from the function and terminal sets, during the tree construction some restrictions have been imposed.

Moreover, a so-called half-and-half method has been chosen to create these random initial structures. This is a compromise between the full and the grow methods which are chosen 50% of times each. The full method generates only full trees, that is the tree path length from any terminal node to the root of the tree is the same. The grow, starting from a node chosen as root, recursively calls itself to produce child trees for any node which needs them. When the initial structures reach the maximum allowed depth all further nodes are restricted to be terminals. This method has been used because it allows to create unbalanced trees with a variety of shapes and sizes.

3 El Niño Anomalies Time Series

The Sea Surface Temperature (SST), monitored from ship reports, buoys and satellite imagery, is an important physical quantity in understanding earth's ocean-atmosphere interactions. The so-called El Niño phenomenon is closely connected with and partially defined by changes in SST. El Niño is characterized by the "*Pacific basin-wide increase in both SSTs in the central and/or eastern equatorial Pacific Ocean and in sea level atmospheric pressure in the western Pacific*" which occurs periodically [24]. Indeed, Rasmusson and Carpenter [25] point out that El Niño is the leading factor in interannual variability of SST off the coast of Peru and Ecuador. Predicting this phenomenon is essential, as El Niño is linked to reduction of fish population in coastal waters and changes in the climate of South America and other global regions, often seriously impacting the economy.

Very often, rather than making reference to actual SST series, researchers deal with SST anomalies series. Any such value represents the departure of a given SST from the long term average temperature.

A rigorous definition [26] states that an El Niño occurs in a region when a positive five-month running mean SST anomaly in the investigated region, exceeding 0.4 K, lasts at least six months. Similarly, there are cold events when the SSTs become unusually cold for at least six months: such events are termed La Niña.

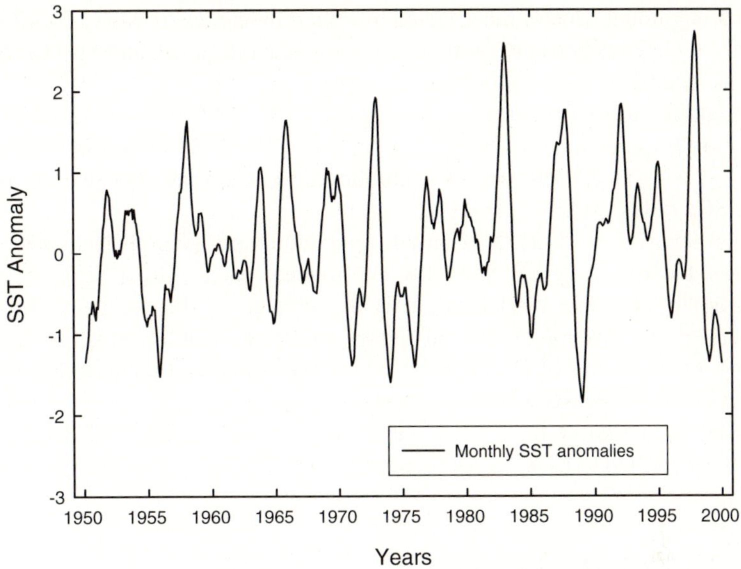

Fig. 1. The El Niño 3.4 SST anomalies original time series

For data sampling of SSTs the equatorial Pacific (between 5°S–5°N) has been divided into a number of regions named El Niño 1, 2, 3, 4, and 3.4 (which encompasses parts of both regions 3 and 4).

In this paper we have decided to take into account the El Niño 3.4 area. This is defined by 120W°–170°W and 5°N–5°S, so it runs along the Equator and is broadly bounded to west by Kiritimati atoll (Kiribati) and to east by Galapagos Islands (Ecuador). For this area literature provides researchers with the monthly series of SST anomalies for the period January 1950–December 1999. The series is publicly available, downloadable at [28], contains 600 values and is reported in Fig. 1.

Forecast of this series has been faced by many researchers with different approaches.

Classical multivariate statistical methods, like for example Multiple Linear Regression (MLR), Principal Component Analysis (PCA), Canonical Correlation Analysis (CCA) and many others, are widely used for data analysis in meteorology and oceanography [27].

Several dynamical models have been applied to predict this series. They are too numerous to be described here. There is an excellent review on them in [29]. Among them we can cite here at least Lamont Simple Coupled Model (LDEO), Australian Bureau of Meteorology Research Center (BMRC) low-order coupled model, University of Oxford intermediate coupled model, Scripps/MPI Hybrid Coupled Model (HMC–3), The Center for Ocean Land Atmosphere Studies (COLA) comprehensive coupled model, NCEP comprehensive coupled model.

Unger and others [30] have approached the problem by combining the forecasts of three input models into a single forecast based on the past behavior of each contributing model. One of the selected models is dynamical (CMP12 NCEP) while the other two are statistical (Constructed Analogue model and Canonical Correlation Analysis model).

Recent advances in Artificial Neural Network (ANN) modeling have led to the nonlinear generalization of CCA. Non-Linear Canonical Correlation Analysis (NLCCA) using an NN approach was introduced by Hsieh and applied to the tropical Pacific SST fields [31].

Tang and others [32] have used a Feedforward Neural Network model. An ensemble of 20 neural networks are trained, each with a different weight initialization. The final model output is the average of the outputs from the 20 members of the ensemble. The advantage of the ensemble model is to reduce variance, or instability of the neural network. By doing so, they have been able to perform several forecasts with respectively 3, 6, 9 and 12 months steps ahead.

As far as we know there is no paper in literature dealing with Evolutionary Algorithms applied to El Niño 3.4 forecast.

4 The Experimental Results

4.1 Parameter Setup and Performance Measures

The El Niño 3.4 SST anomalies series has been divided as it follows: due to the above choice, the former 100 points are used as processing set. Points in the range [101–400] are used to perform training, those in the range [401–588] are used as validation. Finally, the points [589–600] are used to perform prediction. Firstly the value of n has been set equal to 100 and the value of τ has been set equal to 12, so that our system performs 12-month-ahead forecasting. Due to this choice, the first value forecast by our system is that at time $t = 100 + 12 = 112$.

A preliminary set of runs has been carried out to find a good parameter set. The best resulting values are listed in the following. The population size chosen is $3,000$, the maximum number of generations allowed is 350, crossover rate is 0.40, mutation rate is 0.59 and reproduction rate is 0.01. Initial tree depth is in the range [2–6], and maximum tree depth is 30. Initial population is generated by half-and-half method. Tournament selection with size equal to 200 is carried out. Crossover is internal with probability 90%, and external for the remaining 10%. Mutation is equiprobabilistically internal or external. Depth of new subtrees generated by mutation is in the range [0–10].

4.2 Findings

Experiments have been performed on a Sun workstation and require few hours. We have carried out a set of ten runs with the best found parameter values. We consider as the best among those trials the one that gives rise to the best model in terms of lower MSE_v. Results for this best run are reported in Fig. 2 where the training and the validation are shown. It should be noted that values shown have been

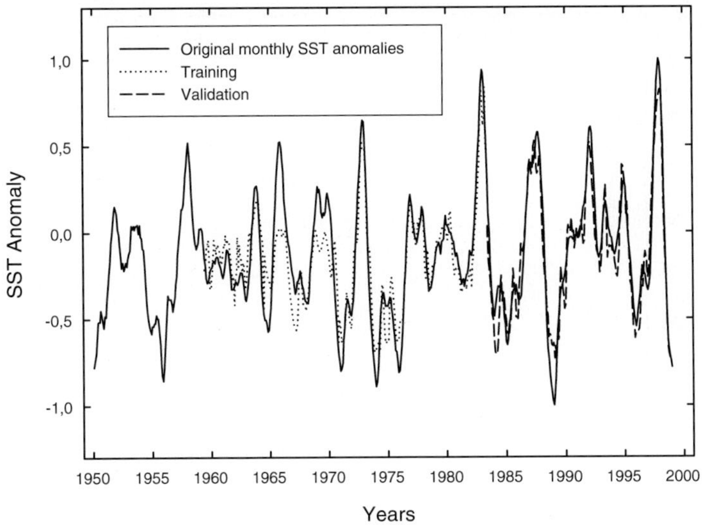

Fig. 2. Training and validation

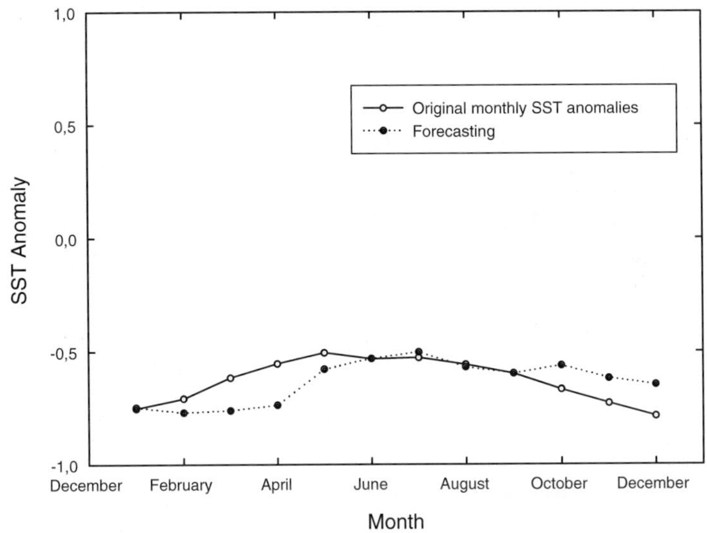

Fig. 3. Forecasting for the twelve months of year 1999

normalized in $[-1.0, 1.0]$. It can be seen that our system models the phenomenon in good accordance with historical data, both hot and cold phases are caught correctly, and all peaks are forecast, though in some cases computed peak heights are lower than real ones. Figure 3 reports the forecasting for the twelve months making up year 1999. Also in this case values are normalized in $[-1.0, 1.0]$. As it can be easily seen,

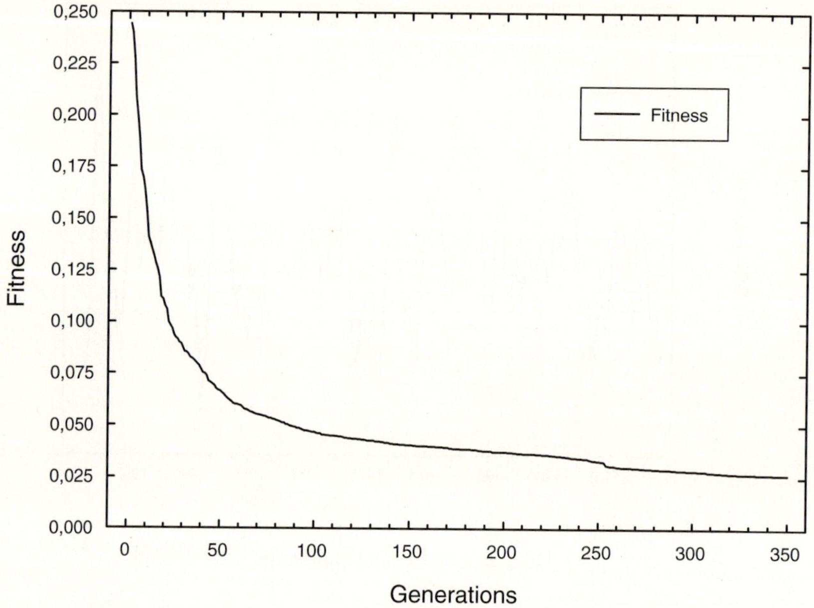

Fig. 4. Evolution of best fitness as a function of number of generations

our forecasting is quite close to actual data: the general shape of this twelve-month time interval is reproduced with good approximation, and can be interpreted as a La Niña phase. Forecast values show a first increasing trend (though initially slower than the real series) followed by a decreasing phase. Values for months from June to September are in excellent agreement with real data. Figure 4 shows the evolution of the best fitness value as a function of the number of generations. The improvement in solution quality is very high during a first phase lasting about 50 generations, though decrease continues satisfactorily in the next phases as well. The values of the errors related to the best model found in this evolution are $MSE_t = 0.026442$, $MSE_v = 0.016818$ and $MSE_p = 0.008829$ for the prediction set.

5 Conclusions and Future Works

In this paper we have presented a GP tool for time series forecasting based on a τ-steps ahead approach. The system has been tested on a publicly available series, that of El Niño 3.4 SST anomalies. Experimental results have demonstrated the effectiveness of the proposed approach in providing good-quality resemblance of the computed series to the real one. Furthermore, good forecasting has been obtained, which guesses the next phase as a La Niña one. This has been confirmed by the real data.

Future work will include the application of the proposed system to other real-world time series in order to further validate the promising results reported in the present paper. Furthermore, a parallel implementation of the GP tool on Multiple Instruction Multiple Data (MIMD) parallel machines will be carried out with the aim to reduce computational times.

References

1. Weigend A, Gershenfeld N (eds) (1994). Time series prediction: forecasting the future and understanding the past, Addison-Wesley, Reading, Massachussetts.
2. Tong H (1990). Nonlinear time series: a dynamical system approach. Oxford University Press.
3. Box G, Jenkins G, Reinsel G (1994). Time series analysis: forecasting and control. Prentice Hall, Englewood Cliffs, New Jersey.
4. Friedman J (1991). Multivariate adaptive regression splines. Annals of Statistics 19:1–142.
5. Mozer N (1993). Neural net for temporal sequence processing. In: Weigend A, Gershenfeld N (eds) Time series prediction: forecasting the future and understanding the past. Addison–Wesley, Reading, Massachussetts.
6. Tsoi A, Back A (1994). Locally recurrent globally feedforward networks: a critical review of architectures. IEEE Trans. on Neural Networks 2:229–239.
7. Koskela T, Lehtokangas M, Saarinen J, Kaski K (1996). Time series prediction with multilayer perceptron, FIR and Elman neural networks. In: World Congress on Neural Networks, INNS Press, 491–496.
8. De Falco I, Iazzetta A, Luongo G, Mazzarella A, Tarantino E (2000). The seismicity in the southern tyrrhenian area and its neural forecasting. Pure and Applied Geophysics 157:343–355.
9. Cybenko G (1989). Approximations by superpositions of a sigmoidal function. Mathematics of Control, Signals and Systems 2:303–314.
10. Angeline P J (1998). Evolving predictors for chaotic time series. In: Rogers S, Fogel D, Bezdek J, Bosacchi B (eds) Proceedings of SPIE: Application and Science of Computational Intelligence, Vol. 3390. Bellingham, WA, USA, 170–180.
11. De Falco I, Della Cioppa A, Iazzetta A, Natale P, Tarantino E (1999). Optimizing neural networks for time series prediction. In: Advances in Soft Computing in Engineering Design and Manufacturing. Springer–Verlag, London.
12. Lee Ki–Youl, Lee Dong–Wook, Sim Kwee–Bo (2000). Evolutionary neural networks for time series prediction based on L–system and DNA coding method. In: Congress on Evolutionary Computation, La Jolla, California, USA, July 16–19. IEEE Press, 1467–1474.
13. Koza J (1992). Genetic programming. On the programming of computers by means of natural selection. MIT Press, Cambridge, MA, USA.
14. Iba H (1993) System identification using structured genetic algorithms. In: Forrest S. (Ed.) Proceedings of the Fifth International Conference on Genetic Algorithms, San Mateo, California. Morgan Kaufmann, 279–286.
15. Oakley H (1994) Two scientific applications of genetic programming: stack filters and nonlinear equation fitting to chaotic data. In: Kinnear K (ed.) Advances in Genetic Programming. MIT Press, Cambridge, MA, USA, 369–389.
16. Mulloy B S, Riolo R L, Savit R S (1996). Dynamics of genetic programming and chaotic time series Prediction. In: Koza J R, Goldberg D E, Fogel D B, Riolo R L (eds), Genetic Programming 1996: Proceedings of First Annual Conference, Stanford University, CA, USA, July 28–31, 166–174.

17. Numata M, Yoshihara I, Yoshizawa M, Abe K (1999). GP-based heart rate prediction for artificial heart control. In: Brave S. and Wu A. S. (Eds) Late Breaking Papers at the 1999 Genetic and Evolutionary Computation Conference, Orlando, Florida, USA, July 13–17, 193–197.
18. Iba H, Nikolaev N I (2000). Genetic programming polynomial models of financial data series. In: Congress on Evolutionary Computation, La Jolla, California, USA, July 16–19. IEEE Press, 1459–1466.
19. Nikolaev N I, Iba H (2001) Regularization approach to inductive genetic programming. IEEE Trans. on Evolutionary Computation, 4:359–375.
20. Duan M, Povinelli R J (2001). Estimating stock price predictability using genetic programming. In: Spector L, Goodman E D, Wu A, Langdon W B, Voigt H M, Gen M, Sen S, Dorigo M, Pezeshk S, Garzon M H, Burke E (eds) Genetic and Evolutionary Computation Conference (GECCO-2001), S. Francisco, California, USA, July 7–11. Morgan Kaufmann, 193–197.
21. Zongker D, Punch W (1995). lilgp version 1.0, Lansing, Michigan State University, GA Research and Application Group. Available at: http://isl.cps.msu.edu/GA/software/lil–gp.
22. Takens F (1981). Detecting strange attractors in turbulence. In: Lecture Notes in Mathematics. Springer–Verlag, Berlin.
23. Jäske H (1996). Prediction of sunspots by genetic programming and their applications. In: Alander J T (ed), Proceedings of Second Nordic Workshp on Genetic Algorithms, Vaasa, Finland, 19–23 August, 79–88.
24. Trenberth K E, (1996) El Niño definition. CLIVAR – Exchanges, 3:6-8.
25. Rasmusson E M, Carpenter T H (1982) Variations in tropical sea surface temperature and surface wind fields. Associated with the Southern Oscillation/El Niño, Monthly Weather Review, Vol. 110, 354-384.
26. Trenberth K E (1997), The definition of El Niño. Bullettin of the American Meteorological Soc., Vol. 78, 2771–2777.
27. von Storch H, Zwiers F W (1999) Statistical Analysis in Climate Research. Cambridge University Press, 484.
28. El Niño 3.4 SST index series.
Available at: http://www.cgd.ucar.edu/cas/catalog/climind/Nino_3_3.4_indices.html.
29. Barnston A G, Glantz M H, He Y (1998) Predictive skill of statistical and dynamical climate models in SST forecasts during the 1997-98 El Niño episode and the 1998 La Niña onset. Bulletin of the American Meteorological Soc., 217–243.
30. Unger D, Barnston A, van den Dool H, Kousky V (1996) Consolidated forecasts of tropical pacific SST in Niño 3.4 using one dynamical model and two statistical models. Experimental Long-Lead Bulletin, Vol. 5:1 of the Hardcopy Version.
31. Hsieh W (2000) Nonlinear canonical correlation analysis of the tropical pacific climate variability using a neural network approach. Journal of Climate, 14:2528-2539.
32. Tang B, Hsieh W W, Tangang F T (1998) Neural network model forecasts of the Niño 3.4 sea surface temperature. Experimental Long-Lead Forecast Bulletin, Vol. 7:1.

Evolutionary Optimization of Parametric Models: the Test Case of Combustion in a Diesel Engine

M. Farina[1], N. Cesario[2], D. Ruggiero[2], and P. Amato[2]

[1] STMicroelectronics, SST Corporate R&D, Via C. Olivetti, 2, 20041, Agrate(MI), Italy, `marco.farina@st.com`
[2] STMicroelectronics, SST Corporate R&D Via Remo De Feo 1, 80022 Arzano (NA), Italy `davide.ruggiero@st.com, nicola.cesario@st.com, paolo.amato@st.com`

Summary. The tuning of parameter values in parametric modelling can be viewed as an optimization problem where the outcome of the optimal model is to be as similar as possible to the experimental data. We give a general formulation of the problem with different fitness function definitions, both in terms of single-objective and multi-objective evolutionary optimization. As a test case we show results on a parametric model of combustion in a combustion chamber of a diesel engine. Such an optimal model will therefore be used for a controller design. The outputs (pressure inside combustion chamber versus rotation angle) of the resulting optimal models are compared to experimental data. Results of different optimization runs with Differential Evolution (DE) and Evolution Strategy (ES) as search algorithms and with different fitness definitions are compared.

1 Introduction

Parametric models are widely used in many technical and scientific areas when an analytical or numerical model is partially unknown or computationally too expensive. As evident some general laws governing the system must be known in order to write prototype parametric equation and leave to the optimization the choice and tuning of such parameters. If the governing laws are unknown and only a linguistic knowledge on the system is available, the field is open for fuzzy and neuro-fuzzy models.

As a first example the parametric model optimization of a mechanical device is described in [1] where excellent results are obtained in terms of similarity between measured data and optimal model output. A binary-coded GA was used with a non-trivial definition of fitness, showing the importance of a proper fitness definition for satisfactory results. A second example is the parametric model optimization of a dynamic structure (a bridge, for example) where the stiffness matrix is usually unknown and almost impossible to be measured or numerically computed. A precise parametric model is essential for a satisfactory semi-active control in case of earthquake; on this topic see [2, 3].

M. Farina et al.: *Evolutionary Optimization of Parametric Models: the Test Case of Combustion in a Diesel Engine*, Advances in Soft Computing **1**, 163–176 (2005)
`www.springerlink.com`

Evolutionary based model optimization has many automotive applications. For example, researchers at University of Tubingen, in collaboration with BMW Group Munich, are working on both the reduction the costs related to engine application at test beds and and to the improvement of the engine system calibration, [4, 5]. Complex engine functionality are the answer of Engine Management Systems developers to the legal exhaust guidelines which are getting more and more strict and to the increasing demands for a reduction of the fuel consumption. These engine functions shall optimize the combustion process and thus increase the efficiency of the engines. However, with the increasing number of adjustable engine parameters the complexity of the look-up table tuning, the so-called application of electronic control units becomes larger, too. In these framework evolutionary algorithms have successfully been applied for solving several combinatorial optimization problems in the engine application (e.g. optimization of the test bed schedule, statistical experimental design, look-up table design).

Another important example is a recent work appeared at the Society of Automotive Engineering SAE conference 2003, [6]. The authors describe the basic role of evolutionary algorithms in the optimization of diesel engine emissions and fuel efficiency. Each modern diesel engine design technique (Air Swirl, EGR, Injection Timing and Multiple Injections etc...) can be optimized by evolutionary algorithms, improving the performance considerably.

In the paper we show the application of parametric model optimization via evolutionary techniques to the automotive field for control purposes. The complexity of automotive devices (engine, suspension system...) highly demands the development of simple and poorly time-consuming models (such as parametric, fuzzy or neuro-fuzzy models) to be easily run in a control loop on cheap processors. In the above context the unaffordable complexity of a differential model of combustion is evident; fluid-dynamic partial differential equations coupled with thermodynamic equations would be necessary and it would be impossible to run such equations on a small and cheap processor for automotive control.

2 Problem Setting

The key point for a successful parametric model optimization is the interaction between the optimization expert and the system's expert; such an interaction give rise to a combination of iterative and non-iterative procedures. The flowchart of the entire optimization process is shown in Fig. 1 where system's expert tasks, optimization expert tasks and joined tasks are outlined with different colors. Moreover the iterative process is shown with full line while non-iterative processes are shown with broken lines.

Regarding the iterative optimization process, an essential point for a satisfactory result is the proper fitness definition in our experience, this is even more important than the choice of the optimization algorithm. The outcome of a parametric model may be of different typology, a single value, a 1-dimensional relationship between two variables or more generally an N-dimensional relationship between $N+1$

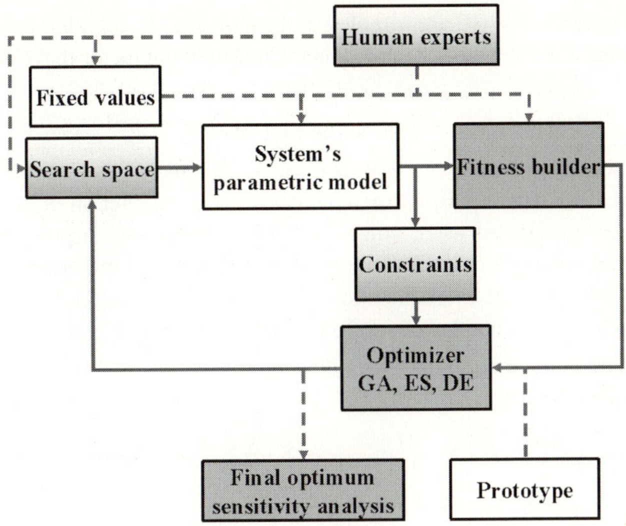

Fig. 1. Flowchart of the optimization strategy with the following meaning of lines and colors: *plain line*: iterative process, *broken line*: human expert intervention, *gray box*: optimization expert task, *white box*: model expert task. Several optimizers are implemented among which ES, DE and Genetic Algorithms (GA).

variables. We consider here the second case. In the test case we will present later the two variables are an angle θ (crank-angle) and a pressure $p(\theta)$.

2.1 Objective Function Definition

From a general point of view a proper fitness definition has to express properly the similarity between the experimental N-dimensional relationship and the model outcome and proper functional norms have to be used for evaluating differences between N-dimensional functions. When considering the one dimensional case. several different functional norms can be used for measuring the gap between the experimental pressure $p_e(\theta)$ and the output of the parametric model $p_s(\theta, X)$ at the current values of optimization parameters $X \in \Omega$ (parameters space). The choice of such gap measuring strategy is highly dependent on the shape of the model output (oscillatory, smooth, ...) and on what kind of similarity is the most important (uniform convergence, no oscillatory behavior, coincidence on some key points, ...). For a general problem's formulation we consider the following set of possible objective functions to be eventually combined once the above information are given by the system's expert.

$$f_1(\mathbf{X}) = \max_{i=n_1,\ldots,n_2} \frac{|p_s(\theta_i) - p_e(\theta_i)|}{P_n}, \qquad f_2(\mathbf{X}) = \sqrt{\frac{\sum_{i=n_1}^{n_2} \frac{(p_s(\theta_i)-p_e(\theta_i))^2}{P_n^2}}{n_2 - n_1}}.$$

where both $p_s(\theta)$ and $p_e(\theta)$ are sampled in n points $(\theta_1,\ldots,\theta_n)$, the indexes n_1 and n_2 are such that $1 \le n_1 < n_2 \le n$, and P_n is the normalizing factor that can be defined either as $P_n^{(1)} = \max_{i=1,\ldots,n} p_e(\theta_i)$ or $P_n^{(2)}(\theta) = p_e(\theta)$. For generality we consider two different normalization strategies, the first one ($P_n = P_n^{(1)}$) does not depend on θ while the second one ($P_n = P_n^{(2)}(\theta)$) is θ-dependent and it increases importance of similarity on areas where $p(\theta)$ is small. Generally a small interval $[n_1, n_2]$ of sample points may be considered when fitness is to be defined. Such a procedure may be needed when similarity on only one interval of $p(\theta)$ curve is important.

If no information are available about the most proper similarity function to be used, the following general formulation can be used where one combination of weights w_j and comparison windows values $n_1^{(M)}, n_2^{(M)}$ are automatically looked for by the optimizer.

$$\begin{cases} \min\limits_{\mathbf{X},\mathbf{w},\mathbf{n}_1,\mathbf{n}_2} = \Sigma_{k=1}^{M-1} w_j f_k\left(\mathbf{X}, n_1^{(k)}, n_2^{(k)}\right) + \left(1 - \Sigma_{k=1}^{M-1} w_k\right) f_M\left(\mathbf{X}, n_1^{(M)}, n_2^{(M)}\right) \\ \Sigma_{k=1}^{M-1} w_k \le 1 \quad w_k \le 1 \quad k = 1,\ldots,M-1 \\ \mathbf{X} \in \Omega \quad 1 \le n_1^{(k)} \le n_2^{(k)} \le n \quad k = 1,\ldots,M \end{cases} \tag{1}$$

We point out that, although the above equations where written for a situation where the model's outcome is a 1-D relation between two parameters, similar equations expressing the same ideas can be written for the general case of $N-D$ relations among $N+1$ parameters.

2.2 The Problem from the Point of View of Multi-Objective Optimization

In some cases the requirements for similarity of data and model output expressed by the various functions 1 are contradictory. As an example a model output may be globally quite similar to the data (the mean square error is low) but it may be hard to have at the same time a very small error on some crucial point. Such a situation may give rise to the following multi-objective optimization problem:

$$\begin{cases} \min\limits_{\mathbf{X},\mathbf{n}_1,\mathbf{n}_2} = \left(f_k\left(\mathbf{X}, n_1^{(k)}, n_2^{(k)}\right)\right) \quad k = 1,\ldots,M \\ \mathbf{X} \in \Omega \quad 1 \le n_1^{(k)} \le n_2^{(k)} \le n \quad k = 1,\ldots,M \end{cases} \tag{2}$$

One possible solution of the parametric model optimization problem can be to choose two potentially conflicting requirements such as global MSE (Mean Square Error) and local MSE and solve the multi-objective optimization problem approximating the Pareto optimal front and after that choose the best compromise; this choice may be done after a visual analysis of outcomes of Pareto optimal models.

3 Modelling the Combustion Chamber in a Diesel Engine

A model of combustion in a diesel engine requires knowledge and sub-models of several complex processes: air dynamic in the cylinder, fuel pulverization and vaporization, mix of the two fluids, chemical kinetic setting combustion premixed and diffusive phases. From a general point of view two different typologies of combustion models are usually developed and used: multidimensional and thermodynamic models. The first one aims at forecasting all details of fluids dynamics processes in engine cylinders. The second one, through empirical correlations and through the first principle of thermodynamics, aims at briefly describing the several processes related to combustion. In the simplest approach the fluid is considered to be uniform in temperature, pressure and chemical composition, so that the processes in the combustion chamber only depends on time (that it only depends on the crank angle). This model is known as "single zone" combustion model.

The most important step in the "single zone" combustion model is the evaluation of the *heat release law*. When neglecting mass flows through boundary surface of combustion chamber, heat released by chemical reactions ($\frac{dQ_b}{d\theta}$) is equal to: $\frac{dQ_b}{d\theta} = \frac{dE}{d\theta} + \frac{dL}{d\theta} + \frac{dQ_r}{d\theta}$ where $\frac{dE}{d\theta}$ represent the system internal energy variation, $\frac{dL}{d\theta}$ the power exchanged with the environment by the piston and $\frac{dQ_r}{d\theta}$ the heat leakage caused by the contact of fluid with the chamber chilled sides. Treating the fluid as a perfect gas with a mean temperature $\tilde{T}$, we have $E = mc_v\tilde{T}$. Consequently, the following equation holds:

$$\frac{dE}{d\theta} = mc_v\frac{d\tilde{T}}{d\theta}, \quad \frac{dL}{d\theta} = p\frac{dV}{d\theta}, \quad T = \frac{pV}{mR}, \quad \frac{dT}{d\theta} = \frac{p}{mR}\frac{dV}{d\theta} + \frac{V}{mR}\frac{dp}{d\theta} \tag{3}$$

Finally, combining the (3) each other we obtain the following equation for the *heat release law*:

$$\frac{dQ_b}{d\theta} = \left[\frac{c_v}{R} + 1\right]p\frac{dV}{d\theta} + \frac{c_v}{R}V\frac{dp}{d\theta} + \frac{dQ_r}{d\theta} \tag{4}$$

Based on (4), the Mechatronics team developed a theoretical model to forecast the pressure cycle inside the combustion chamber of a single injection diesel engine. This model has 3 external parameters (describing the engine working point) and 9 input parameters (among which RPM – Rounds Per Minutes, combustion-start, temperature, loss factor, etc.)[1]. The Fig. 2 shows the pressure values versus crank angle θ associated to a given set of external and input parameters. The test case engine is a Common Rail four strokes mono-cylinder diesel engine with the following features: Displacement = 224 [cm^3], Bore = 69 [mm], Piston race = 60 [mm], Compression ratio = 17.5:1, the Fuel supply is a Common Rail direct injection, Max power = 3.5 [kW] at 3600 RPM and maximum torque = 10.5 [Nm] at 2200 RPM.

The aim of the work is the tuning (optimization) of the model parameters to achieve a "good fitting" of the experimental data of the engine pressure cycle. Here "good fitting" has a twofold meaning. One side, the approximation has to ensure the global error to be less than 5% of the max value of experimental data. On the other side, the fitting is asked to have a small error (about 2% of the max value of

[1] Since this theoretical pressure model is a key item of a patent proposal, we are restrained from giving its detailed description.

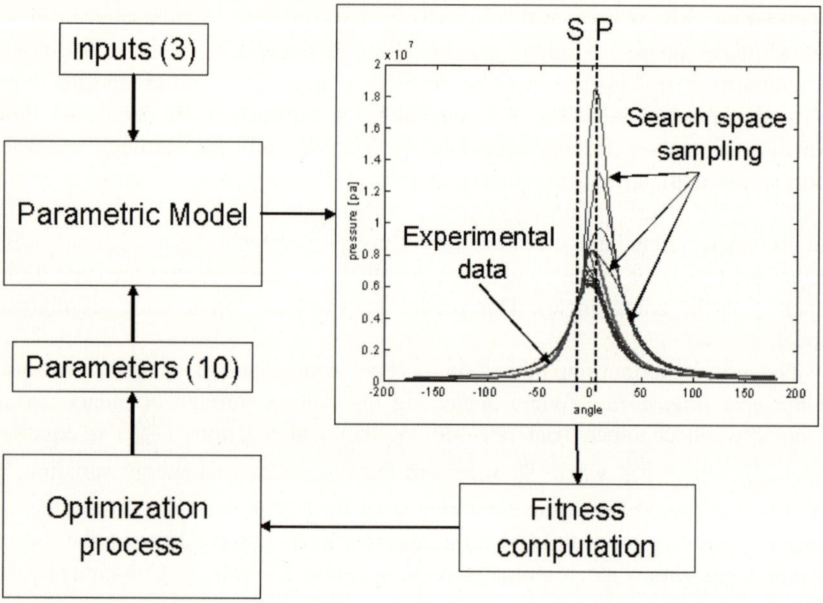

Fig. 2. Data flow for pressure cycle model and for parameter optimization. P: peak pressure point, S: Start of combustion point

experimental data $P_n^{(1)}$) on some given crank angle ranges, and in details: around the combustion starting point (where $-50 \leq \theta \leq -25$) and around the the pressure cycle peak (where $-10 \leq \theta \leq 15$). These tho objectives are in conflict and the eventual multiobjective treating of the problem is thus justified.

4 Optimization Algorithms

One drawback of a model optimization procedure is that each objective function call from the optimizer requires the full run of the model itself with parameters corresponding to the current solution to be evaluated. This may give a relevant burden from the computational point of view. For this reason we have decided to consider only ES and DE and not GA for comparing results.

The two differential evolution DE1 and DE2 algorithms are described in [7, 8, 9] and for a 9-variables real-values box-constrained problem as our test case is a number of 30–50 individuals is sufficiently high for convergence in about 250 iterations; this means approximately 10^4 model runs.

In this work two distinct evolution strategies have been used: $(1+1)$-ES and $(1+9)$-ES [10]. This type of evolution strategies doesn't comprise any selection and recombination. They are the simplest form of the parallel search methods. At each step, all population individuals generate offspring by an asexual reproduction (also referred to as mutation [11]).

Let $\mathcal{P}_t$ be the population at t-th iteration, X_t be an individual in $\mathcal{P}_t$ and J be the number of genes of each individual. For offspring generation we used the following formula:

$$\text{off}(X_t(j)) = X_t(j) + D_t(X_t(j))\sqrt{-2\lg(r_1)}\cos(2\pi r_2) \tag{5}$$

where off is the generative function, $X_t(j)$ is the j-th gene of the individual X at t-th iteration (with $j = 1,\ldots,J$), D_t is the covariance matrix at t-th iteration (whose dimension is $|\mathcal{P}_t| \times J$) which describes the search step of each gene (that is each model parameter) of all individuals and r_1, r_2 are two random number belong to range $[0,1]$.

In both strategies, optimum search involves an updating procedure of the search step D_t based on the "simulated annealing" strategy. Offspring generation is controlled by several algorithm parameters: the hypersphere ray update threshold h, the "annealing" threshold a_{th} and the "annealing" scaling factor a_s. Obviously, offspring generation is controlled by parameters constraints in search space too. The start value of covariance matrix D_0 sets the first hypersphere ray to 10% of the range size of each parameter. The parameter h is the number of initial iterations in which the search hypersphere, centered on the current individual, isn't changed. We set this parameter as 40% of the max number of simulation iterations. After the first h iterations, matrix D_t is updated in this way:

$$D_{t+1}(X_{t+1}(j)) = \begin{cases} D_t(X_t(j))/a_s & \text{if } p(X_t) > a_{th} \\ D_t(X_t(j))a_s & \text{otherwise}. \end{cases} \tag{6}$$

where $D_t(X_t(j))$ is the search ray associated to j-th gene of the current individual $X_t, p(X_t)$ is the ratio between the number of the fitness improvements ($f(\text{off}(X_t)) < f(X_t)$ in a minimum search problem, f is the fitness function utilized) in the last h iterations and h. Usually in literature [12] a_{th} takes the value 0.2 and $0.9 \le a_s \le 0.96$.

5 Single-Objective Optimization Results

Table 1 shows the optimal results of the fitness functions f_1 and f_2 obtained by the different evolution strategies and differential evolutions algorithms. Table 2 lists the optimal values of model parameters $\text{Par}_1,\ldots,\text{Par}_9$ for each optimization strategy.

Table 1. Outcomes obtained by $(1+1)$-ES, $(1+9)$-ES, DE1 and DE2 algorithms with different objective functions. These errors are normalized respect to value max of experimental data $P_n^{(1)}$

	$(1+1)$ES	$(1+9)$ES	DE1	DE2	Prot. Mod.
f_1	0.0693	0.0472	0.0087	0.0041	0.1818
f_2	0.0148	0.0122	0.0094	0.0077	0.0716

Table 2. Optimal parameters obtained by (1 + 1)-ES, (1 + 9)-ES, DE1 and DE2 algorithms with different objective functions. Parameters values are compared with parameters obtained by the prototype of the pressure model (Prot. mod.). This table shows the ranges of each parameter too

	(1+1)ES f_1	(1+1)ES f_2	(1+9)ES f_1	(1+9)ES f_2	DE1 f_1	DE1 f_2	DE2 f_1	DE2 f_2	Prot Mod.	Range
Par_1	4.7144	3.5650	3.4130	3.9787	0.0234	0.1216	0.0454	0.0000	3.3	[0, 5.0]
Par_2	0.5079	0.5385	0.4697	0.4906	0.3958	0.3765	0.3811	0.4345	0.8	[0, 2.0]
Par_3	0.0437	0.0325	0.0102	0.0134	0.0564	0.0679	0.0671	0.0721	0.12	[0.01, 0.1]
Par_4	1.3991	1.3813	1.3870	1.3714	1.3563	1.3445	1.3345	1.3400	1.4	[1.35, 1.40]
Par_5	1.4803	1.6451	1.5803	1.5238	1.1452	2.3257	1.1456	2.1784	4.0	[0, 6.0]
Par_6	14.386	12.3838	11.9032	12.4482	12.456	12.3672	11.234	10.7729	7.0	[6.0, 30.0]
Par_7	68.1875	82.8804	79.5597	69.0244	9.998	89.9997	89.997	89.9997	60.0	[10.0, 90.0]
Par_8	559.0053	492.7448	455.5236	428.8541	553.054	456.5853	455.5853	457.3456	863.0	[400, 1000]
Par_9	0.0425	0.0243	0.0223	0.0253	0.0204	0.0198	0.0204	0.0202	0.4	[0, 1.0]

Figure 3 shows outcomes obtained by DE2 algorithm with the fitness function f_2. As can be seen the experimental data and the optimal model outcome are overlapping and the error is very small. Figures 4 and 5 illustrate respectively a

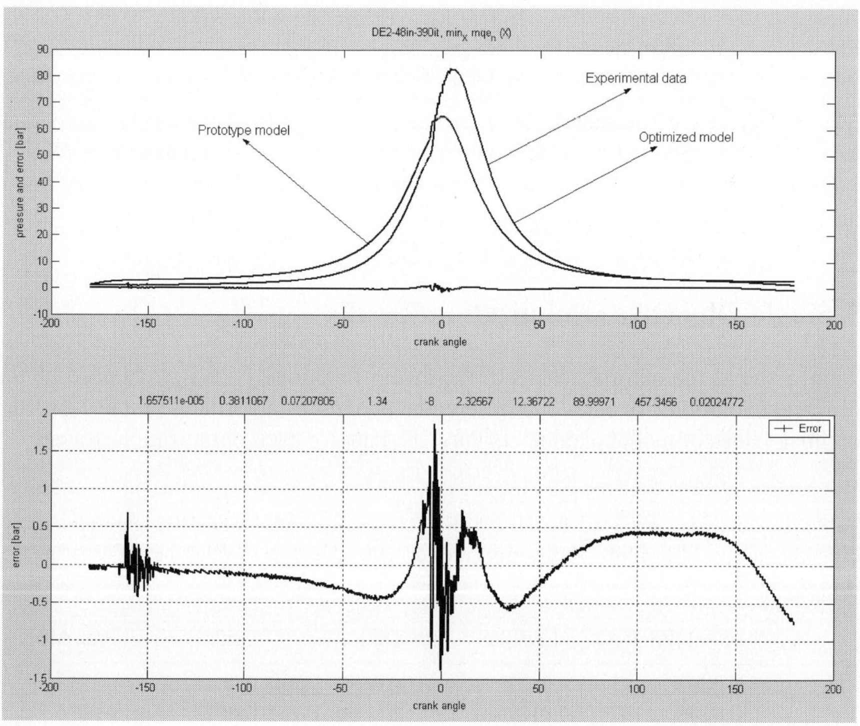

Fig. 3. Pressure curves corresponding to optimized model, data and prototype model (the best model provided by the mechatronics expert) and error for the DE2 optimization

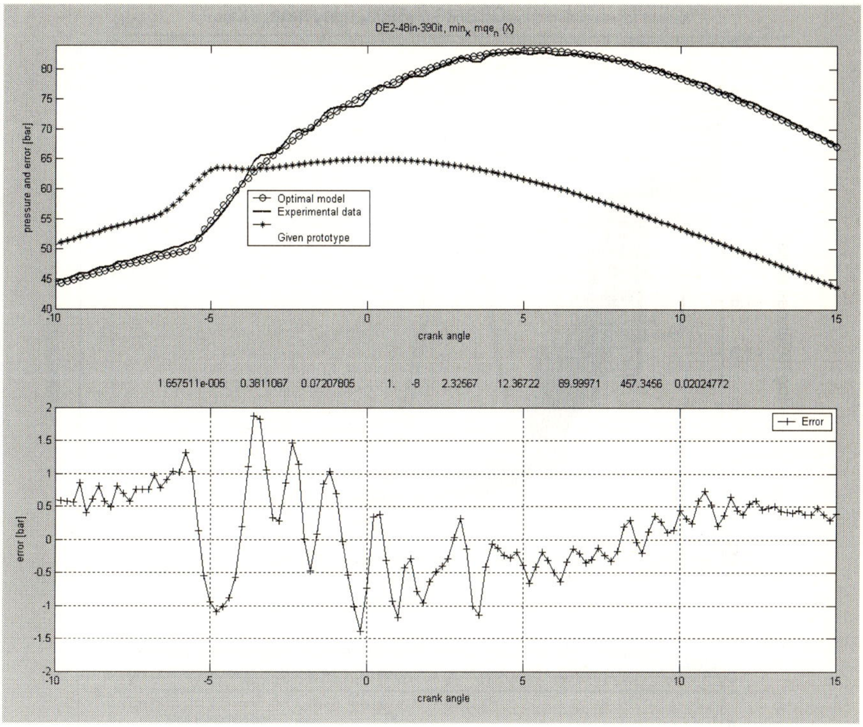

Fig. 4. Zoom of pressure curves corresponding to optimal model and data on the most critical part (peak of the pressure cycle) obtained by DE2 algorithm with the objective function f_2

zoom of the fitting obtained in this case on one of the most critical part of pressure cycle (around peak of the pressure cycle) and the story of the objective function. We point out that, as can be seen from the zoomed figure, the error is of the same size of experimental data oscillations and it is thus a very satisfactory model. Figure 5 shows that the approximation error size is reduced by approximately two magnitude orders. Moreover the pressure curve corresponding to the optimal solution is significantly more similar to the experimental curve than the one corresponding to the "prototype" model given by the mechatronics expert. This means that the optimization process can relevantly improve the solution, exploring areas of the search space that wouldn't have been considered as promising by the mechatronics expert.

At first, it has been made a simulation using $(1 + 9)$-ES algorithm with a weighted fitness function $\frac{f_1+\alpha \cdot f_2}{1+\alpha}$ composed by a uniform error max on the entire domain of the pressure cycle function and by a mean quadratic error on the function "peak" range. In this work we have set α value to 10. The simulation, whose results are showed in Fig. 5 yields objectively the best result obtained by evolution strategies. In fact in this case the error normalized respect to the max value of the experimental data is 0.01.

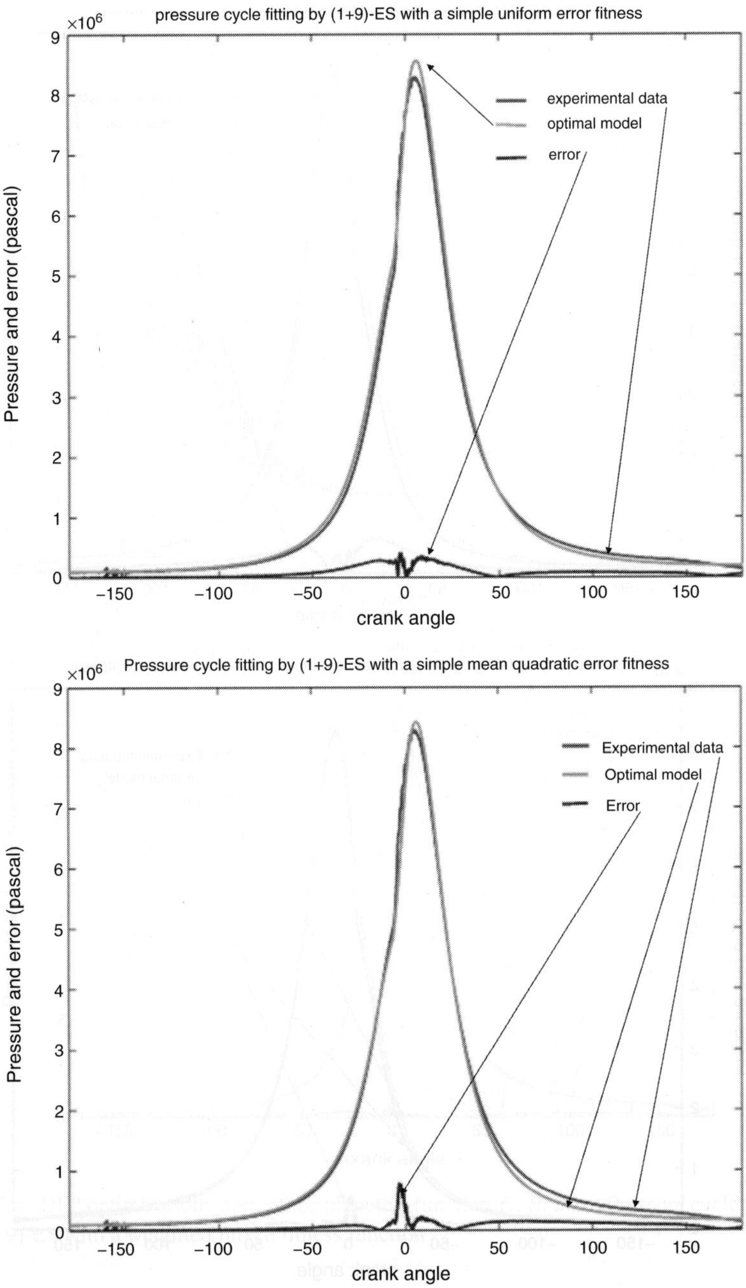

Fig. 7. *Top*: Pressure cycle fitting by (1 + 9)-ES with the uniform error fitness f_1. *Bottom*: Pressure cycle fitting by (1 + 9)-ES with the mean quadratic error fitness f_2

better fitting (according to previous description of the aim of this optimization work) than previous case obtained by the same algorithm but with the mean quadratic error (fitness f_2).

Figures 7 illustrates the result of the pressure cycle fitting obtained by $(1 + 9)$-ES algorithm with the fitness function f_1 (error uniform max), while Fig. 7 illustrates a better fitting (according to previous description of the aim of this optimization work) than previous case obtained by the same algorithm but with the mean quadratic error (fitness f_2). From the results Table 1 and from Figs. 6 and 7 it can be noticed that $(1 + 9)$-ES algorithm yields better outcomes than $(1 + 1)$-ES with both fitness functions f_1 and f_2.

Before going to conclusions we wish to thank Olga Scognamiglio and Giuseppe Palma for the support they gave us with their knowhow on Mechatronics systems.

6 Conclusions

A complete tool for model optimization have been developed. The strategy is based on both algorithm iteration and non-iterative process requiring interactions between the optimization expert and the model's expert. Several evolutionary optimizers have been compared with several objective function definitions.

The key point for such results is the proper choice of both the search algorithm, and the objective function. While the former is usually selected by optimization expert, the latter should be selected through a deep interaction between optimization expert and the system's expert.

The pressure curve corresponding to the optimal solution is significantly more similar (the approximation error size is reduced by approximately two magnitude orders) to the experimental curve than the one corresponding to the "prototype" model given by the mechatronics expert. This means that the optimization process can relevantly improve the solution, exploring areas of the search space that wouldn't have been considered as promising by the mechatronics expert.

References

1. R. Bussola and M. Tiboni. Parametrical identification for a complex mathematical model of indexing cam mechanism by means of a genetic algorithm. *Proceedings of eurogen 99, May 30 – June 3 1999, Jyvaskyla, Finland* .
2. F. J. Molina, P. Peregon, and G. Verzeletti. Time-domain identification from seismic pseudodynamic test results on civil engineering specimens. *Proceedings of Second International Conference on Identification in Engineering Systems, University of Wales Swansee, March 29–31 1999.*
3. F. J. Molina and P. Peregon. Identification of the damping properties of the walls of the safe program. *European Commission, Joint Research Centre, Ispra, Technical Note No. I.98.35* .

4. Alexander Mitterer Thomas Fleischhauer Frank Zuber-Goos Jan Poland, Kosmas Knödler and Andreas Zell. Evolutionary search for smooth maps in motor control unit calibration. *In: K. Steinhöfel (Ed.), Stochastic Algorithms: Foundations and Applications (LNCS 2264).*
5. Alexander Mitterer Andreas Zell Kosmas Knödler, Jan Poland. Memetic algorithms for combinatorial optimization problems in the calibration of modern combustion engines. *In W. B. Langdon et al. (Eds.), Proceedings of the Genetic and Evolutionary Computation Conference (GECCO-2002), Morgan Kaufmann, New York/NY.*
6. H. Hiroyasu and H. Miao. Genetic algorithms optimization of diesel engine emissions and fuel efficiency with air swirl, egr, injection timing and multiple injections. *SAE 2003.*
7. D. Corne, M. Dorigo, and F. Glover. *New Ideas in Optimization*. McGraw-Hill Education, New York, 1999.
8. R. Storn and K. Price. Differential evolution – a simple evolution strategy for fast optimization. *Dr. Dobb's Journal, 22(4)*, pp. 18–24, 1997.
9. R. Storn and K. Price. Differential evolution – a simple and efficient heuristic for global optimization over continuous spaces. *Journal of Global Optimization, 11(4), Kluwer Academic Publishers*, pp. 341–359, 1997.
10. Hans-George Beyer. Towards a theory of "evolution strategies". some asymptotical results from the $(1 \stackrel{+}{,} \lambda)$-theory. 1992.
11. L. Kallel, B. Naudts, and A. Rogers. *Theoretical aspects of evolutionary computing*. Springer-Verlag, New York, 2001.
12. Hans-George Beyer. *The Theory of Evolution Strategy*. Springer-Verlag, New York, 2001.

Part V

Fuzzy Systems and Neural Networks

Rule Learning and Extraction Using a Hybrid Neural Network: A Case Study on Fault Detection and Diagnosis

Shing Chiang Tan[1] and Chee Peng Lim[2]

[1] Faculty of Information Science & Technology, Multimedia University, Jalan Ayer Keroh Lama, 75450 Melaka, Malaysia
sctan@mmu.edu.my

[2] School of Electrical & Electronic Engineering, University of Science Malaysia, Engineering Campus, Nibong Tebal, 14300 Seberang Prai Selatan, Penang, Malaysia
cplim@usm.my

Abstract. A hybrid network, based on the integration of Fuzzy ARTMAP (FAM) and the Rectangular Basis Function Network (RecBFN), is proposed for rule learning and extraction problems. The underlying idea for such integration is that FAM operates as a classifier to cluster data samples based on similarity, while the RecBFN acts as a "compressor" to extract and refine knowledge learned by the trained FAM network. The hybrid network is capable of classifying data samples incrementally as well as of acquiring rules directly from data samples for explaining its predictions. To evaluate the effectiveness of the hybrid network, it is applied to a fault detection and diagnosis task by using a set of real sensor data collected from a Circulating Water (CW) system in a power generation plant. The rules extracted from the network are analyzed and discussed, and are found to be in agreement with experts' opinions used in maintaining the CW system.

1 Introduction

Rule learning and extraction is an important area in neural network (NN) research. Although NN models have exhibited a number of attractive features for fault detection and diagnosis (FDD) applications, there is no single panacea in NN approaches that always yield the best performance on FDD problems. This is particularly true for real problems in large-scale, complex processes in today's modern industrial environments [1]. To circumvent the problem, two key limitations of NN models for FDD tasks are focused in our work. First most NN algorithms need a long training time in response to a large database and cannot perform incremental learning in real time. The ability of NN models in absorbing knowledge continuously and autonomously without corrupting or forgetting previously acquired knowledge is necessary, for instance, in handling on-line learning problems in non-stationary environments. Second, trained NN models often operate as a "black box" which

Shing Chiang Tan and Chee Peng Lim: *Rule Learning and Extraction Using a Hybrid Neural Network: A Case Study on Fault Detection and Diagnosis*, Advances in Soft Computing **1**, 179–191 (2005)
www.springerlink.com

reveals little or no information of the system they represent [2]. Knowledge learned is concealed within a large number of connections (weights), in which domain users can hardly understand and utilize the embedded knowledge to assist in decision-making.

In this paper, we propose a hybrid network, based on Fuzzy ARTMAP (FAM) [3] and the Rectangular Basis Function Network (RecBFN) [4], for undertaking condition monitoring and FDD problems. FAM is a supervised learning architecture of the Adaptive Resonance Theory (ART) [5, 6] family that combines neural network and fuzzy set theory into a common framework. FAM has the ability to undertake the stability-plasticity dilemma, i.e. how a learning system is able to protect useful historical data from corruption (stability) while simultaneously learning new data (plasticity) [5]. It is thus able to learn continuously in a changing data environment and to acquire knowledge in situ while simultaneously providing useful predictions. To enhance its applicability to condition monitoring and FDD tasks, we have endowed FAM with an explanatory facility for its predictions via a rule extraction algorithm. Specifically, we combine FAM and the RecBFN to form a hybrid rule-based adaptive network that is capable of classifying fault patterns as well as of acquiring rules directly from data for explaining its predictions. To evaluate the effectiveness of the proposed system, we have collected a set of real sensor data from a power generation plant to develop health signature for FDD of a crucial process in the plant.

2 The FAM-RecBFN: A Hybrid Network

The hybrid network between FAM and the RecBFN, known as the FAM-RecBFN, is formed in a cascaded mode. The underlying idea for such integration is that FAM works as a classifier to cluster data samples based on similarity, while the RecBFN acts as a "compressor" which groups and refines knowledge from the trained FAM network. The FAM-RecBFN inherits several advantages from its predecessors. First, it requires no prior handcrafting of network architecture. Instead, its network size is built incrementally pertaining to the process/machine under scrutiny via data samples. Second, it is a model of network that overcomes the stability-plasticity dilemma and is potentially applicable to real-time problems. Besides, it has fewer user-defined parameters that need tuning which, in turn, makes rule extraction almost an automated manner. Furthermore, the extracted rules are less restrictive and conflict-free, and they can be expressed in the *If-Then* format that is explicit to domain users. The operation of the FAM-RecBFN is as follows.

2.1 Fuzzy ARTMAP (FAM)

FAM consists of a pair of Fuzzy ART [7] modules, ART_a and ART_b, interconnected by a map field. Figure 1 shows the FAM network structure. Each ART module comprises three layers of nodes: $F_0^a(F_0^b)$ is the normalization layer in which an M-dimensional input vector, $\boldsymbol{a}$, is complement-coded [3] to a $2M$-dimensional

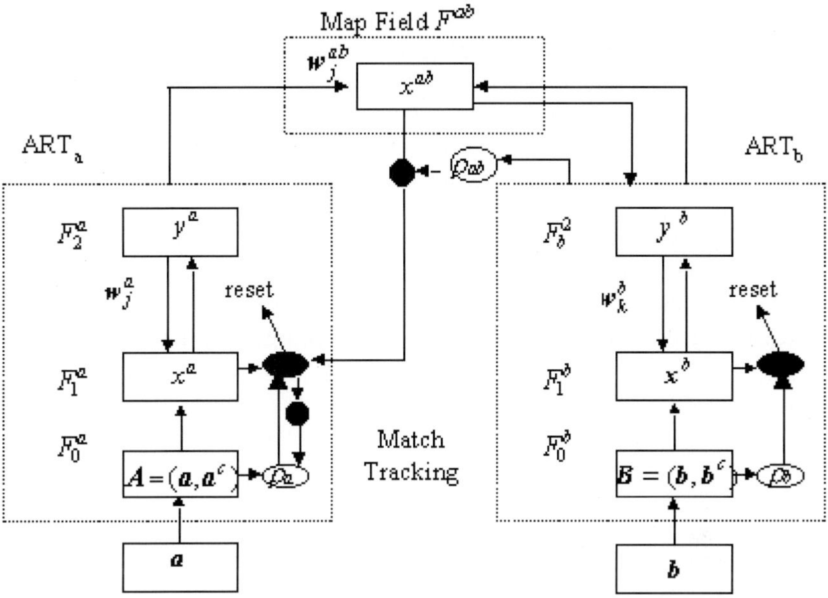

Fig. 1. The architecture of Fuzzy ARTMAP

vector, $A = (\boldsymbol{a},\boldsymbol{a}^c) \equiv (a_1,\ldots,a_M,1-a_1,\ldots,1-a_M)$; $F_1^a(F_1^b)$ is the input layer which receives the complement-coded input vectors; $F_2^a(F_2^b)$ is the recognition layer which is a dynamic layer that encodes prototypes of input patterns and that allows the creation of new nodes when necessary.

The key feature of FAM is in the inclusion of a novelty detector, i.e., the vigilance test, to measure against a threshold (vigilance parameter, ρ) the similarity between the prototype patterns stored, $\boldsymbol{w}_J$, in the network and the input pattern $\boldsymbol{A}$, i.e., $\frac{|\boldsymbol{A}\wedge\boldsymbol{w}_J|}{|\boldsymbol{A}|} \geq \rho$ [3]. When the match criterion is satisfied, learning takes place according to $\boldsymbol{w}_J^{(\text{new})} = \beta(\boldsymbol{A}\wedge\boldsymbol{w}_J^{(\text{old})}) + (1-\beta)\boldsymbol{w}_J^{(\text{old})}$, where $\beta \in [0,1]$ is the learning rate. When the match criterion is not satisfied, a new node is created, and the input is coded as its prototype pattern. As a result, the number of nodes grows with time, subject to the novelty criterion, in an attempt to learn a good network configuration autonomously and on-line. As different tasks demand different network structures, this learning approach thus avoids the need to specify a pre-defined static network size, or to re-train the network off-line.

During supervised learning, ART$_a$ and ART$_b$, respectively, receive a stream of input and target patterns. FAM does not directly associate input patterns at ART$_a$ to target patterns at ART$_b$. Rather, input patterns are first classified into prototypical category clusters before being linked with their target outputs via a map field. At each input pattern presentation, this map field establishes a link from the winning category prototype in F_2^a to the target output in F_2^b. This association is used, during testing, to recall a prediction when an input pattern is presented to ART$_a$.

As pointed out in [3], the orders of input patterns presented to FAM in the training phase can affect the formation of cluster prototypes, hence the network performance. As a result, an ordering algorithm [8] has been proposed to identify a fixed order of training samples that should be presented to FAM in order to improve the performance. In this work, FAM with the ordering algorithm is employed. The ordering algorithm comprises three stages. In the first stage, the input pattern that maximizes the sum $\sum_{i=1}^{M} |\boldsymbol{A}_{M+i} - \boldsymbol{A}_i|$ is selected as the first center, and is removed from the training set S_T to avoid re-selection in subsequent stages. In the second stage, the next $(n_{\text{clust}} - 1)$ input patterns are selected as centers (in which n_{clust} is the number of distinct classes) through an Euclidean-based Max-Min clustering algorithm [9], and they are discarded from S_T. In the third stage, the presentation orders for the remaining input patterns in S_T are determined by finding the minimum of the Euclidean distances between these patterns and the n_{clust} centers. The whole procedure in the third stage is repeated until all orders of the input pattern presentation for the training phase of FAM have been identified.

2.2 The Rectangular Basis Function Network (RecBFN)

The RecBFN is a feedforward network, with an input layer, one hidden layer, and an output layer. Its architecture, as shown in Fig. 2, is similar to the RBF network. Nevertheless, they are different in two aspects: the activation functions of the hidden nodes and the propagation rule.

The input layer is fully connected to the hidden layer. The hidden (RecBF) nodes are dynamically created when necessarily. Each hidden node, which corresponds to a rule, comprises a center-vector $\boldsymbol{r}$ and two sets of widths: a core region (inner rectangle) λ_i^+, λ_i^- and a support region (outer rectangle) $\Lambda_i^+, \Lambda_i^-, (1 \leq i \leq n$ where n denotes the input dimension). Figure 3 illustrates an instance of the RecBF node in a two-dimensional feature space. Each RecBF node is linked to one output node with a non-zero weight, and each output node corresponds to one class.

The RecBFN employs a fast and constructive training scheme known as the Dynamic Decay Adjustment (DDA) algorithm [4, 10, 11]. The DDA algorithm

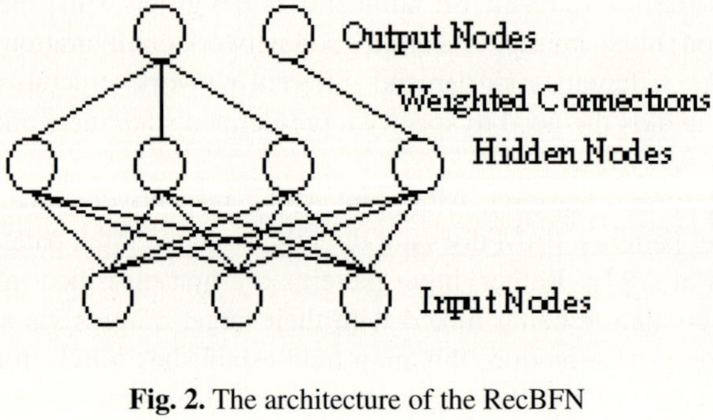

Fig. 2. The architecture of the RecBFN

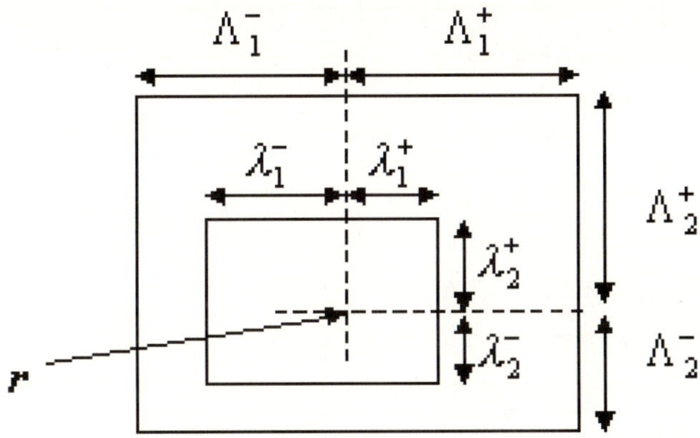

Fig. 3. A RecBF node in a two-dimensional feature space

comprises three steps that create new prototypes when necessary and adjusts the core and support regions of the existing prototypes. The steps are described as follows:

- **cover**: if a new training pattern lies inside the support region of an already existing prototype, its core-region is extended to cover the new pattern;
- **commit**: if a new training instance is not covered by a prototype of the correct class, a new hidden node will be created. The hidden node is assigned with a reference vector that is the same as the new training instance. Meanwhile, the widths of the hidden node will be chosen as large as possible, without running into conflict with the already existing prototypes. This, consecutively, requires the execution of a shrinking procedure towards the existing prototypes of other classes.
- **shrink**: if a new pattern is incorrectly classified by an existing prototype of conflicting class, the widths of this prototype will be reduced (shrunk) so as to solve the conflict.

2.3 The FAM-RecBFN

To effectively form the hybrid FAM-RecBFN, a center estimation procedure is introduced in FAM. This is because the original weight vectors encode the maximum and minimum vertices of a hyper-rectangle which do not reflect the centroids of the pattern clusters represented by each ART_a prototype node. A new set of center weight vectors, $\boldsymbol{w}_j^{a-c}$, is introduced in ART_a to represent cluster centroids [12]. The center weight vectors propagate from the input layer to the hidden layer. If $\boldsymbol{w}_j^{a-c}$ is inside the support region of an already existing hidden node, its core region is extended to cover the pattern. Otherwise, a new hidden node is created to include the center weight vector $\boldsymbol{r}$. The reference vector of the hidden node is the same as the center weight vector. The widths of the prototype are determined using a shrinking algorithm, against the references of already existing hidden nodes of conflicting

Proc Shrink$\left(x = w_j^{a-c}\right)$

//compute new Λ for three cases:

$$\Lambda_{best,k} = \min\{|r_k - x_k| : \forall 1 \le j \le n, j \ne k :$$

$$\left(\frac{\Lambda_k - |r_k - x_k|}{\Lambda_k} \le \frac{\Lambda_j - |r_j - x_j|}{\Lambda_j}\right) \wedge \left(\Lambda_k \ge \varepsilon_{k,min}\right) : \Lambda_j, \Lambda_k < \infty\}$$

$$\Lambda_{max,l} = \max\{|r_l - x_l| : \Lambda_l = \infty\}$$

$$\Lambda_{min,m} = \min\{|r_m - x_m| : \forall 1 \le j \le n, j \ne m :$$

$$\frac{\Lambda_m - |r_m - x_m|}{\Lambda_m} \le \frac{\Lambda_j - |r_j - x_j|}{\Lambda_j} : \Lambda_m, \Lambda_j < \infty\}$$

IF Λ_{best} does exist **THEN**

$\Lambda_k = \Lambda_{best,k}$

ELSEIF $\Lambda_{max,l} \ge \Lambda_{min,m}$ **THEN**

$\Lambda_l = \Lambda_{max,l}$

ELSE

$\Lambda_m = \Lambda_{min,m}$

ENDIF

ENDPROC

Fig. 4. The shrinking procedure

class. Three cases of width shrinking have been proposed (Fig. 4). If an existing finite dimension can be shrunk without shrinking below a preset minimum width, ε_{min}, the one with the smallest loss in volume will be chosen $(\Lambda_{best,k})$. If this is not the case, either one of the remaining infinite dimensions will be shrunk $(\Lambda_{max,l})$, or if this would result in a new width smaller than ε_{min}, one of the existing infinite dimensions $(\Lambda_{min,m})$ will be shrunk below ε_{min}. In fact, the inclusion of ε_{min} to the DDA algorithm and the formation of asymmetric rectangles ease rule extraction with soft boundaries, which are important and useful in suppressing the effect of rule rigidity.

3 Application of the FAM-RecBFN to Fault Detection and Diagnosis

3.1 The Circulating Water (CW) System

Applicability of the FAM-RecBFN to FDD tasks is investigated using a set of real sensor data collected from an electricity power generation plant in Penang, Malaysia.

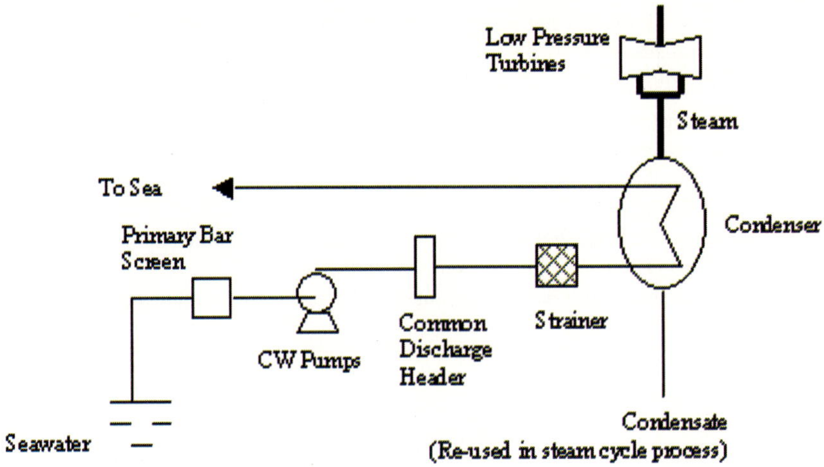

Fig. 5. The Circulating Water system

The system being monitored in the power station is the Circulating Water (CW) system. The main function of the CW system is to supply a sufficient and continuous amount of cooling water to the main turbine condenser to condense steam from the turbine exhaust and other steam flows into the condensers [13]. Figure 5 depicts a simplified overview of the CW system. The CW system includes all piping and equipment (such as condensers and drum strainer) between seawater intake and the outfall where water is returned to the sea.

The circulating water enters the plant from the sea through a primary bar screen at the pump house. The bar screen is used to prevent large sized debris, such as timber and clumps of seaweed, from entering the CW system. In the pump house, the CW pumps draw seawater from the suction chamber to a common discharge header through a hydraulic discharge valve. From the common discharge header, the seawater flows into the CW inlet culvert through a drum strainer. This drum strainer functions as a filter to remove fine debris, such as shells and seaweed in the seawater. The circulating water flows through the culvert up to the turbine condensers, where the water is used to condense steam being exhausted from the low-pressured turbine. After passing through the condenser, the circulating water exits the condenser through a pair of outlet valves into a concrete outlet culvert, before it is discharged back to the sea through an outfall.

3.2 Network Accuracy

A database was formed on 80M Watts of the targeted power generation. Each record in the database consisted of 12 features comprising temperature and pressure measurements at various inlet and outlet points of the condenser, as well as other important information (as tabulated in Appendix). Four fault states of the CW system have been identified and they are summarized in Table 1.

Table 1. Fault-states classification of the CW system

Class	Indication
1	Heat transfer in the condenser is efficient and there is no significant blockage in the CW system.
2	Heat transfer in the condenser is not efficient and there is no significant blockage in the CW system.
3	Heat transfer in the condenser is efficient and there is significant blockage in the CW system.
4	Heat transfer in the condenser is not efficient, and there is significant blockage in the CW system.

The database was partitioned into a training set and a test set, each with 1000 samples. The hybrid network was trained in the fast learning mode with 10 epochs per training session, and by systematically varying the value of the ART$_a$ baseline vigilance [3], $\bar{\rho}_a$, for low to high values. Table 2 depicts the network performance in terms of test accuracy and the number of F_2^a nodes. The highest test accuracy achieved was 95.70%, with a total number of 18 nodes, at $\bar{\rho}_a = 0.5$. This setting was used in subsequent experiments for rule extraction.

Table 2. Classification performance

$\bar{\rho}_a$	Test Accuracy (%)	Number of Nodes
0.0	95.20	15
0.1	95.20	15
0.2	95.20	15
0.3	95.20	15
0.4	95.20	15
0.5	95.70	18
0.6	94.70	20
0.7	94.00	27
0.8	94.70	41
0.9	88.70	119

3.3 Rule Extraction

The centroid of each prototype was retrieved from the trained FAM network and used in the rule extraction phase. A total number of 18 prototypes were obtained, and rule extraction ensued with a predefined minimum width $\varepsilon_{\min}$ of 0.1, for each dimension of the input attributes. Nine rules were extracted, four for Class 1, three for Class 2, one for Class 3, and one for Class 4. Quantization, as proposed in [14], was conducted in which each rule was quantized into 5 levels of linguistic representation

Table 3. Rules extracted from the FAM-RecBFN (1-*very low*; 2-*low*; 3-*medium*; 4-*high*; and 5-*very high*)

No.	Class	Parameter 1	2	3	4	5	6	7	8	9	10	11	12
1	1	1	1	2	1	3	1	2	3	2	2	5	1
2	1	2	1	1	3	3	2	1	3	3	2	3	1
3	1	2	2	2	3	4	3	2	3	4	3	4	2
4	1	3	2	2	3	3–4	3	2	3	4	2	4	2–3
5	2	2	2	1	3–4	4	3	2	2–3	2–3	1–2	3–4	2
6	2	2–3	2	2	3–4	3–4	3	2	3	4	2–3	4	2–3
7	2	3	3	2	3	4	3	2	3	5	3	4	4
8	3	1–3	1–2	2–3	3	3	2–3	1–2	2–3	3–4	1–3	1	1–4
9	4	2–3	2–3	1–2	3–4	3–4	2–3	2	2–3	1–3	1–2	1	2–4

to ease user interpretation (i.e., 1 for *very low*; 2 for *low*; 3 for *medium*; 4 for *high*; and, 5, for *very high*). If compared to the quantization approach in [14], each level of quantization is not necessarily assigned the same interval. Instead, domain experts are free to set the range for each quantization level which conforms to their experience and knowledge. Table 3 shows the extracted rules by considering only their core regions.

3.4 Rule Analysis

With reference to CWOP B (parameter 11) in Table 3, Rule 1 to Rule 7 can be linearly separated from Rules 8 and 9, where in Rules 1 to 7 CWOP B ≥ 3 and Rules 8 and 9 CWOP B $= 1$. The former rules show insignificant blockage in the CW piping system, whereas the latter rules indicate significant blockage. In accordance with the experts' opinions, the CW system is in a state of significant blockage when CWOP B is very low.

Rules from Class 1, particularly Rules 1 and 2, depict a high efficiency of heat transfer of the condensers. This is indicated by the condenser vacuum, VAC (parameter 12), which is very low (i.e., 1). On the other hand, Rule 7 from Class 2 and Rule 9 from Class 4, show poor heat transfer in the condensers, as the condenser vacuum (VAC) is raised from above level 2. The finding is further supported by high scores of quantization levels in LPT A and LPT B (parameters 1 and 2) in Rules 7 and 9, compared to Rules 1 and 2, i.e.,

- In Rule 7, LPT A and LPT B are 3 while Rule 9, both range from 2 to 3.
- In Rule 1, LPT A and LPT B are 1 while in Rule 2, they are 2 and 1, respectively.

Rules 7 and 9 have differences in several aspects. Rule 7 shows a state of inefficient heat transfer in the CW system that has no significant blockage, whereas Rule 9, on the other hand, has significant blockage. Note that CWOP A and CWOP B (parameters 9 and 11) in Rule 7 have quantization levels 5 and 4, respectively, whereas for Rule 9, both scores are from 1 to 3, and 1, respectively. This is the finding that differentiates the two rules.

Most of the rules derived are from Classes 1 and 2. At a glance, Rule 5 from Class 2 is not much different from the rules from Class 1. They, however, do have differences, as follows.

- Rule 5 is the condition where the amount of power generated, GEN (parameter 3), is at level 1. If Rule 5 is to be compared with Rule 2 (which has GEN = 1), one can find that LPT B in Rule 5 is quantized at a higher level (levels 2 and 1, respectively in Rules 5 and 2). Besides, CWIT A, CWIT B, CWOT A, and CWOT B levels in Rule 5 (i.e., from 3 to 4, 4, 3, and 2, respectively) are generally higher than those in Rule 2 (i.e., 3, 3, 2, and 1, respectively).
- If Rule 5 is to be compared with Rule 3, in which Rule 3 has the same condition as Rule 5 in LPT A and LPT B, Rule 3 is able to produce more amount of power. This finding is indicated by GEN in Rule 3 (i.e., 2) has a higher quantization level than that in Rule 5 (i.e., 1). The same finding can be observed in Rules 4 and 5. The underlying reason is, Rule 4 is supported by higher levels of CWIP A and CWIP B (i.e., 3, and 2, respectively), and in addition, by higher levels of CWOP A and CWOP B (i.e., 4 and 4, respectively). Rule 5, however, has a lower range of quantization levels in CWIP A (i.e., from 2 to 3), CWIP B (i.e., from 1 to 2), CWOP A (i.e., from 2 to 3), and CWOP B (i.e., from 3 to 4).
- If Rule 1 is to be compared with Rule 5, obviously, Rule 1 indicates better heat transfer efficiency than Rule 5. This is due to the former has a higher score in GEN at an expense of lower scores in LPT A, LPT B, CWIT A, and CWIT B. For evidence, the scores of LPT A, LPT B, GEN, CWIT A, CWIT B in Rule 1 are 1, 1, 2, 1, and 3, respectively; whereas the scores of the corresponding parameters in Rule 5, are 2, 2, 1, from 3 to 4, and 4, respectively.

4 Conclusions

For NN models to be practically useful and usable as a decision-support tool in handling complex FDD tasks, two key limitations of NN models need to be solved. The underlying limitations are related to the network learning paradigm (i.e., aspect of adaptability), and the black-box phenomenon on how the network reaches a conclusion/prediction given an incoming sensor data. To cope with these limitations, the network should have an ability that learns new data incrementally. An incremental learning network should not suffer from the so-called catastrophic forgetting and neither has to be re-trained with a batch of new and old data. In addition, the network should have an ability to extract knowledge embedded in the learned network and to provide users with reasoning/justification upon a predicted outcome.

In this paper, a hybrid network, known as the FAM-RecBFN, that inherits advantages of FAM and the RecBFN and that overcomes the aforementioned limitations, has been proposed. To evaluate the effectiveness of the FAM-RecBFN, a case study on fault detection and diagnosis has been conducted. Specifically, a set of sensor data pertaining to a CW system from a power generation plant has been collected. The classification results and the extracted rules have been analyzed and discussed. The extracted rules have exhibited a high degree of interpretability, and they are in line with the opinions of the domain experts in maintaining the CW system. The capability of rule interpretability with a substantially high predictive accuracy is particularly beneficial when the system is applied to real-world problems, such as decision support in safety critical applications like power plants.

Further work will be directed to conducting comparative experiments, based on the CW as well as other benchmark data sets, using the FAM-RecBFN and other machine learning techniques, e.g., decision trees such as C4.5 [15], neural-fuzzy or fuzzy-neural systems such as the NEFCLASS system [16, 17] and the EFuNN [18], and the KBANN-based techniques [19]. In particular, the aspect of rule quality (e.g. accuracy, fidelity, consistency and comprehensibility), in addition to rule interpretability, will be investigated. Moreover, it would be interesting to investigate how FAM and the RecBFN could be hybridized as an entity, rather than as in the cascaded mode in the current study. For this, a new network architecture, which extracts conflict-free rectangles from data empirically, will be devised and developed.

Acknowledgements

The authors gratefully acknowledges the research grants provided by University of Science Malaysia, and the Ministry of Science, Technology, and the Environment Malaysia (No. 06-02-05-8002 & 04-02-05-0010) that have in part resulted in this article. The effort and precious time of staff at TNB Prai Power Generation Plant are very much appreciated.

References

1. Mylaraswamy D, Venkatasubramanium V (1997) A hybrid framework for large scale process fault diagnosis. Computers and Chemical Engineering 21:S935–940
2. Altug S, Chow M-Y, Trussell HJ (1999) Fuzzy inference systems implemented on neural architectures for motor fault detection and diagnosis. IEEE Trans on Industrial Electronics 46:1069–1079
3. Carpenter GA, Grossberg S, Markuzon N, Reynolds JH, Rosen DB (1992) Fuzzy ARTMAP: A neural network architecture for incremental supervised learning of analog multidimensional maps. IEEE Trans on Neural Networks 3:698–712
4. Berthod MR, Huber K (1995) From radial to rectangular basis function: a new approach for rule learning from large datasets. Internal Report 15–95 University of Karlsruhe
5. Carpenter GA, Grossberg S (1987) A massively parallel architecture for a self-organizing neural pattern recognition machine. Computer Vision, Graphics and Image Processing 37:54–115

6. Carpenter GA, Grossberg S (1987) ART 2: Stable self-organisation of pattern recognition codes for analogue input patterns. Applied Optics 26:4919–4930
7. Carpenter GA, Grossberg S, Rosen DB (1991) Fuzzy ART: Fast stable learning and categorization of analogue patterns by an adaptive resonance system. Neural Networks 4:759–771
8. Dagher I, Geogiopoulos M, Bebis G (1999) An ordering algorithm for pattern presentation in fuzzy ARTMAP that tends to improve generalization performance. IEEE Trans on Neural Networks 10:768–778
9. Tou JT, Gonzalez RC (1976) Pattern Recognition Principles. Addison-Wesley, Reading, MA
10. Berthod MR, Huber K (1995) Building Precise Classifiers With Automatic Rule Extraction. In Proc. of Int. Conf. on Neural Networks 3:1263–1268
11. Berthod MR, Diamond J (1995) Boosting the performance of RBF networks with dynamic decay adjustment. In: Tesauro, G., Touretzky, D. S., Leen, T. K. (eds.) Advances in Neural Information Processing Systems, vol. 7. MIT Press, Cambridge MA
12. Lim CP, Harrison RF (1997) An incremental adaptive network for on-line supervised learning and probability estimation. Neural Networks 10:925–939
13. TNB (1999) System description and operating procedures (volume 14). TNB Prai Power Station Stage 3.
14. Carpenter GA, Tan AH (1995) Rule extraction: from neural architecture to symbolic representation. Connection Science 7:3–27
15. Quinlan JR (1993) C4.5: Programs for machine learning. Morgan Kaufmann, Los Altos, CA
16. Nauck D, Kruse R (1997) A neuro-fuzzy method to learn fuzzy classification rules from data. Fuzzy Sets and Systems 89:277–288
17. Nauck D, Kruse R (1999) Obtaining interpretable fuzzy classification rules from medical data. Artificial Intelligence in Medicine 16:49–169
18. Kasabov N (2002) Evolving connectionist systems: methods and applications in bioinformatics, brain study and intelligent machines. Springer, London
19. Towell G, Shavlik J (1994) Knowledge-based artificial neural networks. Artificial Intelligence 69 or 70:119–165

Appendix

Table 4. Abbreviations of the CW System parameters

No.	Parameter	Description
1.	LPT A	Low Pressure Cylinder Exhaust Temperature A
2.	LPT B	Low Pressure Cylinder Exhaust Temperature B
3.	GEN	Generator
4.	CWIT A	Condenser Circulating Water Inlet Temperature A
5.	CWIT B	Condenser Circulating Water Inlet Temperature B
6.	CWOT A	Condenser Circulating Water Outlet Temperature A
7.	CWOT B	Condenser Circulating Water Outlet Temperature B
8.	CWIP A	Condenser Circulating Water Inlet Pressure A
9.	CWOP A	Condenser Circulating Water Outlet Pressure A
10.	CWIP B	Condenser Circulating Water Inlet Pressure B
11.	CWOP B	Condenser Circulating Water Outlet Pressure B
12.	VAC	Condenser vacuum

complex plants especially when the available expert knowledge is not reliable. Therefore, it is important to explore adaptation schemes for fuzzy controllers in order to ensure good control performance over a wide range of operating conditions of the plant in spite of a bad expert knowledge [4, 5]. There is no formal method to determine accurately the parameters of the fuzzy controller (membership functions, rules, fuzzy variables). Therefore, an alternative to improve the performance of a fuzzy control system is to combine the fuzzy controller with an adaptation technique to overcome the rule base limitations and to tune the controller.

The current paper is organized as follows. First, the fuzzy PID control system is described. Four adaptation techniques combined with the fuzzy PID controller are presented. Numerical results on a nonlinear plant are shown. Finally, some conclusions and future directions are given.

2 Static Fuzzy PID Control Schemes

As alternatives to overcome the classical control limitations, several modern control techniques (adaptive control, optimal control and predictive control) were developed. However, each advanced control design conception is based on hard mathematical formalisms and it is not an easy task where each control engineer is not prepared to tune in industrial conditions [4].

Another way to deal with complex systems is to use intelligent control systems, which try to mimic the human behavior and do not need any mathematical model. One of the most used intelligent control methods is the fuzzy control. Fuzzy controllers can be divided in three different stages: fuzzification, rule base evaluation, defuzzification (Fig. 1).

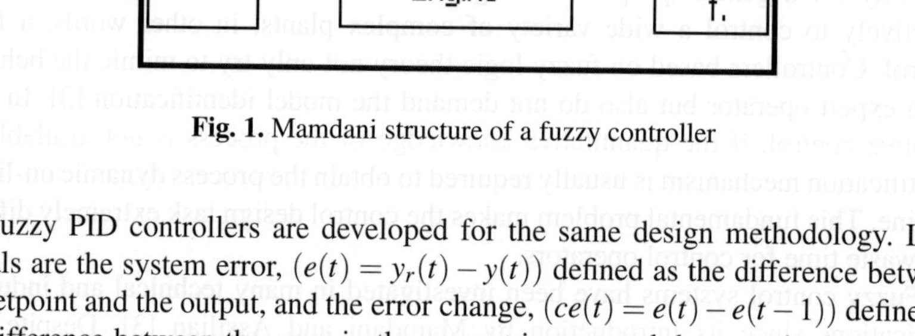

Fig. 1. Mamdani structure of a fuzzy controller

Fuzzy PID controllers are developed for the same design methodology. Input signals are the system error, $(e(t) = y_r(t) - y(t))$ defined as the difference between the setpoint and the output, and the error change, $(ce(t) = e(t) - e(t-1))$ defined as the difference between the error in the current instant and the error in the previous instant. The fuzzification procedure is accomplished with the inputs being distributed

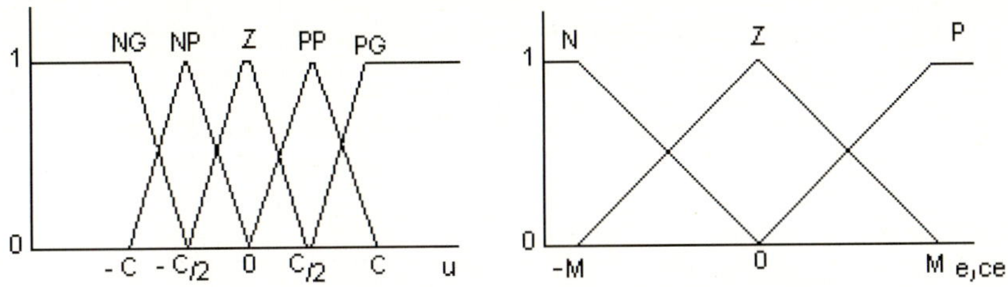

Fig. 2. Membership functions of output and input variables

in three fuzzy variables (negative, zero and positive), with membership functions in triangular and trapezoidal shapes [6]. The control signal is calculated from three (negative, zero and positive) or five fuzzy variables (negative big, negative small, zero, positive small and positive big) depending on the control structure (Fig. 2).

Next, nine different control conceptions for fuzzy PID systems are reviewed, in order to implement closed-loop systems, such as (Fig. 3): (a) FPD+I – a fuzzy PD part is added to a conventional integrator, (b) FPD+(Fk)I – a fuzzy PD is summed to an integrator with the gains calculated with the fuzzy logic, (c) FPD+FPI – two controllers, fuzzy PD and fuzzy PI, are summed and the difference between both is the control variable u_d for PD and u_i for the PI, (d) FPD+FPI (derived of the PD) – if the difference between PI and PD controllers just lays on the control term, other form of obtaining the controller can be using the same base of rules and to calculate the term u_i as the integral of the output of PD, (e) FPD+FI – control signal is obtained by using the isolated integral part of the proportional term, obtaining a fuzzy PD controller added to a fuzzy integrator, (f) FPI+D – a fuzzy PI controller is added to a modified conventional derivative to avoid the derivative kick in the setpoint change [7], (g) FI+FPD Incremental – the controller fuzzy I is added to an incremental fuzzy PD, (h) FP+I+D – the part *P* is calculated in a fuzzy way and adds the derivative and conventional integral parts, (i) FP+FI+FD – all the control part are fuzzy, however, design parameters are calculated in an independent way in order to reduce the rule base and the computational effort. Fuzzy PID controllers of Fig. 3 were proposed in the control literature by: (i) (a–e) – [8]; (ii) (f) – [9]; (iii) (g) – [10] (iv) (h) – [11]; (v) (i) – [12]. In all simulation essays, the controllers are tuned by trial-and-error.

3 Adaptive Fuzzy PID Control Schemes

The use of a large operational range when controlling a nonlinear process with a conventional PID fuzzy controller may lead to an undesirable performance due to a time varying dynamic in the plant. Also, the rules obtained can not fit to all operating points in a satisfactory fashion. To overcome these problems, it is necessary to add some kind of adaptation mechanism [4, 5, 13, 14, 15]. Some of this adaptation structures aim to change the tuning parameters of the controller, while others perform

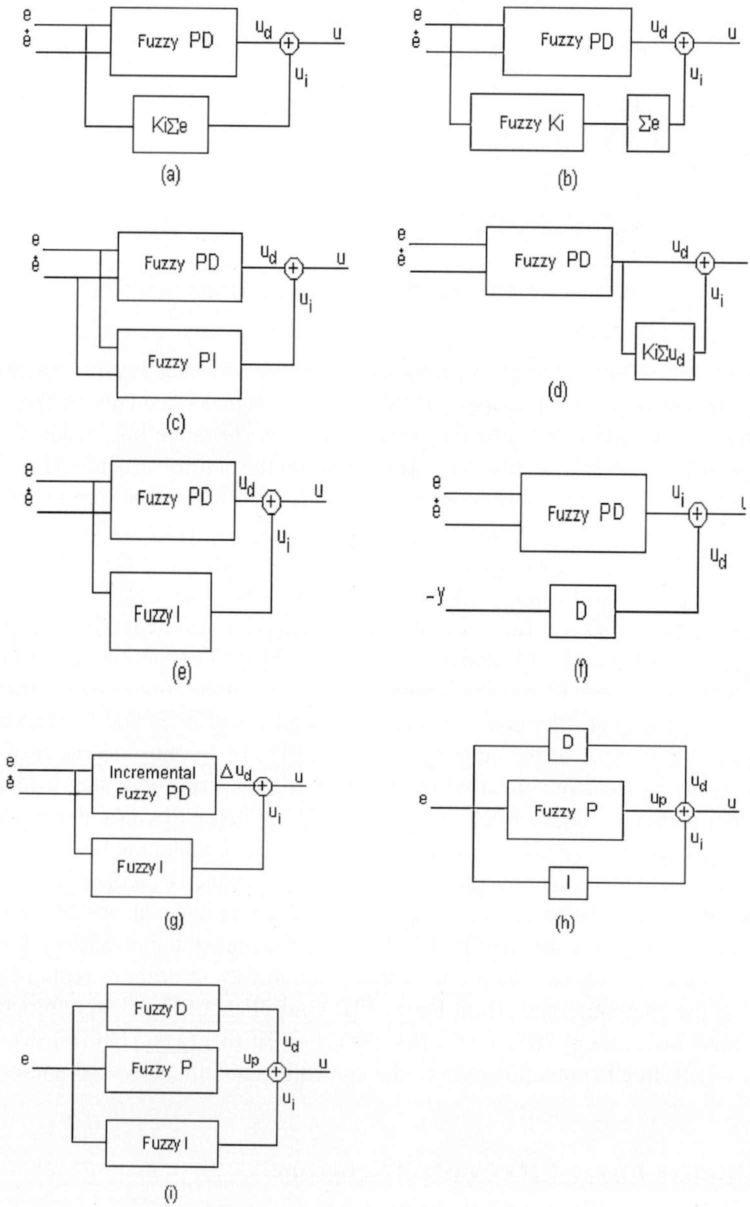

Fig. 3. Design diagrams of fuzzy PID controllers

a compensation on the control signal without act directly in the controller. In this paper both approaches are shown.

All adaptive control structures, presented in this paper, have a similar pattern and are based on a model reference (working as a setpoint filter and smoothing set point changes). The adaptation is performed by a mechanism that evaluates the process output and the desired closed-loop behavior of the output (model reference output).

4 Direct Adaptation

Figure 4 shows the first adaptive fuzzy control structure that uses the recursive least squares algorithm (RLS) to optimize the membership functions of the control variable.

The main idea is to use the standard least squares algorithm to compute values to the center of mass of the membership functions that lead the control variable as close as possible to the desired control (u_{dc}). The desired control is the control signal that gives the best system response to a defined setpoint. Rashid and Heger [15] used a linearizing control to determine u_{dc}, but to find an adequate desired control model to a nonlinear or complex process is not a trivial task. In order to ensure a smooth response of the control system and to overcome the difficulty of selecting the desired control, the equation (1) has been implemented for simulation and real-time applications.

$$u_{dc}(t) = \lambda . |e(t)| \, y_{mr}(t) \tag{1}$$

It is important to observe that the modification performed by the adaptation mechanism shifts the membership functions, as shown in Fig. 5. Accordingly to the way that the adaptation is performed, it may correspond to a rule base modification, which gives a desired dynamic structure.

Figure 6 shows the adaptive fuzzy control structure of Layne and Passino [14] which implements a mechanism of adaptation to optimize the PID tuning parameters. The purpose of the adaptation block is to modify the centers of mass of the membership functions, the difference is the way this is performed. While in the former method the values are calculated at every time sample in this structure the

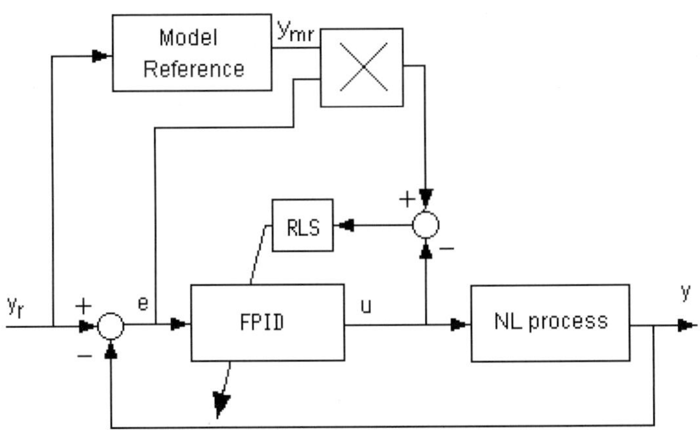

Fig. 4. Adaptive FPID with least squares technique

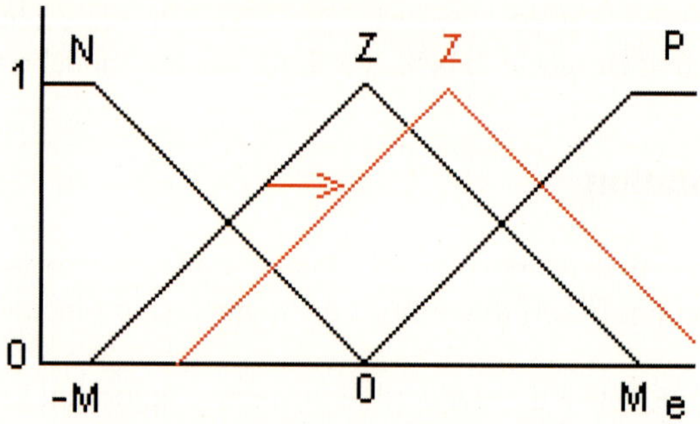

Fig. 5. Membership function shifting

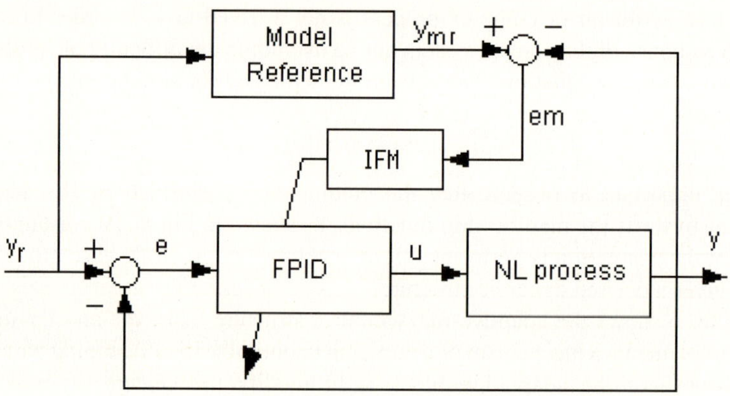

Fig. 6. Adaptive FPID with fuzzy internal model control

mechanism of adaptation computes little values to be added or subtracted to provide a smooth adaptation.

The adaptation mechanism is a fuzzy controller that minimizes the error between the system output and the output of the model reference. This auxiliary controller is also called inverse fuzzy model (IFM) because presents a behavior that is approximately the opposite of the plant, although it does not require the identification of the process model [14].

Indirect Adaptation

Another way to perform adaptation is to add a complementary control signal without changing the controller parameters. Features of these control techniques over the direct adaptation are: (1) small complexity for designing, (2) ease of implementation, and (3) small computational effort. The indirect adaptation is good when the

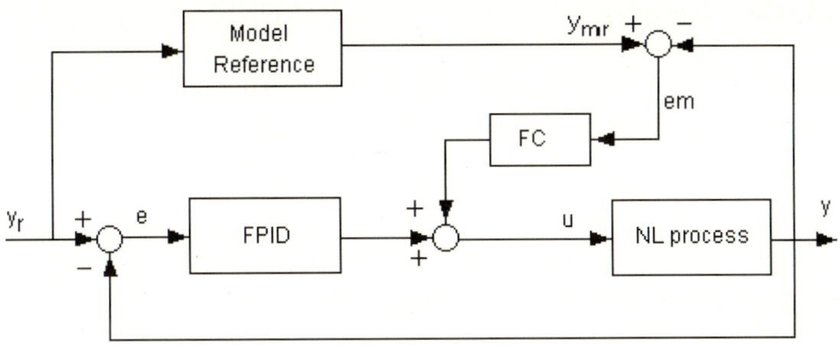

Fig. 7. Adaptive FPID with fuzzy compensation

operational conditions are changing, because it does not modify the control structure, only compensates the control signal when needed.

Minh and Hoang [5] proposed an adaptive fuzzy control design that compares the system output to a model reference output. When the responses do not match, the fuzzy compensation mechanism is activated to bring the system output to the desired point. When the operational conditions are changing, an error appears between $y(t)$ and $y_{mr}(t)$, generating a compensation on the control signal. Figure 7 shows the implemented controller with a PID controller on the adaptation mechanism to perform adaptation.

Finally, the last adaptive fuzzy technique is basically a modification of the former. It replaces the fuzzy controller presented in the main loop for a conventional PID controller (Fig. 8). This approach aims to make the controller design easier, allowing to tune the controller by using any of the numerous design methods presented in the classical control literature and reducing the computational effort [1]. It is necessary to tune the conventional PID to an operating point (usually at the middle of the operational range) and the fuzzy compensation mechanism adapts the control signal to optimize the system response to match the model reference dynamic. The adaptation mechanism is based on the Minh and Hoang controller [5].

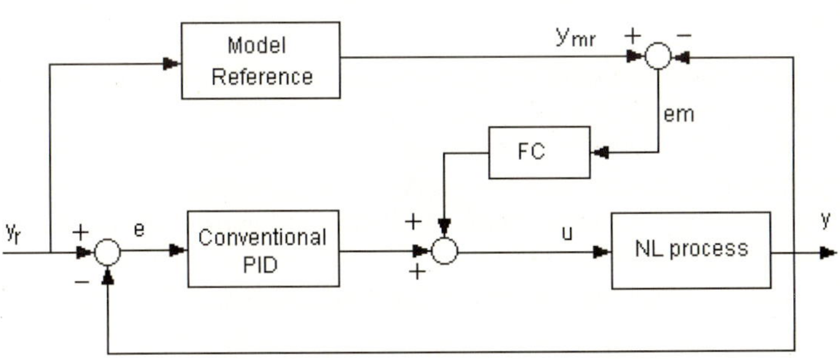

Fig. 8. Adaptive PID with fuzzy compensation

5 Simulation Results

In order to verify the efficiency of adaptive fuzzy PID control structures, several simulations are performed on the control of a nonlinear system (2).

$$\dot{y}(t) = -y(t) + \sin^2(\sqrt{|y(t)|}) + u(t) \tag{2}$$

The nonlinear behavior of the plant is illustrated in Fig. 9, showing the variation of the static gain when the operational point is changing.

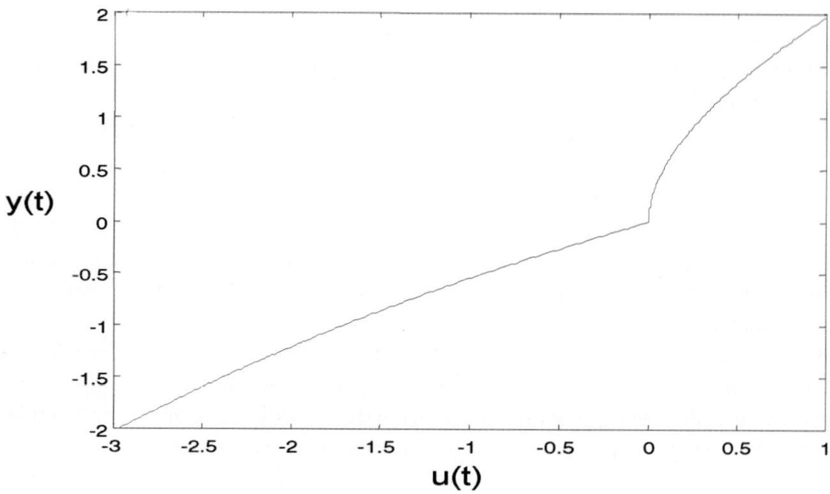

Fig. 9. Nonlinear characteristic of the plant

Simulation tests are performed with a sampling time of 0.5 sec. Servo behavior is assessed by setpoint variations, changing from 0 to -2 at 0 sec and from -2 to 2 at 30 sec. For the regulatory analysis, a load disturbance of 100% is applied at 60 sec.

Fine tuning for the controllers was done experimentally and depended on the dynamic of the plant. In Figs. 11–14, all controllers showed a more conservative control signal and a better response than the non-adaptive fuzzy PID (Fig. 10), which had a small oscillatory behavior on the steady state phase. Adaptive fuzzy PID controllers, from Figs. 12 and 13, provided faster settling time and smaller overshoot than controllers from Figs. 11 and 14.

In order to compare the behavior of each adaptive fuzzy PID control system, according to setpoint tracking and control energy, the following performance indices were used: (1) sum of the quadratic error, calculated by $J_e = (1/N)\sum_{t=1}^{N}[e(t)]^2$, and (2) sum of the quadratic increment control, calculated by $J_u = (1/N)\sum_{t=1}^{N}[u(t) - u(t-1)]^2$. As shown in Table 1, the indices confirmed some important characteristics

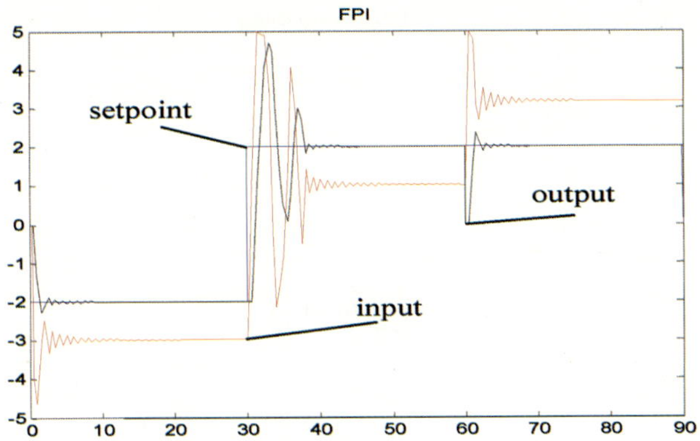

Fig. 10. Non-adaptive FPID

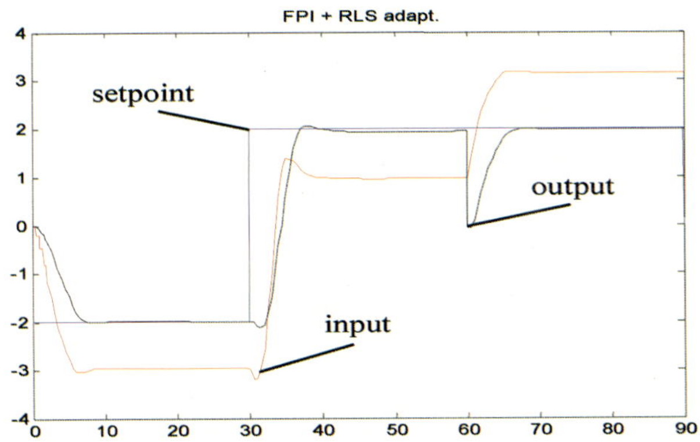

Fig. 11. FPID + RLS adaptation

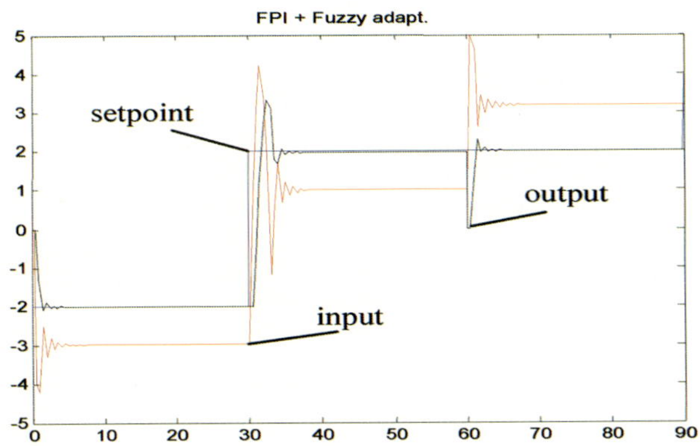

Fig. 12. FPID + Fuzzy adaptation

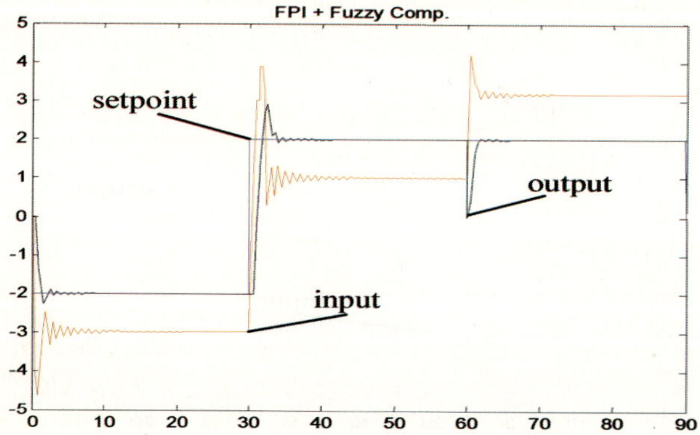

Fig. 13. FPID + Fuzzy compensation

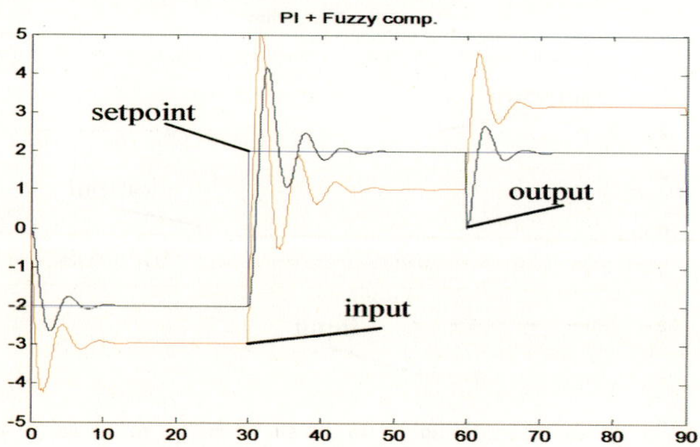

Fig. 14. PID + Fuzzy compensation

Table 1. Indices for error and control signals

Controller Dynamic	J_e	J_u
FPID (Non-Adaptive)	0.4718	0.6050
FPID+RLS Adaptation	0.9667	0.0258
FPID+Fuzzy Adaptation	0.3296	0.4476
FPID+Fuzzy Compensator	0.2977	0.3629
PID+Fuzzy Compensator	0.3368	0.2394

of the closed-loop dynamic. The best performance was obtained from the indirect adaptation (compensation techniques), which, beside the better indices had the advantage to be computationally lighter (did not require excessive computation time).

6 Conclusion

In this paper, several design procedures presented in the process control literature for the PID controller, based on fuzzy control systems, are reviewed.

In order to make the fuzzy logic control less dependent on the quality of the expert knowledge, four techniques for improving the fuzzy PID controllers performance, by adding some kind of adaptation feature when facing nonlinear processes, were presented.

From simulation results, it was possible to show that all four adaptive controllers had better responses than the FPID controller. Adaptive fuzzy PID controllers had a smooth response and a more conservative control action than the non-adaptive fuzzy PID controller.

As a future work, the next step is to assess the adaptive fuzzy PID system on a nonlinear experimental setup. Other fuzzy control systems combined with advanced control techniques, such as, auto-tuning, minimum variance and predictive strategies are also some future considerations.

References

1. K. J. Åström and T. Hägglund, "The future of PID control", *Control Engineering Practice*, Vol. 9, pp. 1163–1175, 2001.
2. K. J. Åström and B. Wittenmark, *Adaptive Control*, Addison-Wesley, New York, 1995.
3. E. Mamdani and S. Assilian, "An experiment in linguistic synthesis of a fuzzy logic controller", *Int. Journal of Man-Machine Studies,* Vol. 7, No. 1, pp. 1–13, 1975.
4. F. Mrad and G. Deeb, "Experimental comparative analysis of adaptive fuzzy logic controllers", *IEEE Trans. on Control Systems Technology,* Vol. 10, pp. 250–255, 2002.
5. T. C. Minh and L. H. Hoang, "Model reference adaptive fuzzy controller and fuzzy estimator for high performance induction motor drives", *Proc. Annu. Meet. IEEE Ind. Applicat. Soc.,* pp. 380–387, 1996.
6. Y. F. Li and C. C. Lau, "Development of fuzzy algorithms for servo systems", *IEEE Control Systems Magazine*, April, pp. 65–72, 1989.
7. L. S. Coelho and A. A. R. Coelho, "An experimental and comparative study of PID control structures", *Advances in Soft Computing: Engineering Design and Manufacturing*, Springer, London, pp. 147–159, 1999.
8. D. P. Kwok, P. Tam, C. K. Li and P. Wang. "Linguistic PID controllers", *11th IFAC World Congress*, Tallin, Estonia, USSR, Vol. 7, pp. 192–197, 1990.
9. H. X. Li and H. B. Gatland, "Enhanced methods of fuzzy logic control", *Proceedings of FUZZY- IEEE/IFES'95*, Yokohama, Japan, Vol. 1, pp. 331–336, 1995.
10. H. A. Malki, D. Misir, D. Feigenspan and G. Chen, "Fuzzy PID control of a flexible-joint robot arm with uncertainties from time-varying loads", *IEEE Transactions on Control Systems Technology*, Vol. 5, pp. 371–378, 1997.
11. W. Li, "Design of a hybrid fuzzy logic proportional plus conventional integral-derivative controller", *IEEE Transactions on Fuzzy Systems*, Vol. 6, pp. 449–463, 1998.
12. M. Golob, "Decomposed fuzzy proportional-integral-derivative controllers", *Applied Soft Computing*, Vol. 18, pp. 1–14, 2001.
13. Y. S. Kung and C. Liaw, "A fuzzy controller improving a linear model following controller for motor drives", *IEEE Trans. Fuzzy Syst.,* Vol. 2, pp. 194–202, 1994.

14. J. R. Layne and K. M. Passino, "Fuzzy model reference learning control", *Journal of Intelligent and Fuzzy Systems,* Vol. 4, pp. 33–47, 1996.
15. N. K. A. Rashid and A. S. Heger, "Tuning of fuzzy logic controllers by parameter estimation method", *Fuzzy Logic and Control: Software and Hardware Applications*, Prentice-Hall, pp. 374–392, 1993.

Wavelet Neural Networks and Its Applications in Chaotic Systems Identification

Leandro dos Santos Coelho[1] and Roger Calixto[2]

[1] Pontifícia Universidade Católica do Paraná, Grupo Produtrônica, Programa de Pós-Graduação em Engenharia de Produção e Sistemas, Rua Imaculada Conceição, 1155, CEP 80215-901, Curitiba, PR – Brazil
leandro.coelho@pucpr.br

[2] Universidade Tuiuti do Paraná, Campus Torres, Faculdades de Ciências Exatas e de Tecnologia, Av. Comendador Franco, 1860, CEP 80215-090, Curitiba, PR, Brazil

Abstract. The combination of wavelets with neural networks can hopefully remedy each others weaknesses, resulting in wavelet based neural network capable of handling system identification problems of a moderately large dimension. A wavelet based neural network is a nonlinear regression structure that represents nonlinear mappings as the superposition of dilated and translated versions of a function, which is found both in the space and frequency domains. In this paper, a wavelet-based neural network is introduced for the nonlinear identification of dynamic systems with chaotic behavior (chaotic time series). The structure of the wavelet based neural network is similar to that of radial basis function neural networks, except that here the activation function of the hidden nodes is replaced by wavelet functions. The proposed wavelet-based neural network is evaluated on two case studies: (i) the Hénon map, and (ii) the Rössler system. Simulation results demonstrate the accuracy and the reliability of the proposed identification methodology based on a wavelet based neural network.

1 Introduction

In recent years, great interest has been devoted to dynamic chaotic systems, since they cover the theories of bifurcations, variable linking, temporal series forecasting, non-linear systems and control systems, amongst others [1, 2, 3].

In various cases the apparent randomness in time series may be due to the chaotic behavior of a nonlinear system (more deterministic) and the sensitivity of the chaotic system to the initial conditions deny the possibility of forecasting of time systems for a long prediction horizon.

Among the many possible methodologies for the identification of nonlinear systems and/or forecasting of time series are neural networks. One type of neural network with interesting properties for this application is the wavelet neural network (WNN) or wavenets. WNNs are a class of neural networks with an intermediate layer in which the base functions are made up of a family of wavelet functions. WNNs have

Leandro dos Santos Coelho and Roger Calixto: *Wavelet Neural Networks and Its Applications in Chaotic Systems Identification*, Advances in Soft Computing **1**, 205–217 (2005)
www.springerlink.com

properties similar to radial basis neural networks certainly an efficient mathematical structure for nonlinear mapping.

This paper presents the mathematical fundamentals and the use of WNNs in two case studies of identifying systems with chaotic behavior. The use of WNNs was motivated by the promissory results seen in papers of non-linear dynamic systems identification using these neural networks [4, 5, 6, 7, 8, 9, 10, 11, 12, 13, 14].

The remaining of this paper is organized in the following manner. Section 2 presents the background and an algorithm for conception of WNNs. The application of WNNs in non-linear identification is presented in Sect. 3. Numerical simulations and the results of applying WNNs are shown in Sect. 4. Finally, in Sect. 5, the conclusions and the perspectives of future research are discussed.

2 Wavelet Neural Networks (WNNs)

Wavelets offer a well-structured mathematical base for signal representation [10, 11]. WNNs, on the other hand, are built into universal function approximators [7] for structural systems identification (gray-box mathematical models). WNN's, as in other neural networks, are used to represent, in a precise way, the input/output of dynamic systems through the learning of training data. Representation of nonlinear mapping through WNNs depends on: (i) proper structure selection of the WNN; (ii) choice of the activation function of the intermediate layer, or in other words, the wavenet base function; (iii) choice of parameters for translation and dilatation (scale); and (iv) the capacity of the learning algorithm to determine the proper parameters for the selected WNN structure.

A WNN can adjust the parameters of the wavelet representation of a dynamic system. However a pre-processing data procedure and a priori knowledge of the parameters of the project are necessary. In the approach used in this paper the WNN structure is adjusted through a Gram-Schmidt [12] orthogonalization method.

The investigation of WNN structures, as mentioned in this paper, reveals a resemblance to three layer radial basis neural networks (input, intermediate, and output layers).

In the case of a problem with N_i problems, a multidimensional wavelet network can be considered. A simple choice would be the multiplication of N_i one-dimensional wavelets for each input, using the following equations:

$$\Phi_j(x) = \prod_{k=1}^{N_i} \phi(z_{jk}) \tag{1}$$

and

$$z_{jk} = \frac{x(k) - m_{jk}}{d_{jk}} \tag{2}$$

where m_j and d_j are translation and dilatation vectors respectively. In this study the WNN's are represented by the equation:

$$\hat{y} = \psi(x) = \sum_{j=1}^{N_w} c_j \Phi_j(x) + a_0 + \sum_{j=1}^{N_i} a_k x(k) \tag{3}$$

Equation (3) can be seen as a neural network with N_i inputs, a hidden layer of N_w wavelets of N_i dimensions with polarization term (bias), a_0, and one linear neuron as output. A WNN can be represented by the diagram in Fig. 1. In this paper, a first derived Gaussian function was chosen, $\phi(x) = \pm x \exp\left(\pm 0,5x^2\right)$, with a mother wavelet.

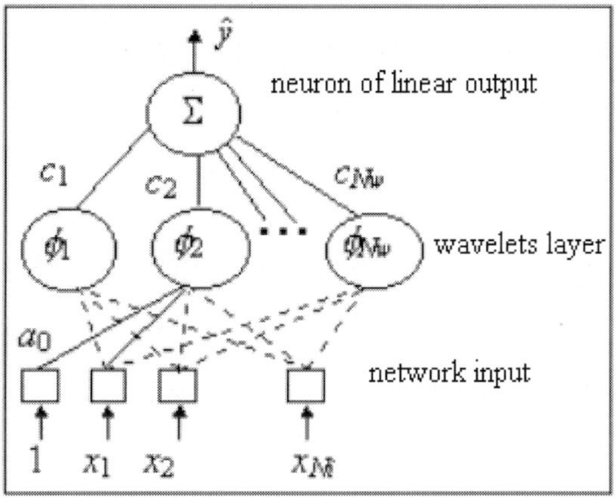

Fig. 1. A wavelet neural network structure

In the next section two case studies of chaotic systems are presented in which the WNNs are applied to the nonlinear identification of the dynamic systems.

3 Nonlinear Identification using WNNs

3.1 Case Study 1: Hénon Map

The French astronomer M. Hénon, in 1975, showed an interesting phenomenon (Hénon maps) present in Poincaré maps created from differential equations. The Hénon [13] map is expressed by the following equation:

$$x(t+1) = 1 - a\left[x(t)\right]^2 + 0,3x(t-1)\ . \tag{4}$$

It can be observed that, depending on the chosen value of parameter a, the system shows different behaviors, anywhere from a periodic regimen to a deterministic chaotic behavior. In this case the dynamic system is sensitive to small perturbations making the mathematic model analysis a much more complex task to represent in a Hénon map.

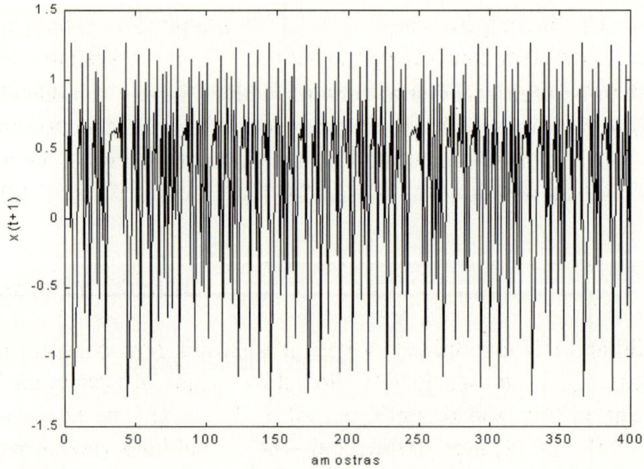

Fig. 2. Data used for the Hénon map forecast through of WNNs

In this paper a chaotic time series, made up of 400 values generated from the initial values $x(0) = 0.1$ and $x(1) = 0.3$ with the parameter $a = 1.4$, was considered. The simulations were conceived to use 200 samples for the estimation phase and 200 for the validation phase of the model obtained by the WNN for the Hénon begin (see Fig. 2).

3.2 Case Study 2: Rössler System

The German scientist O. Rössler [14] proposed a chaotic attractor based on nonlinear differential equations. The Rössler equations are [15]:

$$\frac{dx}{dt} = -y - z\,, \tag{5}$$

$$\frac{dy}{dt} = x + ay\,, \tag{6}$$

$$\frac{dz}{dt} = b + (x - c)z\,, \tag{7}$$

which exhibit a chaotic behavior for (a,b,c) = (0.36; 0.40; 4.50). The chaotic attraction for this case is presented in Fig. 3. The corresponding dynamic behavior of the *x*-*y*, *y*-*z*, and *z*-*x* planes are also presented in Fig. 4.

A set of 20000 samples is used in the identification of this system, which are presented in Fig. 5, where the objective is to identify the dynamic behavior of the *x* coordinate of the Rössler system.

Since (4)–(6) represent a continuous dynamic system instead of a discrete system it's necessary to use a numeric integration method. In this case a 4th order Runge-Kutta method was chosen with a 0.005 step and null initial conditions. The experiment was conditioned to use 10000 samples for the estimation phase and 10000 for the mathematic model validation phase obtained through the WNN.

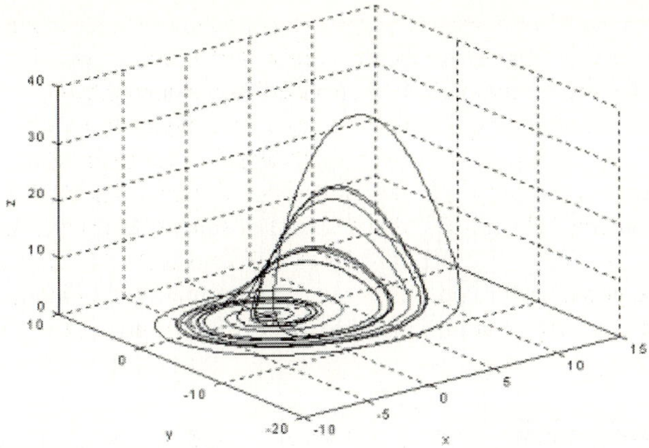

Fig. 3. Rössler system attractor for $(a, b, c) = (0.36, 0.40, 4.50)$

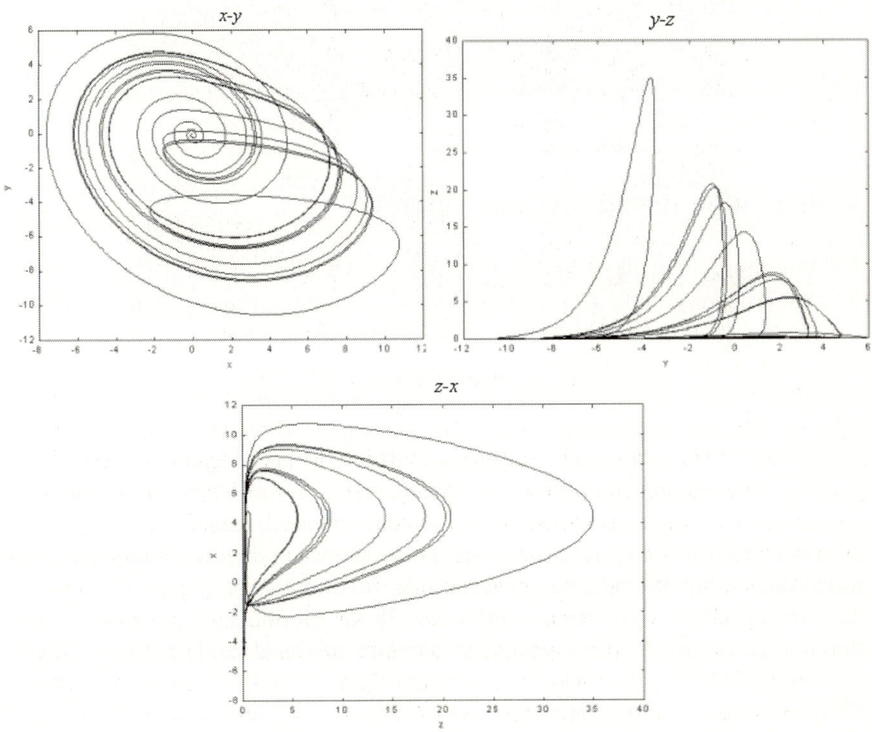

Fig. 4. Dynamic behavior of Rössler system

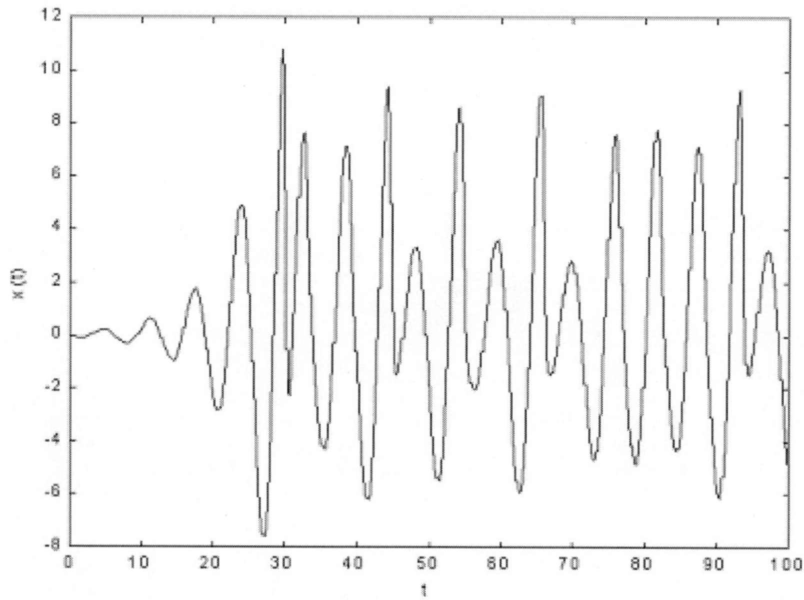

Fig. 5. Time series of the x coordinate used for the Rössler system identification for (a,b,c) = (0.36; 0.40; 4.50) case

4 Simulation Results of Case Studies

The procedure for analysis of the results from the nonlinear identification is divided into the following steps: (i) obtain the input/output data from the dynamic system, (ii) WNN structure choice (number of wavelets, scale and search space), (iii) define which optimization algorithm to use on the WNN (in this case a stochastic gradient algorithm [4] with 100 training epochs for case study 1, and 15 for case study 2), (iv) calculate the parameter estimation of the mathematical model (estimation phase), and (v) obtain the validation of the mathematical model (validation phase).

The performance criteria for the means square error, MSE, is calculated by the sum of the square error difference between the real output, $x(t+1)$, and the desired output $\hat{x}(t+1)$, divided by number of samples in the dynamic system being identified. The WNNs results are shown in Tables 1 and 2 respectively.

The estimated initial number of wavelets for the hidden layer was obtained through a procedure based upon the Akaike criteria [16]. The Akaike criteria measures the residual variance in relation to the number of adjustable parameters. With this in mind an ortogonalization, as proposed in [17], using Gram-Schmidt [12] combined with a decision based on final prediction criteria (FPEC) of Akaike.

The results for identification through WNNs were promising for the identification of both case studies here discussed. Figures 6 and 9 show the best results obtained with WNNs for identification of Hénon maps (see Table 1). The best result was given

Table 1. WNN results for 1st case study (Hénon map)

WNN Inputs	Wavelets in Hidden Layer	MSE (Estimated)	MSE (Validated)
$x(t-1)$	4	3.717×10^{-2}	4.132×10^{-2}
$x(t-1) .. x(t-2)$	16	6.591×10^{-6}	5.759×10^{-3}
$x(t-1) .. x(t-3)$	19	1.636×10^{-6}	5.276×10^{-3}
$x(t-1) .. x(t-4)$	8	1.609×10^{-4}	1.669×10^{-2}
$x(t-1) .. x(t-5)$	2	1.401×10^{-4}	3.705×10^{-2}

Table 2. Results obtained through of WNN for 2nd case study (Rössler system)

WNN Inputs	Wavelets in Hidden Layer	MSE (Estimated)	MSE (Validated)
$x(t-1)$	2	5.437×10^{-4}	6.057×10^{-4}
$x(t-1) .. x(t-2)$	29	2.339×10^{-8}	1.726×10^{-8}
$x(t-1) .. x(t-3)$	35	8.727×10^{-12}	1.672×10^{-12}

by the WNN input array $[x(t-1)\ x(t-2)\ x(t-3)]$ using 19 wavelets (also called *wavelons* according to the convention given by [4] and [9]).

In regards to the best results from the estimation phase, an average error of 1.2103×10^{-6} was obtained with a mean error of -4.620×10^{-3}, and a maximum error of 5.337×10^{-3} was obtained by the WNN. However, in the validation phase, upon verifying generalization capacity of the WNN, a maximum error of $1{,}763 \times 10^{-1}$; a mean of $4{,}023 \times 10^{-2}$ and average of $1{,}613 \times 10^{-7}$ were obtained.

The maximum error obtained was considered acceptable and it was noted, through the average error in both estimation and validation phases, that the identification procedure of the WNN retained a descorrelation of error data (white noise feature).

Figures 10–13 show the best results obtained from the WNN for Rössler identification, as seen in Table 2.

The best results for the WNN were from the input $[x(t-1)\ x(t-2)\ x(t-3)]$ with 35 wavelets in the hidden layer. In regards to the best preliminary results in the estimation phase, a maximum error of 3.100×10^{-6}, mean error of 1.238×10^{-7} and average error of 1.372×10^{-7} was obtained. However, in the validation phase, upon verifying the generalization capacity of the network, a maximum error of 9.608×10^{-6}, mean error of 1.280×10^{-8} and average error of 2.901×10^{-8} were observed. The maximum error obtained was considered acceptable and it was noted, through the average error in both estimation and validation, that the identification procedure of the WNN retained a descorrelation error sign (white noise or Gaussian distribution) similar to that in case study 1. This leads to the conclusion that WNNs are robust when used for forecasting time series. However, some questions remain. The solution of the problem of structural selection present in WNNs depends upon the given values of dilations and translations of the wavelets. In this case there is

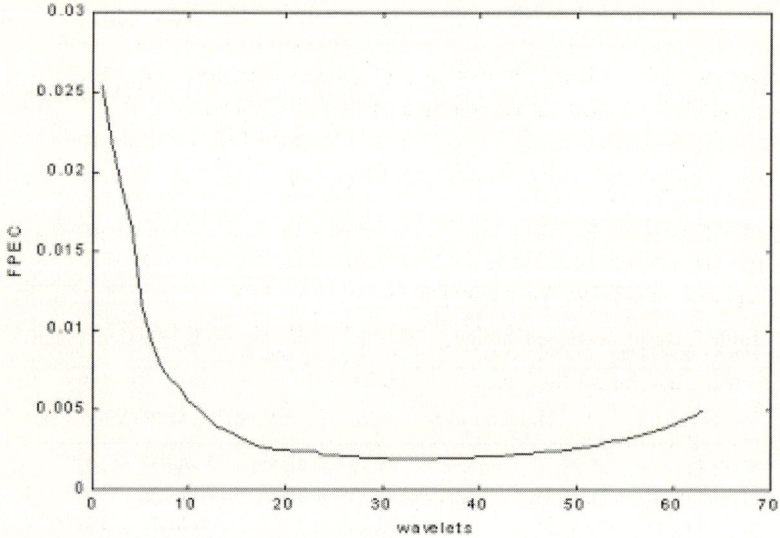

Fig. 6. Results for the FPEC of Akaike for initial network configuration for the identification of Hénon map (19 wavelets in hidden layer)

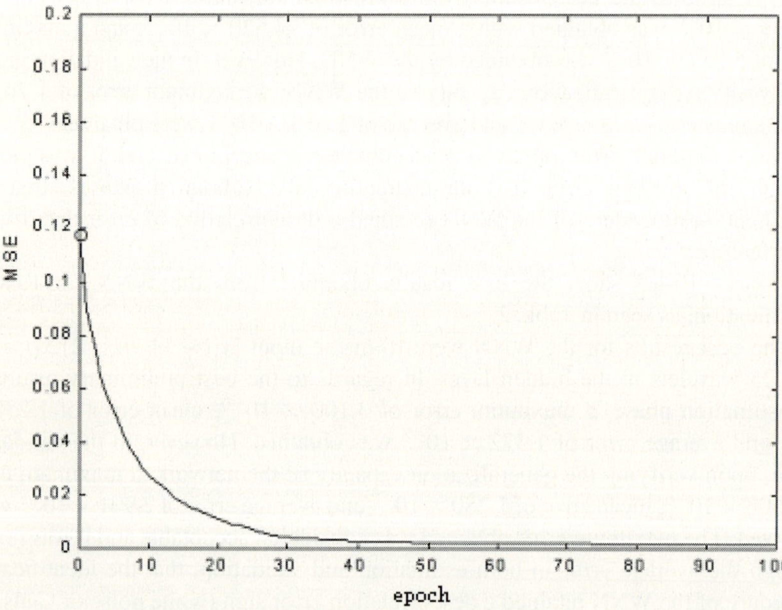

Fig. 7. Convergence of the Hénon map using WNNs (19 wavelets in hidden layer)

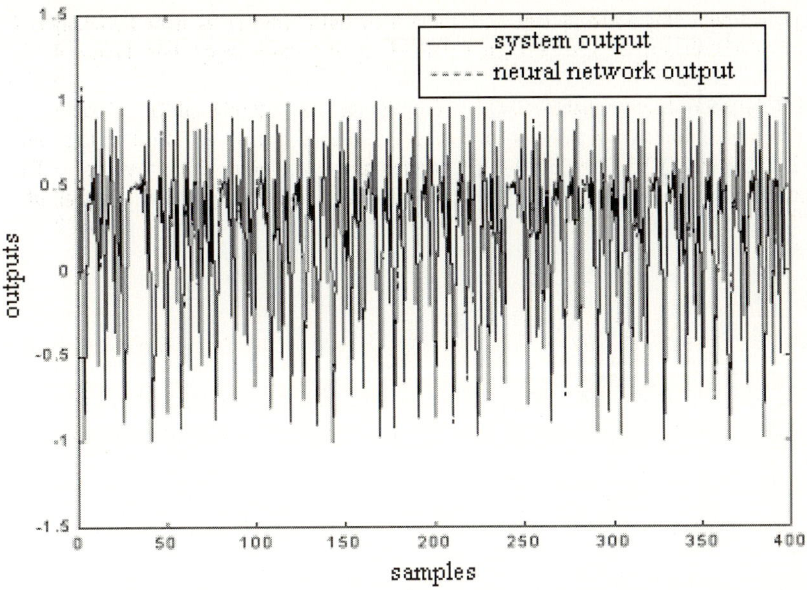

Fig. 8. Identification results for the Hénon map (19 wavelets in hidden layer)

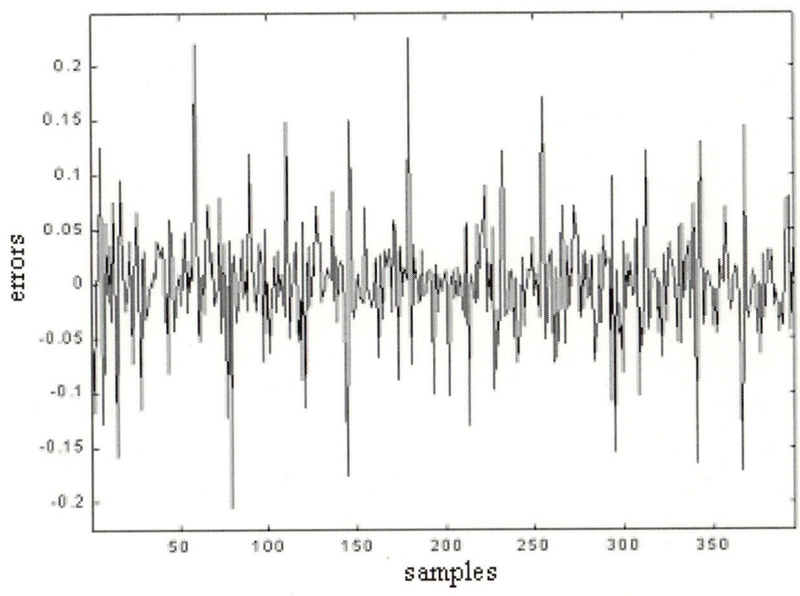

Fig. 9. Error for the Hénon map (19 wavelets in hidden layer)

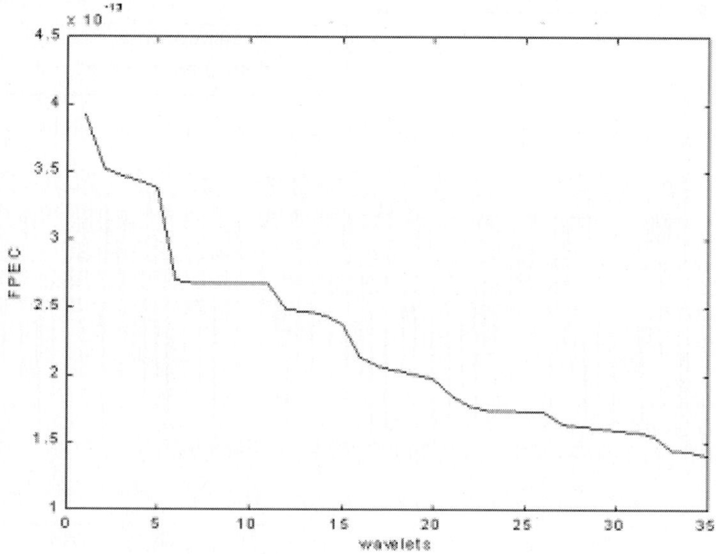

Fig. 10. Results for the FPEC of Akaike for initial network configuration for Rössler system identification (32 wavelets in hidden layer)

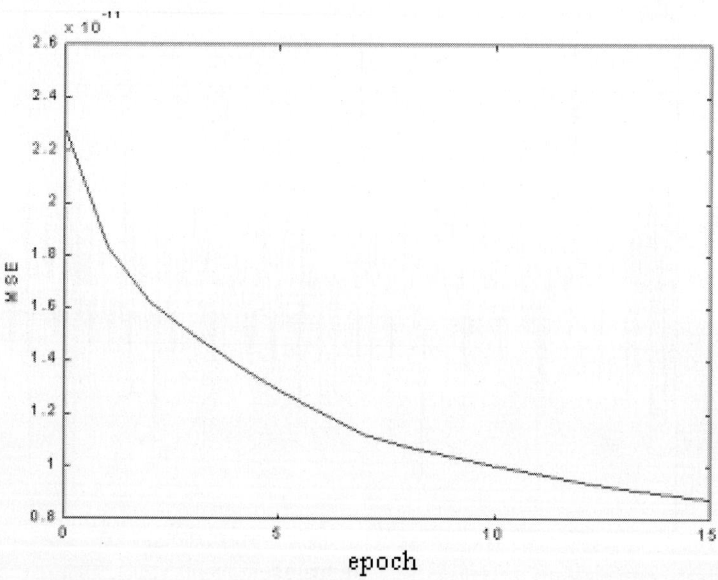

Fig. 11. Convergence of the Rössler system using WNNs (32 wavelets in hidden layer)

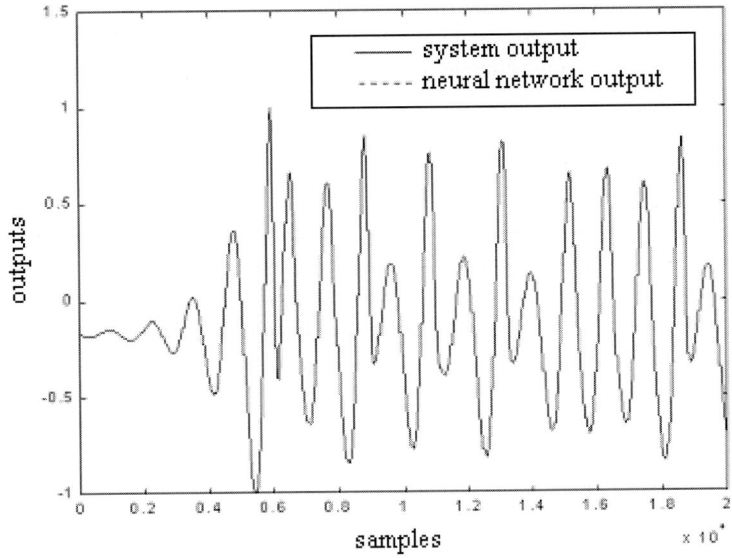

Fig. 12. Rössler system results (32 wavelets in hidden layer)

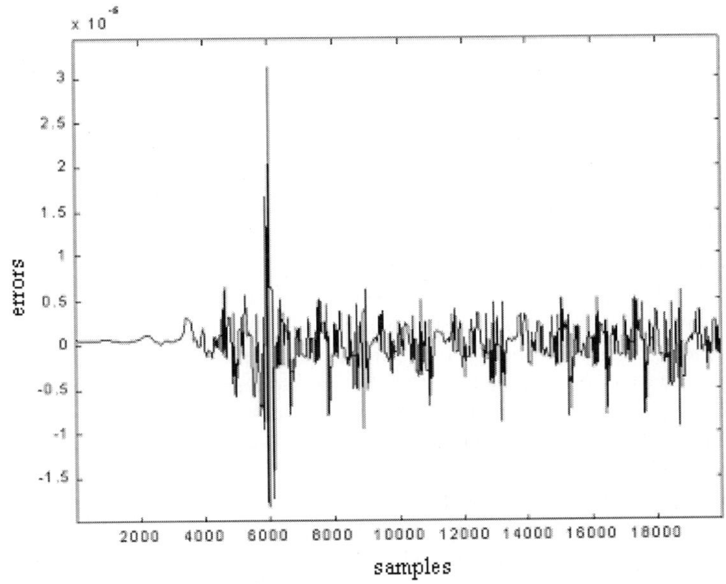

Fig. 13. Rössler system error (32 wavelets in hidden layer)

an underlying need for a deeper study of the relationship between the number of selected wavelets (computational complexity) and the quality of the WNNs answer to forecasting the dynamic chaotic behavior of a time serie.

5 Conclusions

WNNs have advantages due to the high resolution of the wavelets and the learning potential of the feedforward neural networks. Besides the fact that WNNs show advantages due to their capacity of approximation of non-linear functions along with quick and efficient learning, giving the possibility of precise results while offering efficient solutions for multidimensional approximations. These are the characteristics which motivated the writing of this paper, to show the fundamentals and applications of WNNs for the learning of dynamic chaotic systems identification.

Two case studies were analyzed: the Hénon map and the Rössler system. The simulation results show that WNNs learned fast, converged fast and had good approximation to the chaotic dynamics of each study. Future studies will treat the analysis of crosscorrelation algorithms for WNNs with time variant parameters.

References

1. Brown, R.; Berezdivin, R.; Chua, L. O. (2001) Chaos and complexity, *International Journal of Bifurcation and Chaos* 11(1), pp. 19–26
2. Chen, G. (ed.) (1999) Controlling chaos and bifurcations in engineering systems, CRC Press, Boca Raton, FL
3. Wang, J.; Wang, X. (1999) A global control of polynomial chaotic systems, *International Journal of Control* 72(10), pp. 911–918
4. Zhang, Q.; Benveniste, A. (1992) Wavelet networks, *IEEE Transactions on Neural Networks* 3(6), pp. 889–898
5. Szu, H. H.; Yang, X.-Y.; Sheng, Y. (1993) Neural network and wavelet transform for scale-invariant data classification, *Physica Review E* 48(2), pp. 1497–1501
6. Bakshi, B. R.; Stephanopoulos, G. (1993) Wave-net: a multiresolution, hierarchical neural network with localized learning, *AIChE Journal* 39(1), pp. 57–81
7. Kreinovich, V.; Sirisaengtaksin, O.; Cabrera, S. (1994) Wavelet neural networks are asymptotically optimal approximators for functions of one variable, *IEEE Int. Conf. on Neural Networks*, Orlando, FL, pp. 299–304
8. Chen, C. H.; Lee, G. G. (1996) Multiresolution wavelet analysis based feature extraction for neural network classification, *IEEE International Conference on Neural Networks*, Washington, DC, USA, vol. 3, pp. 1416–1421
9. Zhang, Q (1997) Using wavelet network in nonparametric estimation, *IEEE Transactions on Neural Networks* 8(2), pp. 227–236
10. Daubechies, I. (1990) The wavelet transform, time-frequency localization and signal analysis, *IEEE Transactions on Information Theory* 36(5), pp. 961–1005
11. Nikolaou, M.; Vuthandam, P. (1998) FIR model identification: parsimony through kernel compression with wavelets, *AIChE Journal* 44(1), pp. 141–150

12. Golub, G. H.; Van Loan, C. F. (1983) *Matrix computations*, Baltimore: The John Hopkins University Press
13. Hénon, M. (1976) A two dimensional mapping with strange attractor, *Communications in Mathematical Physics* 50, pp. 69–77
14. Rössler, O. E. (1976) An equation for continuous chaos, *Physical Letters* 35A, pp. 397–398
15. Alligood, K. T.; Sauer, T. D.; Yorke, J. A. (1996) *Chaos: an introduction to dynamical systems*, Springer, London, UK
16. Johansson, R. (1993) *System modeling and identification.* Prentice-Hall, Englewood Cliffs, NJ
17. Hull, J. R.; Pendse, H. P. (1997) A neural network algorithm using wavelets and auto regressive inputs for system identification, *International Conference on Neural Networks*, Houston, TX, Vol. 2, pp. 728–732

Fuzzy Specializations and Aggregation Operator Design in Competence – Based Human Resource Selection

Miguel-Ángel Sicilia[1], Elena García-Barriocanal[1], and Rafael Alcalde[2]

[1] Computer Science Department, University of Alcalá, Ctra. Barcelona km. 33.600–28871, Alcalá de Henares, Madrid (Spain)
{msicilia, elena.garciab}@uah.es
[2] R&D Department, Cátenon Human Resource Selection, Goya 115 – 28009 Madrid (Spain)
ralcalde@catenon.es

Summary. The central component of most knowledge-based *Human Resource Management* (HRM) systems is a model of the actual or required knowledge and abilities of employees, applicants and job positions. The notion of *competence* has been used in many of them to describe levels of skills and knowledge as applied to concrete work situations. Nonetheless, the imprecise nature of relationships and interactions between competences has been neglected in existing approaches. In this paper, a model for imprecise *gen-spec* and composition relationships between competences is described, aimed at coming up with more detailed and realistic selection processes. A concrete case study is also described, illustrating how the `Hr-Xml` canonical format for competency definition and interchange can be extended to give support to those relationships.

1 Introduction

The technology of *Knowledge-based Systems* (KBS) can be applied to give support to [11] or complement [13] a wide range of *Human Resource Management* (HRM) activities, including human resource planning, recruitment, selection, and staff development. In consequence, application areas for artificial intelligence techniques range from strategic ones, like manpower planning [6], to operational activities like performance appraisal or individualized training. Nonetheless, all of these application areas share a common requirement: the construction and maintenance of some form of knowledge base regarding *human resources* – both employees and job applicants – , with detailed and up-to-date information about the history, skills and abilities of each individual. Such kind of knowledge base requires a *epistemologically adequate* schema – in the sense given by McCarthy [12] – to properly represent the value of each individual, organized around a set of concepts describing difficult to characterize human traits like knowledge, abilities and attitudes. The notion of *competencies* – that form the basis of current industry

Miguel-Ángel Sicilia et al.: *Fuzzy Specializations and Aggregation Operator Design in Competence – Based Human Resource Selection*, Advances in Soft Computing **1**, 219–230 (2005)
www.springerlink.com

proposed standards like `Hr-XML`[1] – can be used as the core structuring criterion for that purpose. Competence is understood as the relation between humans and work tasks in the context of concrete work *roles*, i.e. the concern is not about knowledge and skills by themselves, but about which knowledge and skills are required to perform a specific task in an efficient way [9]. Nonetheless, as competence can be considered as 'skills or knowledge applied to a concrete business objective', the three terms are in some cases used interchangeably.

This competence-oriented view of human resources is considered to form the basis for organizations that are more responsive to its constituency [7]. As a consequence, a number of HRM-related systems have been built that organize its core model around the notion of competence or skill, some of them using ontologies for that purpose [21]. In all of them, some notion of relationship between competencies is used, but they do not capture the diverse forms of interaction between competences that are used by HR consultants in their everyday's work. For example, in `MASEL` the only relationship is that of competence grouping by company role [3], and in `CommOnCV` sector-specific competencies are grouped by position and activity [8]. Some systems use the concept of *skill tree* [1, 20], but the hierarchical relationship embodied in those trees is not clearly defined, and specializations are intermingled with composition relations, thus making difficult to aggregate and obtain overall realistic competence scores.

The focus of this paper is the fact that relationships between competencies or skills inside a HRM system are in many cases of an imprecise nature. For example, the "developing dynamic Web page" skill subsumes more specific ones like "developing `Jsp` pages" and "developing `Php` pages" to some extent, and depending on the kind of HRM process being carried out, a given degree of specificity might be required. Another example is that of competencies that depend on each other like "customer interaction" that may depend to some extent – or may be correlated in some way – with competencies like "customer tracking" and "customer value estimation". These relationships are of a different nature than the more precise groupings found in `CommOnCV`, since the former are of a finer level of granularity. Nonetheless, selection services depend critically on these lower-level definitions, as they ultimately require numerical scores or sorting criteria for candidates that are computed from them. Possibility theory [2] has been applied elsewhere [15] to model compentences and competence matching, but limiting relationships between competences to boundary restrictions of possibility in specializations.

Given that imprecise relationships between competences have not been properly addressed in existing work, in this paper we aim at addressing at least some of them. Concretely, we approach the following two different kinds of relationships (they were identified by HR consultants in an informal analysis session):

(i) Generalization-specialization (*gen-spec*) relationships.
(ii) Aggregation-oriented relationships (compositions).

[1] http://www.hr-xml.org/

Relationships of type (i) essentially provide a mean to deal with competences at different levels of abstraction or specificity that are somewhat related by subsumption, i.e. they appear in the same work context and require related abilities or skills, and only specificity makes them different. For example, "formal writing with word processors'- is a general term that may subsume "formal writing with Microsoft Word". Relationships of type (ii) allow for the description of competences that are present whenever a combination of competences appear. Following the example, in a given context, "formal writing" may be considered an aggregate of "academic paper writing" and "professional writing" (which in turn may be an aggregate of "review writing" and "report writing"). The differences between types (i) and (ii) are subtle in some cases, as will be detailed in the following sections.

The rest of this paper is structured as follows. In Sect. 2, a model for fuzzy relationships of type (i) is described. Section 3 deals with relationships of type (ii), assuming some form of HRM scenario involving aggregation of competencies modelled after the concepts described in the previous section. Section 4 sketches some details about a concrete implementation case study that uses the proposed canonical model of competencies provided by the `Hr-XML` Consortium. Finally, conclusions and future research directions are provided in Sect. 5.

2 Modelling Vague Competence Specializations

In what follows, competences will be considered in the context of selection, which is one of the most common activities in HRM, necessary both in internal project group selection or assessments, skill gap analysis, recruitment and other tasks. It will be assumed also that HR consultants drive concrete selection activities.

In addition, a possibilistic approach following [15] is used, since it provides a realistic framework for modelling human traits, given that it provides a upper probability bound "disconnected" with randomness, as described by Smithson [18]. The process of competence assessment can be summarized as follows: Initially, the score for each competence (in a set C of them) of a given individual is set as "non applicable", indicating that the competence has not been assessed for that individual. After that, a consultant carries out an initial curriculum inspection, assigning a possibility distribution for some of the competencies. Then, subsequent analysis and selection processes will eventually lead to gather additional evidence about some competences, resulting in the progressive definition of the profile.

Following the usual notation, a measure space $(X, \mathcal{B}, \mu)$ is defined by a σ-field $\mathcal{B}$ of subsets of X and a (finite) measure μ defined on $\mathcal{B}$ such that $\mu(X) < \infty$. The function μ is said to be a *possibility measure* if $\mu(\emptyset) = 0, \mu(X) = 1$ and for any countable sequence of subsets S in $\mathcal{B}$, $\mu(\bigcup_{i \in S} s_i) = sup\{\mu(s_i) | i \in S\}$. It should be noted that a possibility measure Π such that $\Pi(A) = 1$ if $A \neq \emptyset$, and $\Pi(\emptyset) = 0$ describes the *least informative* possibility measure. Given a possibility measure Π a possibility distribution $\pi : X \rightarrow [0,1]$ can be defined such that $\pi(x) = \Pi(\{x\})$.

Possibility distributions for a given individual $h \in H$, denoted by $\pi^h_{\tilde{c}}(x), c \in C$ are defined following the usual conventions: (1) $\pi^h_{\tilde{c}}(x) = 0$ means that $\tilde{c} = x$ is impossible,

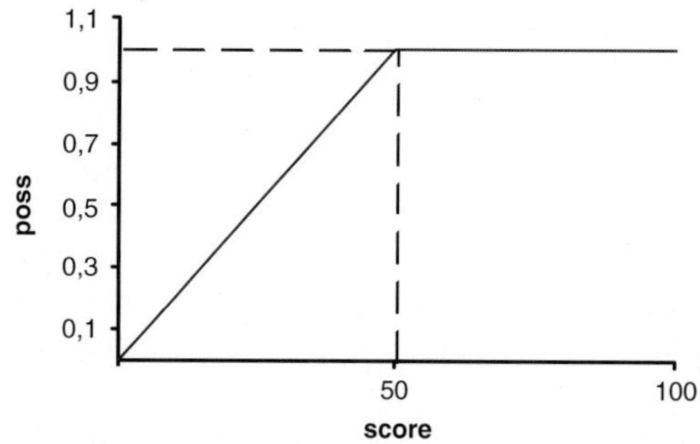

Fig. 1. Example possibility distribution

and (2) $\pi^h_{\tilde{c}}(x) = 1$ means that $\tilde{c} = x$ is possible without any restriction. The domain of values of competencies is normalized in the [1,100] real interval.

Figure 1 depicts an example of possibility distribution for a given competence. The rationale in this case is that of assigning a score obtained from a self-assessment that yielded the value "3". According to the function, it's granted complete possibility to values above that medium point, proceeding in this case in an optimistic way. The selection of the shape is competence and assessment method-dependant, and therefore, consultants are responsible for that decision.

Possibility may be additionally restricted by certifications, grants or experience. For example, the presence of a certification in a programming language or technology can be used to give full possibility for related competences to a certain extent.

The following property regarding *gen-spec* relations between competences must be true for any candidate profile. Given a pair of competencies a and b so that b is an specialization of a:

$$\forall x \ \pi^h_{\tilde{a}}(x) \geq \pi^h_{\tilde{b}}(x) \tag{1}$$

The rationale for (1) is that it's possible that for an individual the more specific competence is known to be not present at all (i.e. that $\pi^h_{\tilde{b}} \equiv 0$), but having that more specific competence entails that the more general one is also present. Despite this restriction, some specialized competences are less differentiated to some of its generalizations than others. This fact should be taken into account when using flexible approaches to selection, since in many practical situations, more specific competences can be *substituted* by more general ones, e.g. a selection searching for individuals with a given level of "Oracle 6i Backup Administration" may also consider individuals with an appropriate level in "Oracle Backup Administration", but not those having a more general competence like "Relational database backup administration". This leads to a concept of degree of substitutability between competences connected by *gen-spec* relations.

The approach to model this kind of imprecise relationships follows the resemblance-relation approach described in [14]. Concretely, we'll denote a generalization relationship between two competences in C as defined in (2).

$$a \succ^d b \quad a,b \in C \tag{2}$$

$$d = \{\phi_i(a,y) \mid a \succ^d y \wedge y \in C\} \tag{3}$$

A discriminator d determines the taxonomic criterion that justifies the relationship, and can be represented in the most general case by a set of predicates (3) – one for each direct specialization – that determines the specific properties of the instances of each subclass. Each of the predicates ϕi characterize one of the subclasses discriminated. Following the example, $product(X)$ and $version(X)$ could represent assertions discriminating backup competences like those mentioned above.

Given a competence, its direct subclasses are divided in disjoint sets (partitions), according to their discriminators. P denotes the set of (local) partitions.

$$P_a = \{p_{(d,a)} \mid a \in C\} \quad \text{where} \quad p_{(d,a)} = \{c \mid c \in C \wedge a \succ^d c\} \tag{4}$$

Given a concrete $p_{(d,a)}$, a notion of *distance* from a to each of its specializations is required to model the concept of substitution mentioned above. We have used *resemblance relations* to model that specialization distance. A resemblance relation R on a crisp domain D is a binary fuzzy relation (5).

$$R : D \times D \rightarrow [0,1] \tag{5}$$

which satisfies reflexive (6) and symmetric (7) properties.

$$R(x,x) = 1 \ \ \forall x \in D \tag{6}$$

$$R(x,y) = R(y,x) \ \ \forall x,y \in D \tag{7}$$

Given this definition, a separate partial resemblance relation R can be obtained locally for each partition of subclasses, so that we operate on a set of relations (8) in the form (9).

$$\Xi_D = \bigcup_{x \in P} R_x \tag{8}$$

$$R_x : p_{(d,c)} \cup \{c\} \times p_{(d,c)} \cup \{c\} \rightarrow [0,1] \tag{9}$$

Relations are labelled partial since they only contain competence-sub-competence relationships, that is, relations are really defined in the form $R_x : \{c\} \times p_{(d,c)} \rightarrow [0,1]$, i.e. from a specified competence to all its specializations that are discriminated by an specific d (although this could easily be extended to siblings). This enables a form of stepwise simple reasoning in which competences at hierarchy level i can be substituted with the closest competence in the $i \pm 1$ level traversing *gen-spec* relations through the different discriminators. The use of these resemblance relations in selection models the notion of substitution, as will be described in Sect. 4. The elicitation method for these relations was based on the techniques described in [14], using HR consultants as experts.

3 Designing Aggregation Schemes for Competencies

Weighted additive approaches like the percentage matching used in `OntoProper` [19] have been used for the process of selecting individuals given a desired profile for a task or job. But these approaches neglect the fact that the presence of one concrete skill or competence may be correlated or be interpreted as being covered by others. This and other forms of interaction between competencies point out that the problem of score aggregation for concrete selection processes require a careful examination of the nature of the skills involved in the process. Since relationships between competencies are seldom precisely defined, fuzzy integrals are good candidates for the design of flexible and interpretable aggregation schemes as described in [4]. Following the possibilistic framework described above, the aggregation of competence levels can be described as an aggregation of possibilities normalized in the [0,1] interval:

$$\Omega : I^n \to I \tag{10}$$

Our first attempt to model these forms of aggregation has been that of designing fuzzy measures that are used as capacities using the Choquet integral (11) as a concrete form of Ω operator.

$$C_v(x) = \sum_{i=1}^{n} x_{(i)}[v(\{j|x_j \geq x_{(i)}\}) - v(\{j|x_j \geq x_{(i+1)}\})] \tag{11}$$

The fuzzy measure v is specific to each aggregated competence and should in the general case be specified by HR consultants (a sample process for the elicitation of those measures in other domain can be found in [16]). A fuzzy measure on a set X is a monotonic (i.e. $v(S) \leq v(T)$ whenever $S \subseteq T$) set function $v : 2^X \to [0,1], v(\emptyset) = 0, v(X) = 1$. A number of interactions between competences can be modelled by fuzzy measures, including substitutiveness, complementarity and preferential dependencies, as described by Marichal [10].

An example is given in Table 1, in which the value of the typical *soft* competence "customer loyalty management" (clm) is obtained by aggregation of four competences[2], namely, "Campaign eligibility criterion setting" (e), "customer portfolio management" (m), "customer segmentation analysis" (s) and "customer loyalty program management" (l).

A positive correlation is assumed between e and s since they share a common collection of required skills to be attainable. This interaction, according to [10], expresses that the marginal contribution of s to every combination of criteria that contains e is strictly less than the marginal contribution of s to the same combination when e is excluded. In addition, m and l are considered as substitutive by the consultants (due to the fact that both reference a common collection of management skills), so that the presence of m or l produces almost the same effect than the presence of both.

[2] The characterization of the competence is actually more complex, but a simplification is described for clarity

Table 1. Example fuzzy measure $v(X)$

$n=1$	$n=2$	$n=3$	$n=4$
$\{e\}\to 0.15$	$\{e,s\}\to 0.5$	$\{e,m,s\}\to 0.9$	$\{e,m,s,l\}\to 1$
$\{m\}\to 0.2$	$\{e,m\}\to 0.65$	$\{e,s,l\}\to 0.95$	
$\{s\}\to 0.5$	$\{e,l\}\to 0.65$	$\{e,l,m\}\to 0.7$	
$\{l\}\to 0.15$	$\{m,s\}\to 0.7$	$\{m,l,s\}\to 0.75$	
	$\{s,l\}\to 0.65$		
	$\{l,m\}\to 0.2$		

It should be noted in Table 1 that the weight of $\{e,m\}$ is only slightly augmented when the substitutive competence l is added to the set, and the set of correlated competences $\{e,s\}$ has a weight of 0.5, which represents a difference of 0.15 from absolute additivity. In more complex cases, mathematical functions are required to model those relationships (and in presence of larger numbers of competences, *k-additivity* must be studied to be able to define v [5]). The requirement for two correlated criteria i and j is that they are subadditive i.e., that $v(\{i,j\}) < v(\{i\})+v(\{j\})$. Two substitutive competences are required to satisfy the relationship expressed in (12), so that the addition of a substitutive criterion have a small effect in the fuzzy measure (having no effect if the criterion are completely interchangeable).

$$v(T) < \left\{ \begin{matrix} v(T\cup i) \\ v(T\cup j) \end{matrix} \right\} \approx v(T\cup\{i,j\})\ ; \quad T\subseteq X-\{i,j\} \tag{12}$$

The results of the Choquet integral for the measure in Table 1 can be compared with those of a simple weighted mean ($\mathcal{W}$) to appreciate the differences (both expressed in a zero to three scale). For example, given the input vectors $x_1 = (2,0,3,3)$ and $x_2 = (2,3,3,3)$, $\triangle\mathcal{W}(x_1,x_2) = 0.85$, while $\triangle\mathcal{W}(x_1,x_2) = 0.2$. This reflects the effects of substitutiveness between m and l. If we consider $x_3 = (3,2,0,2)$ and $x_4 = (3,2,2,2)$, we have $\triangle\mathcal{W}(x_3,x_4) = 0.4$ and $\triangle\mathcal{C}(x_3,x_4) = 0.15$, reflecting the effect of the degree of correlation between e and s.

The just described example illustrates the subtleties and context-dependency of competence composition in the general case, which calls for further studies on this kind of relationship. It should be noted that this form of relationship is different to that described in the previous section. For example, the presence of a given possible level x in competence s does not entail necessarily that the aggregated possible level in *clm* would be over that value.

4 Integrating Vague Relationships and Aggregation Schemes in Canonical Models

The just described competency relationships have been used to build a Web-based tool used as a support system in human resource selection. The tool allows a consultant to query for matching candidates for a given desired profile, and the tool uses the relationships described in Sects. 2 and 3. The overall layout of the tool, showed in Fig. 2 simply allows for the definition of a given required profile (a job profile), and returns the individuals in the database that better match the given profile.

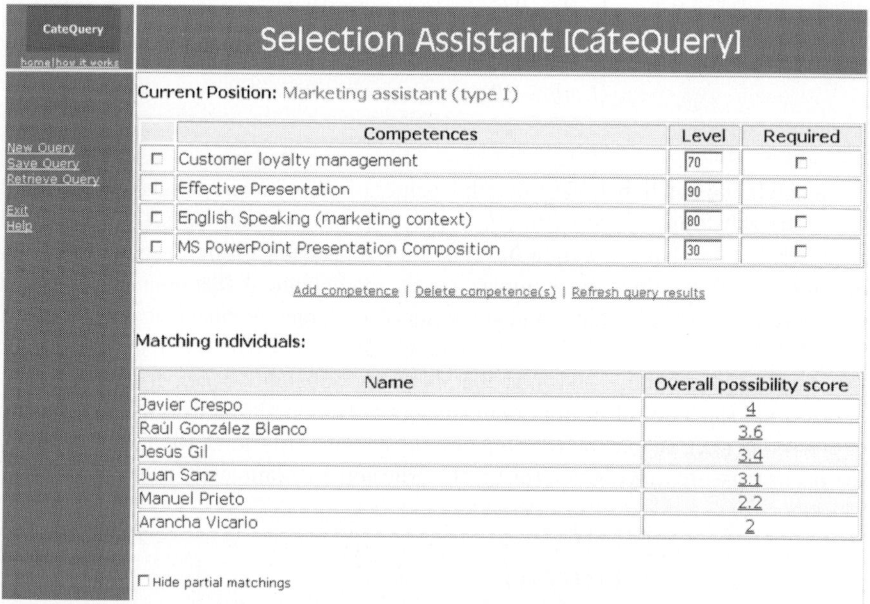

Fig. 2. Overall layout of the candidate-query Web tool

Job positions are described by a set of competency scores that are associated to a specific selection process. A number of explicit scores is specified by the job seeker directly, that are considered as completely certain and *required* requirements. Other scores may be *desirable* but not mandatory requirements. In consequence, a job profile $p \in P$ for a given organization may be characterized as a set of scores:

$$p \equiv REQ_p \cup DES_p \tag{13}$$

$$p \equiv \{r_i\} \cup \{d_j\}, r_i, d_j \in (c,x), c \in C, x \in [1,100] \tag{14}$$

The selection of the better subset of candidates for a given job position is carried out according to the following compatibility formula that represents the possibility of fit of the profile of candidate h to the job position p.

$$POSS_p^h = trunc\left(\prod_{(c_i,s_i)\in REQ_p} \pi_{c_i}^h(s_i)\right) \cdot \sum_{(c_i,s_i)\in DES_p} \pi_{c_i}^h(s_i) \tag{15}$$

The *trunc* function truncates the real value to an integer. This entails that if any of the required scores is below perfect possibility, the overall score becomes zero.

Otherwise, the effect of that required competences is that of multiplying one by the rest of the formula. The *POSS* value can be divided by $|DES_p|$ to have a normalized value for comparing degrees of matching in different selection processes.

When obtaining the possibility grades in expression (15), the relationships described in previous sections are used in the following way:

(i) *Gen-spec* relationships are used to substitute more specific competences with direct generalizations both in *REQ* and *DES*, for individuals in whose profile the concrete competence is not present or does has a lower value than that of the generalized competence.
(ii) Aggregation operators associated to competences (as *clm* in the example of Sect. 3) are used whenever they appear in *REQ* or *DES*, computing the aggregated value from the partial values of sub-competences. This may occur recursively if more than one level of composition occurs.

In Fig. 2, the column "overall possibility score" is shadowed if a relationship of type (i) has been used as substitute. These scores are links that lead to a page providing details on the competences actually used in the computation of the score.

Competences involved in the described process can be described in standardized markup following the `Hr-Xml` conventions. But a number of extensions to the "Competencies 1.0 (Measurable Characteristics)" recommendation are required to describe both resemblance relations and aggregation-oriented compositions:

- The recommendation states that "competencies can be recursive", but a way to differentiate types of relationships is required.
- Each type or relationship must be accompanied by a collection of information elements describing the concrete characteristics of the relationship.

The following example fragment of extended markup illustrates a possible straightforward way of adding those required information items:

```
<?xml version="1.0"?> <Competency name="Communication Skills"
  xmlns:xsi="http://www.w3.org/2001/XMLSchema-instance"
  xsi:noNamespaceSchemaLocation="http://ns.hr-xml.org/
                        Competencies-1_0/Competencies-1_0.xsd"
  xmlns:fuzzy="http://www.dei.inf.uc3m.es/hr/">
  <Competency name="Written Communication Skills">
         <CompetencyEvidence name="WRITTENTEST1-A"
                                  dateOfIncident="1995-01-01"
                                  lastUsed="2000-01-01">
                 <NumericValue minValue="3"
                               maxValue="5"
         description="SEP-equivalent Skill-Level Range">5
                 </NumericValue>
         </CompetencyEvidence>
     <Competency name="Technical Writing Skills"
                        fuzzy:type="specialization">
         <fuzzy:distance value="0.4"/>
         <fuzzy:discriminator name="Kind of writing"/>
```

```
            <CompetencyEvidence name="c1" >
            <!-- etc...-->
        </Competency>
    </Competency>
    <Competency name="Oral Communication Skills">
            <CompetencyEvidence name="ManagerObservation"
                dateOfIncident="1996-01-01"
                lastUsed="2000-01-01">
              <NumericValue minValue="1"
                 maxValue="5"
                 description="Company XYZ Skill Range">5
              </NumericValue>
            </CompetencyEvidence>
     </Competency>

     <!--  etc... -->

    <fuzzy:AggregationScheme file="aggr1.xml" />
    </Competency>
```

In the above fragment, extended elements and attributes are put into the `fuzzy` namespace. By default, competence nesting is interpreted as "aggregation", and the `AggregationScheme` elements is provided an alternative for the official simple weighting scheme of `Hr-Xml`, and points to a separate file. Specialization relationships are marked explicitly by the `type` attribute, and elements `distance` and `discriminator` give details about the characteristics of the relationship.

5 Conclusions and Future Work

Imprecision is an inherent characteristic of both composition and generalization-specialization relationships in competence modelling, which makes necessary the development of richer and more realistic knowledge representations of contextually–situated competences. In this paper, we have proposed resemblance relations to model distance between competences and its specializations, and the use of the Choquet integral as a device to produce interpretable aggregations of competence levels considering interactions between them. Both relationships have also been integrated in a simple scoring possibilistic framework based on previous work [15], that has been used to develop a prototype tool that allows for querying for individuals that match a given job position. The resulting model provides a point of departure for further studies about the structure of competences that are more ambitious than current simple compensatory scoring models like the one described in [19].

The modelling constructs devised as part of our current competency model are fragmentary aspects of the problem of competency modelling, and in consequence, they need to be integrated in a comprehensive framework to give support to knowledge-based applications. Future work should focus on the specifics of each of the different kinds of competence relationships and their interactions to be able to develop a new generation of more realistic HRM tools. It should be noted that here we only deal with specific forms of imprecision, but other forms of imperfect information [17] are also inherent to most HRM systems. For example, *ambiguity* appears whenever they have to deal with conflicting information like assessments coming from peers and managers, and imprecision and uncertainty is also inherent to measurement instruments like questionnaires or indirect tests.

References

1. Benjamins, R., López, J. M. Contreras, J., Casillas, J., Blasco (2002) Skills Management in Knowledge-Intensive Organizations. In: Proceedings of the 13th International Conference on Knowledge Engineering and Knowledge Management. Ontologies and the Semantic Web. Springer Lecture Notes in Computer Science, 2473: 80–95.
2. Dubois, D. and Prade, H. (1988) Possibility Theory. Plenum Press, New York.
3. Garro, A., Palopoli, L. (2002) An XML Multi-agent System for E-learning and Skill Management. In: Proceedings of the Workshops on Agent Technologies, Infrastructures, Tools, and Applications for E-Services. Springer Lecture Notes in Computer Science 2592: 283–294.
4. Grabisch, M. (1997) Fuzzy Integral as a Flexible an Interpretable Tool of Aggregation. In: B. Bouchon–Meunier (ed.): Aggregation and Fusion of Imperfect Information, Studies in Fuzziness and Soft Computing. Physica-Verlag: 51–72.
5. Grabisch, M. (2000) The interaction and Mobius Representations of Fuzzy Measures on Finite Spaces, k-Additive Measures: a Survey. In: M. Grabisch, T. Murofushi, and M. Sugeno (eds.): Fuzzy Measures and Integrals – Theory and Applications. Physica Verlag: 70–93.
6. Grundy, T. (1997) Human Resource Management – a Strategic Approach. Long Range Planning 30(4): 507–517.
7. Hagan, C. M. (1996) The Core Competence Organization: Implications for Human Resource Practices. Human Resource Management Review 6(2): 147–164.
8. Harzallah, M., Leclère, M., Trichet, F. (2002) CommOnCV: Modelling the Competencies Underlying a Curriculum Vitae. In: Proceedings of the 14th International Conference on Software Engineering and Knowledge Engineering. ACM Press: 65–71.
9. Lindgren, R., Stenmark, D., Bergquist, M. and Ljungberg, J. (2001) Rethinking Competence Systems for Innovative Organizations. In: Proceedings of the 9th European Conference on Information Systems: 775–786.
10. Marichal, J. L. (2000) An Axiomatic Approach of the Discrete Choquet Integral as a Tool to Aggregate Interacting Criteria. IEEE Transactions on Fuzzy Systems 8(6): 800–807.
11. Martinsons, M. G. (1997) Human Resource Management Applications of Knowledge-based Systems. International Journal of Information Management 17(1): 35–53.
12. McCarthy, J. L. (1977) Epistemological Problems of Artificial Intelligence. In: Proceedings of the International Joint Conferences on Artificial Intelligence: 1038–1044.
13. O'Donell, S. E. (1990) Expert Systems Provide New Access to Knowledge. HR Magazine 35(9): 30–32.
14. Sicilia, M. A., García, E., Díaz, P., Aedo, I. (2003). Representation of Concept Specialization Distance through Resemblance Relations. In: Benítez, J., Cordón, O., Hoffmann, F., Roy, r. (eds.): Advances in Soft Computing – Engineering, Design and Manufacturing. Springer: 113–129.

Personalizing Information Services for Mobile Users

Wei-Po Lee

Department of Information Management, National University of Kaohsiung, Kaohsiung, Taiwan

Summary. The advances of mobile devices and wireless telecommunication infrastructure have allowed mobile users to conveniently transmit information through a wireless environment. To enhance the information services via the wireless transmission, the WAP Forum was founded and a new protocol was proposed to support wireless applications. However, as in the World Wide Web, the increasing information leads to the problem of information overload. One method to overcome such a problem is to model individual users to provide personalized information services. By this way, only the information predicted as user-interested can reach the end user. This paper presents a multi-agent framework in which a decision tree-based approach is employed to learn a user's preferences. To assess the proposed framework, a mobile phone simulator is used to represent a wireless environment and a series of experiments are conducted. The experimental studies have concentrated on how to deliver appropriate information to the individual user, and on how the system can adapt to a user's most recent preferences. The results and analysis show that based on our framework the WAP-based personalized information services can be successfully achieved.

1 Introduction

In recent years, the advances of mobile devices and wireless telecommunication infrastructure have allowed mobile users to conveniently transmit information through a wireless environment. To enhance the information services via the wireless transmission, the WAP (Wireless Application Protocol) Forum (http://www.wapforum.org) was founded and a new protocol was proposed to support wireless applications. Together with WAP, a specific language WML (Wireless Markup Language) was also invented to describe the information content for mobile access, as HTML is to the Internet and World Wide Web. In this way, content providers can easily and rapidly offer more and more information for mobile users. Though it is convenient to access on-line information via mobile devices, the problem of information overload caused by the increasing digital contents has to be overcome, especially when there is only limited communication bandwidth available.

One way to remedy the above problem is to model the preferences of individual users and then to provide personalized information services. In a wireless

Wei-Po Lee: *Personalizing Information Services for Mobile Users*, Advances in Soft Computing **1**, 233–246 (2005)
www.springerlink.com

broadcasting environment, the service provider can use a personal profile to record the most recent preferences of each individual user, and then to deliver information to a specific user according to his preferences. Similar to the work of information filtering, to personalize information services is to reason a user's preferences by analyzing his personal information, especially the information he has accessed in the past. In this way, only the information considered as user-interested can reach the end user. The user can thus access the information he is interested in whenever he wants, to save the time spent on reading large amount of electronic documents.

In this paper, we describe how we adopt an agent-based methodology to develop a prototype system for personalizing mobile information services. Our system includes four agents: a film agent to gather a user's preferences in different ways; a news agent to collect Internet news for user's identification; a learning agent to analyze and model a user's interests; and an adaptation agent to evaluate the performance of the system and adapt to the user's newly changed interests. To assess our framework, a mobile phone simulator is used in the experiments for two kinds of mobile services: one is about the film, and the other, the news. The experimental results show that our learning agent can model a user's interests in real time and provide sensible information recommendations for mobile users. Furthermore, we also show that even if a user changes his interests occasionally, the learning agent can continuously learn his most up-to-date preferences to ensure the quality of services.

2 Background

To deploy data-oriented applications for wireless terminals, WAP has provided a promising technological solution for mobile information services. WAP specifies an end-to-end communication platform between the mobile terminals and a gateway, and the latter is defined to be a server to convert WAP and HTTP, and vice versa. Different system architectures have been proposed to enable WAP-compliant applications (e.g., [1, 2]), and many information services have been developed for the mobile users (e.g., [3, 4]).

As services created by HTML do not fit well on handheld devices, the content providers have been deploying WAP services by conducting completely new WML-based applications or building mechanisms that can automatically transfer the original HTML pages into WML. In the latter case, when a mobile user asks for the Internet information, the WAP gateway will filter the useless tags and reconstruct the content of the pages, and then convert it into WML pages [5]. However, unlike with the Internet, a wireless environment has some restrictions such as low bandwidth, limited computational resources and power of batteries, etc. Therefore, information services for mobile users must be provided in a more efficient way. An ideal solution is to build a recommendation system that gives information to each individual user according to his own preferences, rather than delivers all information to all users. In this way, the user can save his time spent on reading irrelevant information and the transmission bandwidth can be utilized more efficiently.

Regarding the work of personalization, the most important issue is to construct a computational model for each individual user to predict his preferences for incoming information. In this way, only the information considered as user-interested can reach the end user. Therefore, the work of personalization can be regarded as classification: using the personal information already known to build a model to predict the events unknown. Some machine learning approaches have been applied to construct users' profiles, for example, a statistic-based approach tfidf (term frequency and inverse document frequency, [6]) was used in [7] to build a user's profile for recommending Web pages, and a reinforcement learning method was employed in [8] for book recommendations. Our work presented here differs not only in the learning approach used, but in that we apply the services to the mobile network. Most importantly, our work includes a specially designed mechanism for system adaptation, with which a user's profile can be re-learnt to correct the prediction errors and to meet the user's most recent preferences.

3 Personalizing Mobile Information Services

3.1 System Architecture

The major tasks of a personalized recommender system include collecting a user's personal interests, building a model to describe the information collected, and adapting to the user's most recent interests. In this work, we design and implement an agent-based framework to provide WAP services. To incorporate this personalization framework with a mobile device and to experience the overall system in a wireless environment, we adopt a mobile phone simulator developed by the NOKIA mobile phone company. The simulator is an application software for emulating WAP network; it thus plays both roles of mobile browser and WAP gateway in a simulated environment for mobile information access. Figure 1 shows the typical mobile environment and the roles of the simulator. With the simulator, we do not have to consider the problems of signal translation, encoding/decoding, etc., and can concentrate on the design of new mobile information services. In this work the NOKIA simulator UP.Simulator 4.1.1 is used as the mobile information receiver for the end user, and a WML database is built to work as the Internet service provider. Figure 2 illustrates the system architecture in which the computation for personalization is performed on the server side. As can be seen, three types of agents for information gathering, user modeling, and system adaptation are included. Each agent is responsible for a specific task, and different agents work simultaneously to achieve the overall task in a distributed way.

3.2 Information Gathering

In order to reason the user's preferences, the first task a personalized system has to perform is to collect his personal information. In this work, two kinds of information services are considered. One is to provide the user the information about the daily

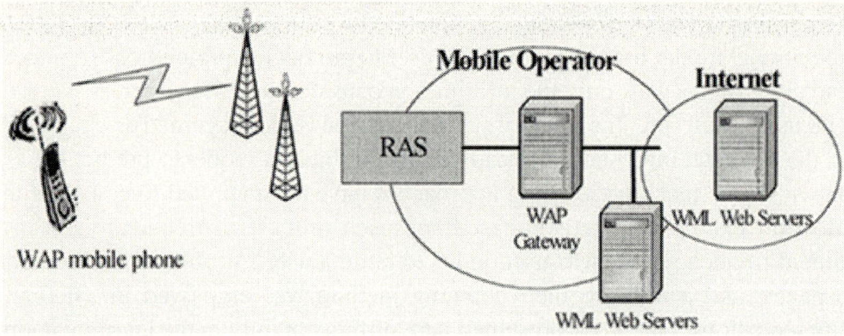

Fig. 1. The typical mobile environment and the mobile phone simulator

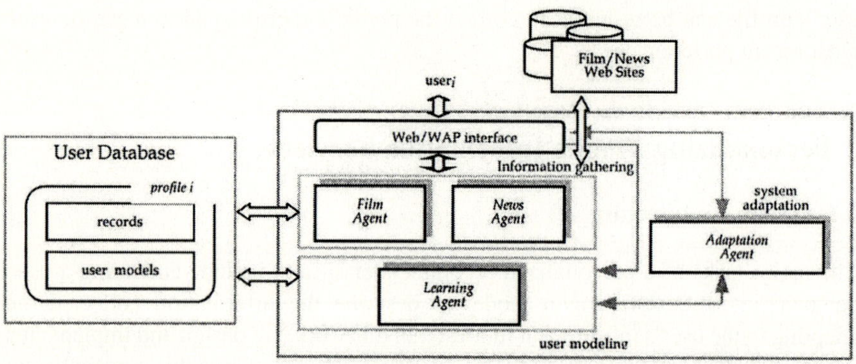

Fig. 2. The typical mobile environment and the mobile phone simulator

TV film programmes that may seem interesting to him. This is for the increasingly numerous digital channels following the invention of the set-top-box and digital TV. The other service is to inform the user whenever there appears a piece of news he may be interested in available on-line. The film agent and news agent shown in Fig. 2 are designed to collect the relevant information respectively. For the film service, the film agent finds the daily film programmes to be broadcasted and extracts the corresponding information (such as genres, actors, introductory notes, etc.) of these films from the on-line resources. Then it suggests a programme list to the individual user according to his personal profile (shown in the experimental section). The user can select an item to read the detailed information provided by the film agent and identifies whether he likes this item or not by the mobile device. Alternatively the film agent can present the user a form (as shown in Fig. 3) that allows him to explicitly point out his interests through the World Wide Web. With the user's feedback, this agent can precisely learn his preferences. For the news service, the news agent in Fig. 2 extracts the on-line daily news from the CNN news site (http://www.cnn.com), re-organizes the news, and presents the news to the user by the interface shown in Fig. 4, according to his profile. As for the films described above, the user can mark the news to indicate his interests by the desktop PCs or

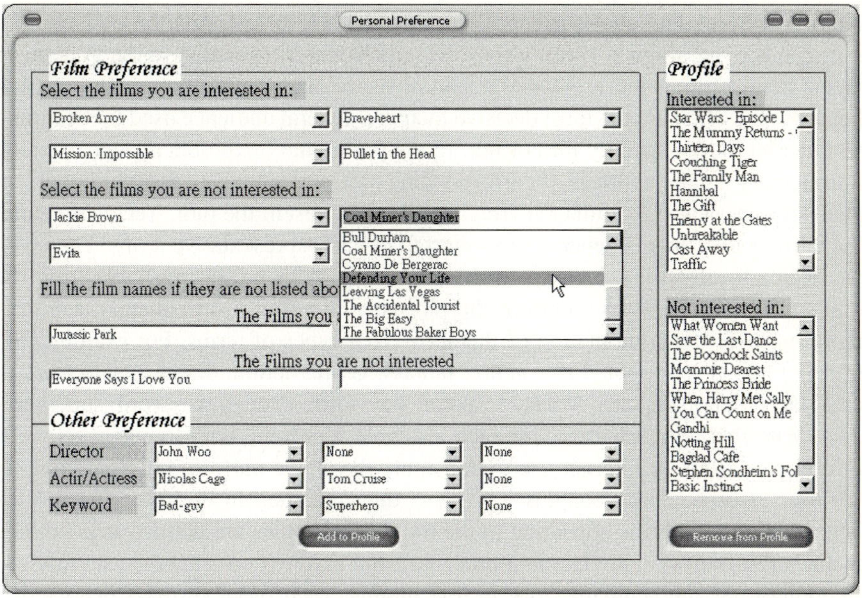

Fig. 3. Information being gathered by the information agent

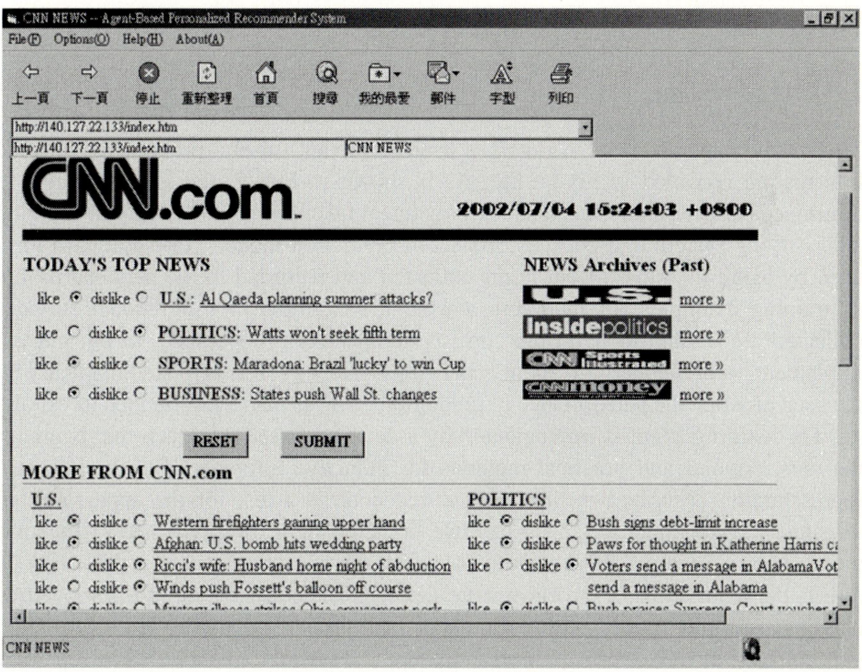

Fig. 4. Interface presented by the news agent for information collection

mobile devices. The above agents both play the role of information collectors; a user can activate them arbitrarily to provide personal information to the system. In this system, a film is represented as a feature vector of genre, director, cast, and plot, because they often are the most decisive factors a general audience used to determine whether he likes a specific film or not. In the preliminary experiments, we used the introductory note of a film as its corresponding plot, and employed the traditional text analysis approach (i.e., tfidf) to extract words to represent the plot. Yet, as is found, the introductory note is generally too short to extract representative terms. Therefore we decide to directly use the keywords associated with a film (available from the on-line film database) to represent the plot part of a film. Consequently, a film is represented as the combination of the above four types of terms. For example, the film Batman is represented as ⟨ genre: fantasy, genre: action, director: Tim Burton, cast: Jack Nicholson, cast: Michael Keaton, keyword: revenge, keyword: superhero, ... ⟩. With this representation, for each available film item, the film agent retrieves information from an Internet site to constitute the above uniform representation. If the user gives feedback for this item, it is then recorded in the user's profile for later training. All terms appearing in the training examples are defined as candidate features (attributes) in the learning procedure and used to construct the user models. In addition, as the standard work in text mining, each piece of news extracted is represented as a word vector in which the feature words are determined by the tfidf method, after the common words are removed and the stemming procedure is performed.

3.3 User Modeling

The second type of agent in Fig. 2 is used to model a user's preference from the information recorded in his profile. As is shown in Fig. 2, the learning agent is developed for user modeling. The learning agent takes the responsibility of building a classifying system that consists of a set of prediction models (or classifiers) for a user, by using the most recent items collected and recorded in the personal profile as training examples. In modeling a user, the examples for individual services collected are trained to build the corresponding models separately. In our current implementation, the learning agent acts in a passive way: it is activated manually by the user, or when the performance of prediction is lower than a pre-defined threshold.

The learning agent is implemented by a decision-based approach that is one of the most popular and practical methods for inductive inference. For the keyword representation described in Sect. 3.2, a tree node is a test for the appearance of a feature keyword and this node has two branches indicating whether a film (or a piece of news) includes this keyword. After a series of attribute testing, the leaf node directs the item to one of the classes pre-defined. The construction of the above tree classifier is based on the measure of a statistical property that is called information gain for each attribute. This property measures how well a given attribute separates the training examples according to their target classification. The details about the decision tree learning can be found in [9].

Though decision trees can perfectly separate training examples by a covering tree, their predictive performance for the test cases is not as good as that in the training phase. This overfitting phenomenon occurs mainly in the situations when the number of training examples is too small to form a representative sample set or when there are noises in the data. As a result, the tree is overly specialized to the training data. To avoid the overfitting cases, we adopt the method of adaptive resampling for multiple trees. This method is to reduce the variance between training and testing cases, and it has been observed that inducing multiple decision trees from the same training set is an efficient approach. In such an iterative approach, training examples are firstly sampled from the original training set and used to construct a tree. Then this tree is given a vote for determining the categories of other items. After that, the training examples are replaced by the newly sampled cases, and a new decision tree is constructed again. This process continues and the category with most votes determines the final answer. It has been shown that when the number of induced decision trees increases, the variance can gradually be reduced and the corresponding prediction performance can be enhanced [10]. In the sampling phase mentioned above, the technique of adaptive resampling usually performs better than others: instead of randomly sampling a training set in each iteration, this technique increases the probabilities to be chosen for those cases previously misclassified [11].

For simplicity, in our current implementation the system only predicts whether a user is interested in a specific item or not, rather than the degree in which he favors the item. In other words, each decision tree shall predict an item into one of the pre-defined classes (user-like or user-dislike) Here, ten decision trees are included, and their voting results are used to determine whether a user likes a specific item or not. That is, an item will be predicted as user-like if it receives at least 5 votes for this class. As the number of news pieces is much larger than the number of films, therefore, instead of directly constructing a single classifier to categorize all news documents, our system decomposes the overall classification task into several subtasks in which an individual classifier is developed respectively for each category of news, for example business, sports, or politics. For each category, the news cases recorded in the profile are used as training examples to build the corresponding classifier. In this way, the personal model for news prediction in a user profile includes several classifiers for different categories of news. Each piece of Internet news extracted is sent to the corresponding classifier according to its category originally assigned by the news site, and the classifier predicts the user's interests for this news. Then only those predicted as user-interested are sent to the end user.

3.4 System Adaptation

With the multiple decision trees learnt, the system can classify films and news into appropriate classes of preference. Yet, in the real world situations a user's interests may change occasionally. In this situation the model built earlier is not able to characterize the most up-to-date interests any more, and consequently the system performance declines. To deal with this situation, the system must continuously track the user's feedback in order to adapt to his most recent interests. The adaptation agent is thus designed for this purpose. Because the prediction errors occur here, the examples used in the previous training phase have to be gradually removed. Under such circumstances, the system uses the misclassified items to replace the oldest training examples in the user's profile and re-build the personal models. As the decision tree learning approach can build new models in real time in our application case, the learning procedure is thus activated for each faulty prediction. In addition to the above case, the user can also arbitrarily add new example or remove the old ones from the profile to derive a new model.

4 Experiments and Results

4.1 The Mobile Environment

In the experiments, we use a NOKIA mobile phone simulator (UP.Simulator 4.1.1) as a wireless platform to perform WAP-based services on a simulated mobile network. Here, the wireless communication is simulated by connecting the WAP-phone to the IP address of the server, as shown in Fig. 5(a). As can be seen in Fig. 5(b), two

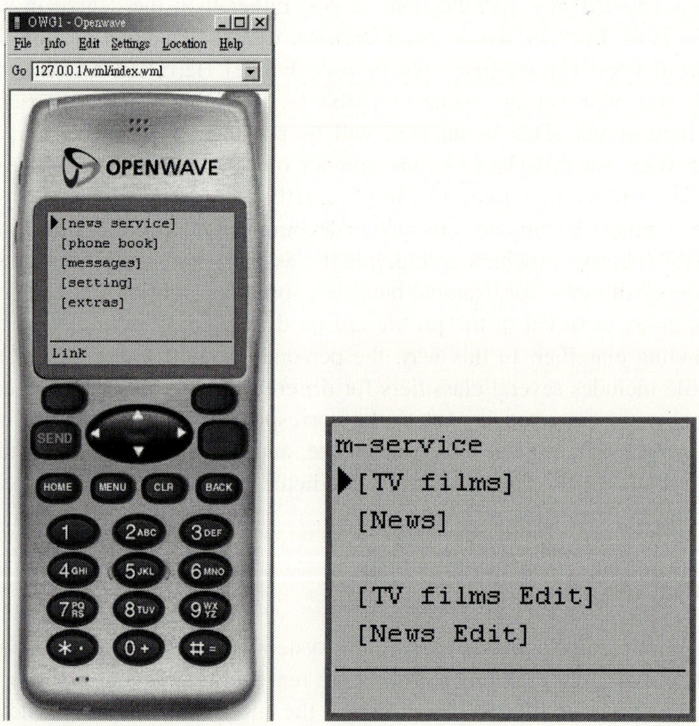

Fig. 5. (**a**) *Left*: The simulated mobile phone; (**b**) *Right*: The services provided

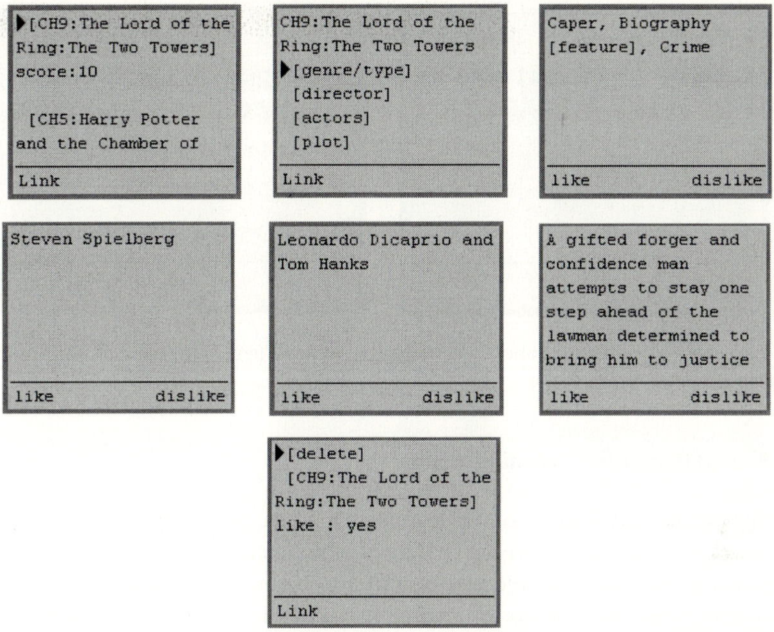

Fig. 6. (**a**) *Upper*: The film programmes predicted as user-like; (**b**) *Middle*: The user can choose a certain programme to view the relevant details; (**c**) *Bottom*: The user can directly remove an item previously added

kinds of mobile services are provided, including the recommendations of daily TV film programmes and the real time news. For the film service, the system presents the user with a list of programmes predicted as user-like, and each programme is associated with its corresponding tree votes as a score (Fig. 6(a)). The user can select a programme to see the relevant information, for instance the genres, directors, actors, and the plot of the film (Fig. 6(b)). After reading the descriptions, he can give his opinion to the system by pressing the like/dislike button. The user's feedback will then be used to modify his personal profile, and the user model within the profile will be adjusted whenever necessary. In addition, the user can also choose the option of editing listed in Fig. 5(b) to directly remove the items previously added (Fig. 6(c)). Similarly, the user can choose the news service, read the titles of news pieces recommended by the service provider (Fig. 7(a)), and select the specific pieces news to further read the details (Fig. 7(b)). Again, he can give his responses to the system via the like/dislike button. The examples illustrated in Fig. 6 and Fig. 7 show that the WAP-based information services have been successfully established in the simulated mobile environment.

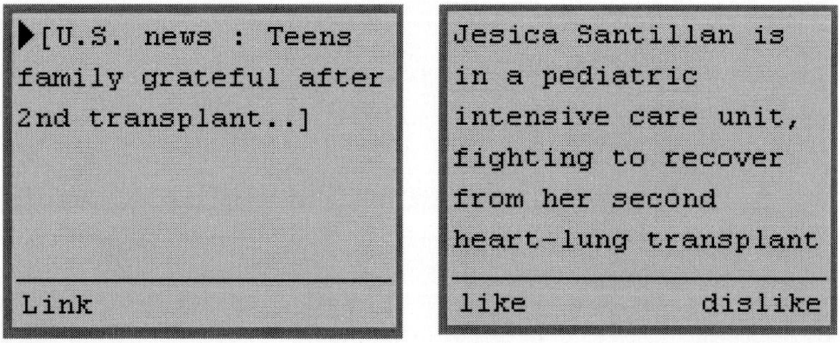

Fig. 7. (**a**) *Left*: The user can choose a certain piece of news to read the content; (**b**) *Right*: The detail of the news selected

4.2 Evaluation of the Learning Agent

To evaluate the performance of our learning agent, we have implemented a decision tree-based mechanism to develop models of preferences for users. In the first set of experiments, we only used the plot of a film to be the representative feature: each film was represented by its corresponding keywords. In this way, a decision tree was constructed from the keywords within the training set. In the training phase, different numbers of examples (including 60, 80, 100, and 120 training examples) were used to develop trees: a half of them were positive examples and the other half, negative. After the training, each tree model was tested by another 40 cases not contained in the training set for further evaluation. Table 1 shows the test results. The test results indicate that in average 73.8 $\sim$ 77.3% accuracy can be obtained. This may be satisfactory as there were only 60 to 120 training examples used in the learning phase and they were distributed between different genres. Though using sparse samples to reason about a user's preferences in a large space could result in a relatively loose user model, it is nevertheless a more realistic situation.

After showing that our learning agent is capable of developing tree classifiers to model a user's preferences, we compared it with the well-known *k*-nearest neighbor method. Three different values of *k* (1, 3, and 5) were used, and for each *k* different numbers of reference examples were provided as in the above decision tree experiments. The *k*-NN-1, *k*-NN-3, and *k*-NN-5 in Table 1 are the results. As can be seen, in our user modeling cases, the decision tree learning outperforms the *k*-nearest neighbor method.

Table 1. A comparison of the decision tree and *k*-nearest neighbor methods

# Training Examples	60	80	100	120
decision tree	73.8	74.7	76.1	77.3
k-NN ($k = 1$)	67.5	67.2	69.3	68.5
k-NN ($k = 3$)	67.2	67.4	70.4	69.4
k-NN ($k = 5$)	66.7	66.8	70.4	68.8

In addition to the representation of plot keywords, we also used the combination of keyword, genre, director, and cast to represent a film for further performance comparison. The results show that with the hybrid representation, the accuracy of prediction can be slightly improved from 77.3% to 78.5% in average. Though there seems to be no significant difference in prediction between the above two representations, we have noticed that the hybrid representations can give more consistent results for different experimental runs. The reason could be that users normally have their own ways to choose films, for example some tend to focus on the plots, but others on the directors. Therefore using the hybrid of different features can mediate various cases and then obtain more stable results.

To investigate the effect of adaptive resampling for multiple trees, we conducted a series of experiments to observe their corresponding efficiency. In the experiments, ten trees were used, and here one additional copy of each misclassified case was added to the original example pool so that its corresponding probability to be chosen as a training example for the next iteration was increased. A set of experiments was conducted in which the hybrid representation was used. The results show that the resampling technique was able to improve the prediction accuracy from 78.5% to 81.1% in average.

The above decision tree approach was also employed to learn classifiers for news categorization. As is analyzed, each classifier here is developed for a specific category of news; it is to predict whether or not a user is interested in a certain piece of newly available news assigned to this category by the news Web site. Four sets of experiments were performed for four categories of news, including business, politics, sport, and U.S. domestic, as expository examples. For each category, different numbers of training cases collected by the new agent were used to construct classifiers. Table 2 shows the prediction performance of 40 test cases. The result is similar to those of decision tree-based text mining work [11]. It shows the feasibility of this approach in news classifications. As the user's models for films and news recommendations are developed by the same approach, the same technique of adaptive resampling can also be applied to news classifications.

4.3 Evaluation of the Adaptation Agent

To evaluate the adaptability of our system, we conduct experiments to examine how the system can adapt to the real world situations. In the case that the recommendation performance declines (i.e., the prediction fault occurring), the user's preferences

Table 2. The results for news categorizations

# Training Examples	60	80	100	120
business news	76.1	77.2	78.6	80.4
politics news	80.5	81.7	81.1	81.8
sport news	79.4	81.4	81.1	81.6
U.S. domestic news	75.0	74.7	77.5	79.3

are re-learnt and his personal model is updated. Here, the strategy is that the case misclassified is used to substitute the training example firstly added to the user's profile. Figure 8(a) shows a simulation example in which a stream of new cases (i.e., film programmes) were used to continuously test the most recent user model. It indicates how the performance of the system changes during the 90 consecutive recommendations for a certain user. In this figure each data point in the upper curve indicates the number of correct cases accumulated from the most recent ten recommendations (representing the first ten programmes recommended to the user), and the gray data point at the lower part of the figure, the correctness of each single prediction. As can be seen, with the mechanism of adaptation the system can keep its performance at a certain level. In this example, the user suddenly changed his preferences into completely different ones at the twentieth step (it was simulated by exchanging the positive and negative cases) and it caused the decline of system performance seriously. To deal with each fault, the adaptation agent resumed the learning agent to build a new model. The new model did not improve the system performance immediately because the number of cases representing the new preferences was not enough to dominate the set of learning examples; consequently the new model could not reflect the new preferences truly. After most of the original training examples were replaced by the new ones, the model learnt could then represent the user's current interests and the system performance was thus improved. Similarly, the learning agent can be resumed to correct the prediction faults for the news service. Figure 8(b) shows the typical simulation result.

5 Conclusions and Future Work

With the coming 3G systems and mobile IP, WAP and similar technologies will continue to play an important role in the development of information services. To mediate the problem of information overload along with the exponentially increasing digital information over the mobile networks, in this paper, we emphasize the importance of providing personalized information services for the mobile users. To achieve the personalization, we present a framework that mainly includes three parts: information gathering, user modeling, and system adaptation. The first part is to collect a user's personal information in different ways; the second part focuses on how the system can know the individual user by reasoning his preferences; and the third part emphasizes the system's adaptation ability in response to the user's feedback. The decision tree approach is employed in user modeling and a series of experiments are conducted to inspect the corresponding performance. Also the k-nearest neighbor method is implemented for comparison. In addition, the system is evaluated to show how it can re-build a user's model in respect to the changed interests. Two services are performed as examples, and the experimental results have shown the promise of our proposed approach.

Based on the work presented, we plan to develop further an intelligent conversion interface to incorporate voice transmission for audio information services. Also it is worthwhile to investigate how to combine the personalization techniques

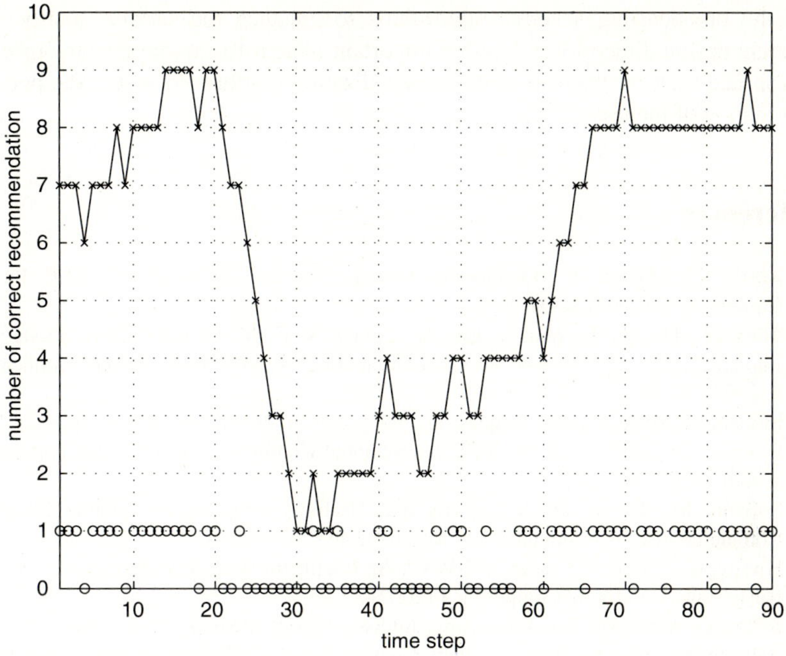

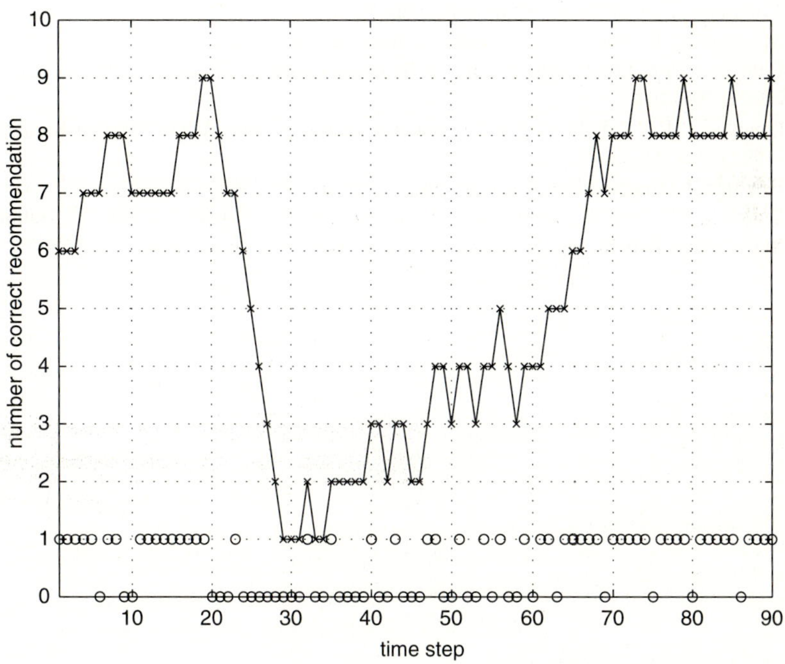

Fig. 8. The examples illustrate how the system performance changes during the consecutive recommendations. In the above figures, the *circles* indicate the correctness for each single recommendation; and the *cross marks*, the accumulated accuracy of ten consecutive recommendations

and the broadcasting schedule algorithms to enhance the service quality from different design dimensions. Another important issue is to explore how to embed an appropriate trust mechanism to the personalization platform to secure the personal information of the users.

References

1. Read, K., Maurer, F.: Developing Mobile Wireless Applications. IEEE Internet Computing. **7** (2003) 81–86
2. Törö, M., Huynh, T., Zhu, J., Liu, K., Leung, V. C. M. CORBA Based Design and Implementation of Universal Personal Computing. Mobile Networks and Applications **8** (2003) 75–86
3. Mohan, M. Voice Enabled Request and Response for Mobile Devices Supporting WAP Protocol. Proceedings of the IEEE International Conference on Vehicular Technology. (2000) 2736–2743
4. Soriano, M., Ponce, D. A Security and Usability Proposal for Mobile Electronic Commerce. IEEE Communications Magazine **40** (2002) 62–67
5. Hwang, G., Tseng, J., Huang, Y. I-WAP: An Intelligent WAP Site Management System. IEEE Trans. on Mobile Computing **1** (2002) 82–95
6. Salton, G. Automatic Text Processing. Addison-Wesley, Reading, MA. (1989)
7. Balabanovic, M., Shoham, Y. Fab: Content-Based Collaborative Recommendation. Communication of the ACM **40** (1997) 66–72
8. Mooney, R. J., Roy, L. Content-Based Book Recommending Using Learning for Text Categorization. Proceedings of the ACM International Conference on Digital Libraries (2000) 195–204
9. Quinlan, J. R. C4.5: Programs for Machine Learning. San Mateo: Morgan Kauffmann. (1993)
10. Breiman, L. Bagging Predictors. Machine Learning **24** (1996) 123–140
11. Weiss, S. M., Apte, C., Dameran, F. J., Johnson, D. E., Oles, F. J., Goetz, T., Hampp, T. Maximizing Text-Mining Performance. IEEE Intelligent Systems **14** (1999) 63–69

Detecting and Verifying Dissimilar Patterns in Unlabelled Data

Manolis Wallace[1,2,*], Phivos Mylonas[3], and Stefanos Kollias[4]

[1] National Technical University of Athens: Image, Video and Multimedia Systems Laboratory
wallace@image.ntua.gr
[2] University of Indianapolis: Athens Campus
[3] National Technical University of Athens: Image, Video and Multimedia Systems Laboratory
fmylonas@image.ntua.gr
[4] National Technical University of Athens: Image, Video and Multimedia Systems Laboratory
stefanos@cs.ntua.gr

Clustering of unlabelled data is a difficult problem with numerous applications in various fields. When input space dimensions are many, the number of distinct patterns in the data is not known a priori, and feature scales are different, then the problem becomes much harder. In this paper we deal with such a problem. Our approach is based on an extension to hierarchical clustering that makes it suitable for data sets with numerous independent features. The results of this initial clustering are refined via a reclassification step. The issue of evaluation of hierarchical clustering methods is also discussed. The performance of the proposed methodology is demonstrated through the application to a synthetic data set and verified through application to a variety of well known machine learning data sets.

1 Introduction

Clustering of data is an problem that is related to numerous scientific and applied fields [6]. Although researchers in the field of *data mining* have worked in this direction for long, and numerous related texts exist in the literature, it is still considered an open issue, as it is difficult to handle in the cases that the data is characterized by numerous measurable features. This is often referred to as the *dimensionality curse*.

Works in the field of *classification* focus in the usage of labelled (characterized) data, also known as *training data*, for the automatic generation of systems that are able to classify (characterize) future data. This classification relies on the similarity

* Corresponding author

Manolis Wallace et al.: *Detecting and Verifying Dissimilar Patterns in Unlabelled Data*, Advances in Soft Computing **1**, 247–258 (2005)
www.springerlink.com

of incoming data to the training data. The main aim is to automatically generate systems that are able to correctly classify incoming data [6].

Typically, in order to pursue such a task, one first needs to detect the *patterns* that underly in the data, and then study the way these patterns relate to meaningful classes. Even when using self-training systems, such as *resource allocating neural networks*, that are able to adapt themselves to the training data, good results may only be achieved when the patterns are known before hand, so that they may be used for proper initialization [5].

Although the tasks of classification and clustering are closely related, an important difference exists among them. While in the task of classification the most important part is the distinction between classes, i.e. the detection of class boundaries, in the task of clustering the most important part is the identification of cluster characteristics. The latter is usually tackled via the selection of cluster representatives and cluster centroids, or via the extraction of (fuzzy) rules [6].

Efficient solutions have been proposed in the literature for both tasks, for the case in which a unique similarity or dissimilarity measure is defined among input data elements [11]. When, on the other hand, multiple independent features characterize data, and thus more than one meaningful similarity or dissimilarity measures can be defined, both tasks become more difficult to handle. A common approach to the problem is the lowering of input dimensions. This may be accomplished by ignoring some of the available features (*feature selection*) [7], or by applying some space transformation [3].

In the case when input features are not independent from each other, a decrease of dimensions is very helpful. On the other hand, when input features are independent, or when the relation among them is not known a priori, which is often the case with real data, a decrease of space dimensions cannot be accomplished without loss of information. Therefore, if the relation among features is not known before hand, and the aim is to detect the patterns that exist in the data, the decrease of dimensions is not possible. Moreover, the differences in measurement scale among different features also tend to disrupt the process of clustering. The difficult problem of initial analysis of data as to properly re-scale features and select which ones to use in the process of clustering is known as *data pre-processing*.

In this work we attempt to tackle detection of patterns in multi-dimensional data that have not been pre-processed, when the count of distinct patterns in the data and the relation among input features are unknown. The proposed algorithm is an extension of agglomerative clustering and is based on a soft selection of features to consider when comparing data. The results of this initial clustering are refined via a reclassification step; this step, although unsupervised, is based on the principles of the Bayes classifier. This step also contributes to the experimental evaluation of the method's efficiency.

The structure of the paper is as follows: in Sect. 2, after a short introduction to agglomerative clustering, we present the main problems that are related to our task. In Sect. 3, we present the proposed method for initial clustering and in Sect. 4 we explain how a Bayes – based classifier can be used to refine, as well as to experimentally verify the efficiency of the algorithm. Finally, in Sect. 5, we present experimental results for the proposed algorithm and in Sect. 6, we present our concluding remarks.

2 Agglomerative Clustering and Related Problems

Most clustering methods belong to either of two general methods, partitioning and hierarchical. Partitioning methods create a crisp or fuzzy clustering of a given data set, but require the number of clusters as input. When the count of patterns that exist in a data set is not known beforehand, partitioning methods are inapplicable; an hierarchical clustering algorithm needs to be applied.

Hierarchical methods are divided into agglomerative and divisive. Of those, the first are the most widely studied and applied, as well as the most robust. Their general structure is as follows [9]:

1. Turn each input element into a singleton, i.e. into a cluster of a single element.
2. For each pair of clusters c_1, c_2 calculate a compatibility indicator $CI(c_1, c_2)$. The *CI* is also referred to as cluster similarity, or dissimilarity, measure.
3. Merge the pair of clusters that have the best *CI*. Depending on whether this is a similarity or a dissimilarity measure, the best indicator could be the maximum or the minimum operator, respectively.
4. Continue at step 2, until the termination criterion is satisfied. The termination criterion most commonly used is the definition of a threshold for the value of the best compatibility indicator.

The two key points that differentiate agglomerative methods from one another, and determine their efficiency, are the compatibility indicator and the termination criterion used. Major drawbacks of agglomerative methods are their high complexity and their susceptibility to errors in the initial steps, that propagate all the way to their final output.

The core of the above generic algorithm is the ability to define a unique compatibility indicator among any pair of clusters. Therefore, when the input space has more than one dimensions, an aggregating distance function, such as Euclidean distance, is typically used as the *CI* [14]. This, of course, is not always meaningful. Cases exist, in which the "context" can change the similarity or dissimilarity measure to be used [13]. For example, two films may be compared based on their topic, or on their directors.

In such cases, a selection of meaningful features needs to be performed, prior to calculating a *CI*. In the example of films, although two films may be similar to one another as far as their content is concerned, two other films may be similar as far as their cast is concerned. In other words, it may not be possible to select a single distance metric, which will apply in all cases, for a given data set. Moreover, one feature might be more important than others, while all of the features are useful, each one to its own degree. In other words, hard (crisp) feature selection is not always possible, either.

3 Soft Feature Selection and Clustering in Multi-Dimensional Spaces

Elements are usually grouped together based on their similarity in a single or a few features. When the total number of features is high, small distances in a small subset of them barely affect the overall distance, when an aggregation of distances in all features is used. Thus, only when the correct subset of features is considered, can elements be compared correctly [4, 10].

In this paper we tackle feature selection based on the following principle: while we expect elements of a given meaningful set to have random distances from one another according to most features, we expect them to have small distances according to the features that relate them. We rely on this difference in distribution of distance values in order to identify the *context* of a set of elements, i.e. the subspace in which the set is best defined.

More formally, let c_1 and c_2 be two clusters of elements. Let also $r_i, i \in N_F$ be the metric that compares the i-th feature, and F the overall count of features (the dimension of the input space). A distance (dissimilarity) measure between the two clusters, when considering just the i-th feature, is given by

$$f_i(c_1, c_2) = \sqrt[\kappa]{\frac{\sum_{a \in c_1, b \in c_2} r_i(a_i, b_i)^\kappa}{|c_1||c_2|}} \tag{1}$$

where e_i is the i-th feature of element e, $|c|$ is the cardinality of cluster c and $\kappa \in R$ is a constant.

The context is a soft selection of features to consider when calculating an overall distance value. We can define it as a fuzzy set x defined on N_F, with a scalar cardinality of one. Then, the overall distance between c_1 and c_2 is calculated as

$$d(c_1, c_2) = \sum_{i \in N_F} x_i(c_1, c_2)^\lambda \cdot f_i(c_1, c_2) \tag{2}$$

where x_i is the degree to which i, and therefore f_i, is included in the context, $i \in N_F$ and $\lambda \in R$ is a constant.

According to the principle presented in the beginning of this paragraph, the features that relate c_1 and c_2 are the ones that produce the smallest distances f_i. Therefore, the "correct" context can be calculated through the solution of an optimization problem, as the context that produces the smallest overall distance.

When $\lambda = 1$ the solution is trivial: the feature that produces the smallest distance is the only one selected. The degree to which it is selected is 1. If more than one features produce the best distance, then they are equally selected, as there is no information as to which should be favored.

When $\lambda \neq 1$ and $\exists i \in N_F : f_i(c_1, c_2) = 0$, then the features for which $f_i(c_1, c_2) = 0$ are the ones the are (equally) selected.

When $\lambda \neq 1$ and $f_i(c_1, c_2) \neq 0 \forall i \in N_F$, then the optimization problem is not trivial and has to be solved. According to the following lemma it can be solved analytically, which means that the optimization problem does not affect the algorithmic complexity of the process:

Lemma 1. *When $\lambda \neq 1$ and $f_i(c_1,c_2) \neq 0 \forall i \in N_F$, then the best context x, and equivalently the best compatibility indicator CI, is given by:*

$$x_F(c_1,c_2) = \frac{1}{\sum_{i \in N_F} \left[\frac{f_F(c_1,c_2)}{f_i(c_1,c_2)}\right]^{\frac{1}{\lambda-1}}} \tag{3}$$

$$x_i(c_1,c_2) = x_F(c_1,c_2) \cdot \left[\frac{f_F(c_1,c_2)}{f_i(c_1,c_2)}\right]^{\frac{1}{\lambda-1}} \tag{4}$$

Proof. We have demanded that the scalar cardinality of the context is one; the optimization problem we have to tackle is constrained. $|x| = 1$ is equivalent to $\sum_{i \in N_F} x_i = 1$. Thus, replacing

$$x_F = 1 - \sum_{i \in N_{F-1}} x_i \tag{5}$$

the minimization of

$$d(c_1,c_2) = x_F(c_1,c_2)^{\lambda} \cdot f_F(c_1,c_2) + \sum_{i \in N_{F-1}} x_i(c_1,c_2)^{\lambda} \cdot f_i(c_1,c_2)$$

is reduced to an unconstrained optimization problem. From (5) we have

$$\frac{\partial x_F(c_1,c_2)}{\partial x_i(c_1,c_2)} = -1 \forall i \in N_{F-1}$$

and thus

$$\frac{\partial \{x_F(c_1,c_2)^{\lambda} \cdot f_F(c_1,c_2)\}}{\partial x_i(c_1,c_2)} = -\lambda \cdot x_F(c_1,c_2)^{\lambda-1} \forall i \in N_{F-1}$$

Easily now, demanding that

$$\frac{\partial d(c_1,c_2)}{\partial x_i(c_1,c_2)} = 0 \forall i \in N_F$$

we have

$$x_i(c_1,c_2) = x_F(c_1,c_2) \cdot \left[\frac{f_N(c_1,c_2)}{f_i(c_1,c_2)}\right]^{1-\lambda} \forall i \in N_{F-1} \tag{6}$$

Combining (6) with (5) we also have

$$x_F(c_1,c_2) = \frac{1}{\sum_{i \in N_F} \left[\frac{f_F(c_1,c_2)}{f_i(c_1,c_2)}\right]^{\frac{1}{\lambda-1}}}$$

□

As λ increases, pairs of clusters that are related by fewer features, and thus have greater values in their contexts, are obviously assigned smaller distances. In order for distances to be usable as compatibility indicators, they need to be unaffected by cluster direction in comparison to the axes. Thus, it is imperative that they are transformed, as to become directly comparable to each other. The following, adjusted, compatibility indicator is used:

$$CI(c_1,c_2) = \frac{d(c_1,c_2)}{x_\lambda(c_1,c_2)} \tag{7}$$

$$x_\lambda(c_1,c_2) = \sum_{i \in N_F} [x_i(c_1,c_2)]^\lambda \tag{8}$$

When features are quantized to a small set of levels, as is often the case with digital data, cases for which $f_i(c_1,c_2) = 0$ are not rare. Especially in the first steps of agglomerative clustering, when clusters are of small size, the best CIs are almost always zero. Since, as we have already mentioned, errors in the initial steps of agglomerative clustering propagate all the way to the final output, it is important to always make the best selection possible for the pair of clusters to merge. Therefore, especially for the case of CIs that are equal to zero, we introduce one more criterion: out of all the pairs for which $CI = 0$, the one that has zero distances for the most features will be selected. In other words, out of all the pairs of similar clusters, the ones that are similar according to the greatest number of features are selected.

As far as the termination criterion is concerned, a threshold on the value of CI can be used, as Lemma 2 guarantees that this is meaningful.

Lemma 2. *The above mentioned CI is non decreasing as we move from one step to the next.*

Proof. Proof is trivial and is omitted for the sake of space. □

This way, the algorithm gradually groups elements together, based on their similarities; for each cluster, a different fuzzy subset of features may be considered for the calculation of similarities. This soft feature selection may also be perceived of as a re-scaling of features, thus making up for the skipped step of data pre-processing.

The average values of features for each cluster form the centroid, i.e. a "virtual" element that is located at the center of the cluster, when all of its elements are placed in the F-dimensional space. Its position may be considered as a description of the feature values of the pattern that this cluster corresponds to. The variances of the values for each feature indicate the importance of each feature for the definition of the cluster; this may be perceived of as an estimation of the radius of the cluster in the direction of each feature.

Assuming that the clustering has produced meaningful groups of elements, the latter may be used for the initialization of an adaptive neural classifier; the fact that clusters are described through center and variance combinations makes the output ideal for the initialization of RBF based networks [12].

4 Refinement and Verification through Bayesian Classification

As stated in Sect. 1, the primary aim of clustering algorithms is not to correctly classify data, but rather to identify the patterns that underly in it. Therefore, "wrong" elements in clusters may be acceptable, as long as the overall cluster correctly describes an existing and meaningful pattern. This implies that feeding labelled data to the algorithm and measuring the classification rate may not be enough to evaluate the actual efficiency of the algorithm.

In order for a clustering algorithm to be truly evaluated, the patterns that are described by the detected clusters need to be extracted and examined. In this work we examine whether detected patterns are meaningful by evaluating a classifier that is created by using them. Out of the numerous classification schemes that exist in the literature we have chosen to work with the Bayesian classifier, although others could have been chosen as well [8].

Specifically, each cluster is considered to describe a distinct class. Furthermore, we assume that all features of members of a class follow a gaussian distribution. Thus, using the centroid and standard deviations of each cluster, we may design the mixture of Gaussians that describe the class. The Bayes classification scheme calculates for each input element a the probabilities $P(p_i/a), i \in N_T$, where T is the count of detected patterns, and classifies a to the pattern for which it has the greatest probability. Probabilities are easily computed by applying the transformation:

$$P(p_i/a) = P(a/p_i)P(p_i)$$

where $P(p_i)$ is equal to the relative cardinality of cluster c_i, i.e.

$$P(p_i) = \frac{|c_i|}{\sum_{j \in N_P} |c_j|}$$

and $P(a/p_i)$ is given as the value of a in the mixture of Gaussians that describe pattern i, i.e.

$$P(a/p_i) = \prod_{j \in N_F} \frac{1}{\sqrt{2\pi} s_{ij}} e^{-\left(\frac{a_j - m_{ij}}{2 s_{ij}}\right)^2}$$

where m_{ij} and s_{ij} are the centroid value and standard deviation for the j-th feature of pattern i and a_j is the j-th feature of element a.

Using this scheme, we may reclassify all data that were used for clustering. If the clustering was successful, i.e. if the detected patterns are meaningful, then this process will refine the classification rate by removing some of the clusters' members that were a result of errors in the initial steps. Thus, this process offers an indication of the clustering's true performance. Moreover, it makes the overall algorithm more robust, as opposed to simple hierarchical clustering, as it is more resilient to errors in the initial steps.

5 Experimental Results

In this section we list some indicative experimental results of the proposed methodology. In Subsect. 5.1 we provide an example of application to a simple synthetic data set, which facilitates the visualization of the algorithm's performance.

Continuing, in Subsect. 5.2 we list results from application to real data sets from the machine learning databases.

5.1 Synthetic Data

In order for the visualization of the synthetic data set to be feasible, we have limited it to two dimensions. Three classes of data were created, using a gaussian random generator. The parameters of the gaussian distributions used for the generation of the data set are presented in Table 1.

Table 1. The parameters for the generation of the synthetic data set

Class	m_1	s_1	m_2	s_2	Elements
A	2	0.5	1	0.1	100
B	1	0.9	3	0.1	100
C	1	0.1	2	0.7	100

The synthetic data set.

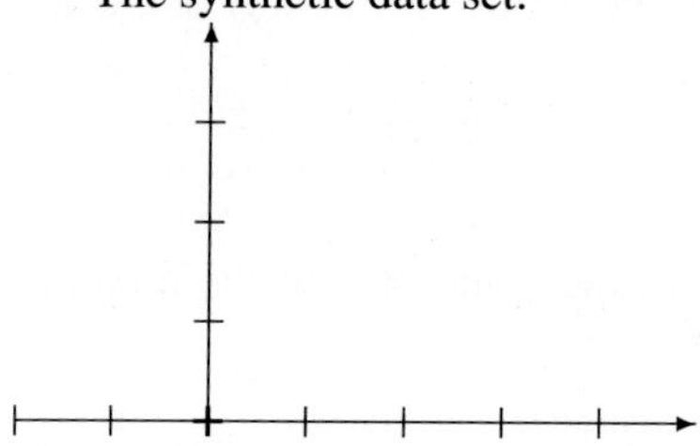

As can be seen from the Table, as well as from the diagram, the three classes are not clearly distinguished from each other, and the subspaces that best characterize each class differ to a great extent. This makes distance aggregation – based approaches inefficient; this is verified in Table 3. The initial classification step produces a classification rate of 91% (assigning each cluster to the class that dominates it), and the reclassification refines this to 95.7%, indicating that the initial step, although has a smaller rate, has correctly identified the underlying patterns.

5.2 Real Data

Iris Data

The iris data set contains 150 elements, characterized by 4 features, that belong to three classes; two of these classes are not linearly separable from each other. This is a relatively easy data set, as the number of clusters in the data is equal to the

Table 2. The clusters produced, for the synthetic data set. Format: (class 1, class 2, class 3)

Method	Cluster 1	Cluster 2	Cluster 3	Classification Rate
Euclidian clus.	(0, 0, 13)	(28, 0, 87)	(72, 100, 0)	66.7%
Initial clus.	(4, 0, 86)	(92, 5, 8)	(4, 95, 6)	91%
Bayesian reclass.	(6, 0, 98)	(94, 5, 2)	(0, 95, 0)	95.7%

Table 3. Classification rates for iris data ($\kappa = \lambda = 2$)

Method	Cluster 1	Cluster 2	Cluster 3	Classification Rate
Initial clus.	(36, 4, 12)	(13, 0, 38)	(1, 46, 0)	80%
Bayesian reclass. 1	(33, 0, 2)	(17, 0, 48)	(0, 50, 0)	87.3%
Bayesian reclass. 2	(35, 0, 0)	(15, 0, 48)	(0, 50, 0)	90%

number of classes. The labels of the elements were not used during clustering and reclassification; there were used, though, for evaluation purposes. Results are shown in Tables 2 and 3.

The considerable refinement that a single step of Bayesian reclassification offers is indicative of the validity of the detected clusters. This observation is supported even more by the fact that recursive application of the reclassification step refines even more the results, even though this step is unsupervised (it does not use element labels).

Wisconsin Breast Cancer Database

The Wisconsin breast cancer database contains 699 elements, which are characterized by the following attributes: clump thickness, uniformity of cell size, uniformity of cell shape, marginal adhesion, single epithelial cell size, bare nuclei, bland chromatin, normal nucleoli, mitoses. All these attributes assume integer values in $[1, 10]$. Elements are also accompanied by an id, and class information; possible classes are benign and malignant. 65.5% of the elements belong to the benign class and 34.5% to the malignant class. 16 elements are incomplete (an attribute is missing) and have been excluded from the database for the application of our algorithm.

This data set, having a greater number of features, is considered to be more difficult than the iris data set. Detailed results acquired using the proposed methodology are available in Tables 4 and 5. It is worth noting that, although the classification rate of the initial clustering procedure are not extremely high, the reclassification step refines it considerably. Furthermore, for the case where $\kappa = \lambda = 5$, the reclassification step classifies every element to one of exactly two clusters, each one almost totally dominated by one class.

This performance is not far from that of trained classification systems that utilize the same dataset; a classification rate of 97% is reported in [2]. This is indicative of

The EMBASSI system [2] was born by a very big consortium occupied in defining the technologies and their ergonomics requirement to implement an intelligent shop assistant to facilitate user purchasing and information retrieval. These objectives are pursued by multi-modal interaction: common text dialogs as well as speech recognition devices are used to sense user requests whilst a 3D face is the front-end of the system. The agent is able to send its response also via classical multimedia content (video-clips, hyperlinks, etc.).

In [3] an agent is described which is not just able to be animated but also to answer to users based on emotions modeled on the Five Factor Model (FFM) of personality [4] and implemented using Bayesian Belief Networks. Moreover the Alice chatterbot is used in order to let the web agent process and generate responses into the classical textual form.

We consider "creative" those literary systems that generate composition characterized by a certain degree of *novelty* and *quality*, as in [5]. Here we present two examples of these systems, both dealing with poetic production: Poevolve and McGonagall.

The system Poevolve [6] takes its name from "poetry" and "evolve", for is an attempt to generate limericks using evolutionary algorithms. In this working prototype the user can set some parameters for the genetic algorithm which guides the evolution of a population of limericks with restricted lexicon.

Quality is assured by a recurrent neural network trained on the ratings done by 160 students. This network assigns fitness values to individuals of the population of limericks. A the same time, novelty is assured by genetic operators: mutation and crossover.

Another system is McGonagall, developed by Manurung [7], which uses evolutionary techniques. Poetry generation is considered as an evolutionary exploration of all possible texts.

Genetic operators work at different representation levels, while fitness takes into account semantics, rhetorics, and phonetics. In such a way, the system features an integrated architecture, because of the interaction between text planning and surface realization.

Manurung motivates this architectural design recalling that poetry combines form and content.

Unfortunately, it was impossible to evaluate experimental results of these two systems, due to the lack of examples.

The remainder of our paper is organised as follows: Section 2 gives an overview of the system, Sect. 3 details the client side of the UPoet application whilst Sect. 4 explains the server side. Section 5 shows our experimental results while future work is drawn out in Sect. 6.

2 The UPoet System Architecture

Our UPoet application was born integrating the **YODA** generative module with a part of the S.A.M.I.R. **(Scenographic Agents Mimic Intelligent Reasoning)** [8]

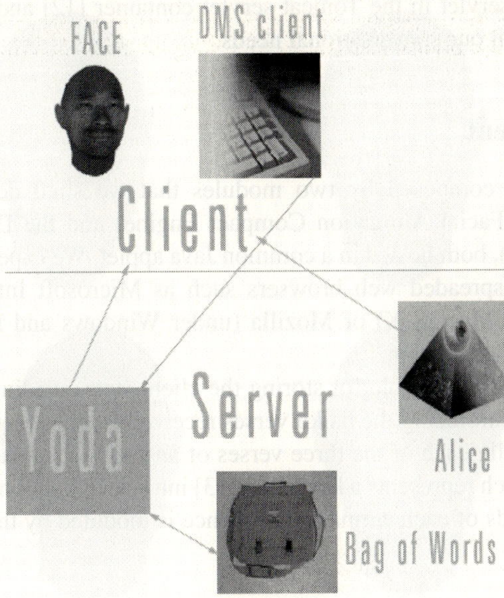

Fig. 1. The UPoet Architecture

system: A 3D intelligent character, resembling the appereance of a zen monk, acts as an animated front-end to the haiku, some sort of minimalist japanese poetic compositions [9], generated by the YODA module. This original virtual poet can be embedded into any kind of web applications or in a stand-alone fashion, just for the sake of entertainment.

A general picture of the application is given in Fig. 1: It is split between a client, a Java applet, enclosing the animated 3D face and a minimal interface to let users chat with the system (the DMS client) and a small HTTP layer to ensure synchronized communication with the server. Basically the user can chat with UPoet, receiving visual (i.e. animation of the 3D face) and textual responses but he can also request the generation of a haiku: At this time a message is issued to the server which receives the user request about the subject of the haiku. The server will try, using the YODA module, to generate a context dependant haiku: It means that the words and the meaning of the composition will be as related as possible to the suggestions given by its user.

The server is composed not only by the YODA module but also by a small set of Java classes acting as a filter on the suggestions given by the user, removing not so useful words for the generation phase (i.e. articles, conjunctions etc.). A list of these *useless* words is mantained in an apposite bag-of-words [10].

The Alice chatterbot [11], used to give responses to users when not in text-generation mode, completes the server side set of modules and represents the core of the server side part of the Dialogue Management System. Alice was slightly modified

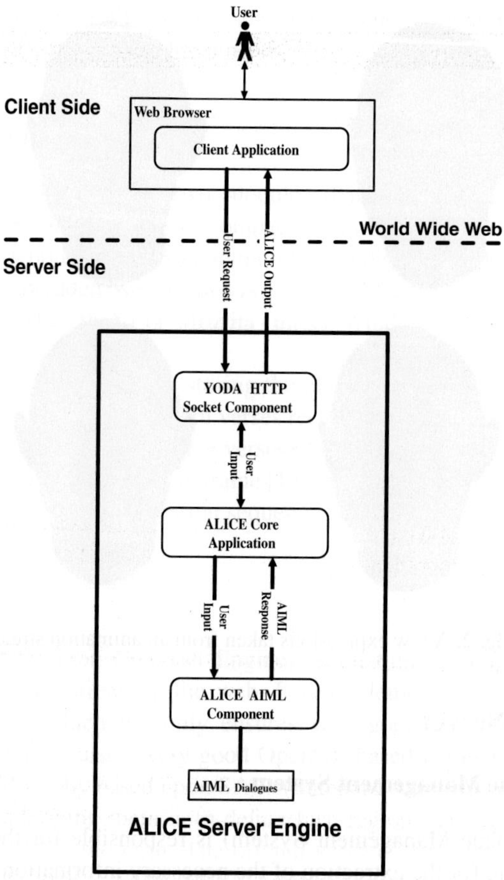

Fig. 3. The DMS Architecture

```
<pattern>I WOULD NOT *</pattern>
<template> No one says you have to. </template>
</category>
```

The <category> tag indicates an AIML category, the basic unit of chat robot knowledge.

The category has a <pattern> and a <template>. The pattern in the first example is the wildcard symbol "*" that matches any input. In general the pattern is a combination of text and wildcards as in the second example ("I WOULD NOT *"). The template is the bot's answer associated to the user input detected by the pattern. Just the text "Hello!" in the first example or "No one says you have to." in the second example.

ALICE has been fully integrated in UPoet as a Java Servlet and all the knowledge of the system has been stored in the AIML files, containing all the patterns matching user input.

Dialogues data are exchanged through simple built-in classes handling the classical HTTP sockets communication.

4.2 YODA

YODA is a prototype of creative literary system, developed in [16]. Technically, YODA is a natural language generation system, which is also creative and interactive, as we shall explain later.

YODA is developed in such a way that the process of generation should proceed as follows:

- *text planning*, i.e., selection of knowledge and output structuring;
- *surface realization*, i.e., conversion of information blocks in grammatically (well-formed) sentences.

Text planning is dependent on the kind of our desired output, while surface realization depends mainly on the language to use; in our case, the English language.

To demonstrate YODA features, we chose to generate *haiku*. This short poetic form originated in Japan during 17th century, as starting verses of a collective composition, *renga*: usually, the first verses of renga contain an indication of the period in which the renga is composed. These first verses split to become haiku; the indication of the period is given in term of natural events, pertaining a specific season.

In Japanese language a haiku is made up of 17 *onji*, "phonetical symbols" which have no exact equivalent in English language. Currently is being used a transposition which maps one onji to one English syllable. These 17 syllables are distributed on 3 verses, respectively of 5, 7, and 5 syllables.

YODA is interactive in the sense that the user gives suggestions to generate the haiku. Suggestions are given in terms of words, related or not to seasons.

YODA is creative in the sense that elaborates "creatively" these suggestions, in order to select a set of words to use in the haiku. This set is semantically related to the suggestions. The aim is to make the user feel that YODA has taken into account his suggestions.

4.3 YODA Architecture

YODA is constituted by several components: the `User Interface`, the `Surface Realizer`, the `Text Planner`, the library `JWNL 1.2.1`, and the lexical resource `WordNet 1.7.1` as shown in Fig. 4.

The component `WordNet 1.7.1 database` is a lexical database of English language. A short description which can be found on the main site of WordNet [17].

The great advantage of WordNet over other lexical databases and dictionaries is the possibility to go across large semantic networks. From the standpoint of a natural language generation system, this fact implies that the system has access to a resource that's not "flat", like an ordinary browsable dictionary.

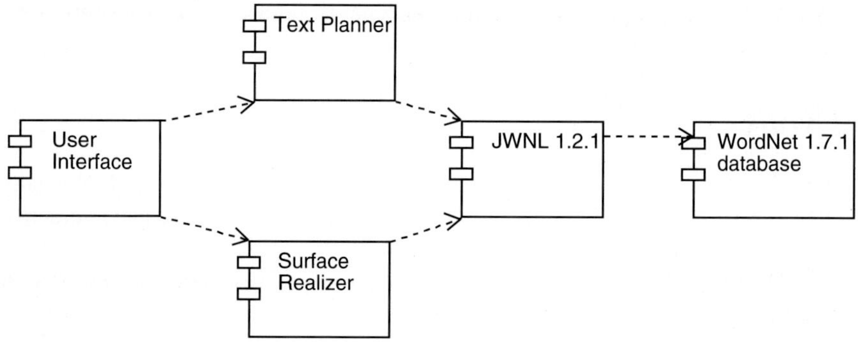

Fig. 4. YODA Architecture

The fundamental unit of WordNet is the *concept*. Every concept has associated lemmas, which constitute all together a *synonyms set*, a.k.a. *synset*. Pointers between synset implement semantic relationships between concepts. This organization leads to hierarchical structures which are navigable following pointers. For example, nouns are structured in trees, while adjectives in clusters.

YODA is written in Java, so it is necessary the use of a library to access WordNet from Java code. YODA uses the component `JWNL 1.2.1`, i.e. the *Java WordNet Library 1.2.1* developed as a SourceForge project, available at [18].

This library gives access to all the informations of WordNet, but in a format that is too verbose for a natural language generation system like YODA. The library is wrapped, to get a format more suitable for our purposes.

Moreover, JWNL gives the possibility to serialize the entire WordNet dictionary. Doing so, during the startup phase YODA loads the dictionary and every access is done in memory, rather than from disk. In a client-server context, like ours, the dictionary is loaded just one time on the server, shortening users waiting.

The component called `User Interface` makes possible the user to insert suggestions. Here "user" stands for a human or a system giving suggestions. In the stand-alone version of YODA, the (human) user inserts suggestions at the prompt, while in UPoet the user (from the point of view of YODA) is the client application.

The `Text Planner` concerns the semantics. Instead of generating text with random (or pseudo-random) words, like many other systems do, YODA chooses the words to use following the user's suggestions.

Suggestions are not isolated words, but linked to other nodes in a semantic network. The heuristics guiding the process of text planning exploit the structure of WordNet.

The `Surface Realizer` takes care of the morphological and syntactical aspects of text generation. YODA is able to get the stem of nouns and verbs. Moreover, it can declinate nouns and conjugate verbs, in order to obtain agreement during the process of generation of the text. In fact, text is generated starting from single sentences. Each sentence is generated independently from others, using words chosen by the `Text`

`Planner` component. During the process of haiku generation, syllable counting is performed.

For the syntactic aspect, YODA uses a CFG modeled upon *sentence frames*. In WordNet, every verb has one or more corresponding *sentence frame*, each indicating the correct use of the verb (e.g., if the verb is transitive or intransitive). This information makes possible the generation of well-formed sentences.

To summarize the process of haiku generation, we briefly explain the steps involved:

- the user inserts one or more words, taken by YODA as suggestions;
- for each term of this set, it is determined the suitable part(s)-of-speech;
- from this set are obtained (via WordNet) the terms effectively being used for the generation: rather than use terms inserted by the user, YODA selects (for example) synonyms; if a term is related to a particular season, then selection is guided by the appropriate season;
- if the number of words found is not yet enough for generation, then some other words are chosen randomly;
- eventually, the surface realizer generates sentences that take into account user's suggestions, while fitting haiku structure.

4.4 Haiku and Facial Expressions

As Higginson states [9], a haiku is a way to share feelings: UPoet shares emotions with the user by means of haiku and facial expressions.

The 3D agent changes its facial expression according to the content of the haiku generated by YODA. For the realization of this feature, we follow the theory of Ekman and Friesen [19] on facial expression. This theory claims that every facial expression is given by a combination of six fundamental expressions, corresponding to six emotions: anger, disgust, fear, happiness, sadness, and surprise.

UPoet manages the mapping between text and expression our 3D facial animation engine (F.A.C.E.) and our natural language generation system (YODA), synchronizing text and expression on the client-side of the application.

The mechanism used for the mapping from words to expression is based on a metric distance defined in terms of semantic relationships in WordNet. Distances are calculated between words used in the haiku and words corresponding to the six emotions. Then, for every verse of the haiku, YODA passes text and corresponding facial expression.

YODA receives suggestions from the user via a filtering mechanism based on a *bag of words*. This bag contains words that should not be accepted as suggestions by YODA: articles, conjunctions, determiners, pronouns, prepositions, and other words. Every other word which passes this filter is passed to YODA as a suggestion.

3. S. Kshirsagar and N. Magnenat-Thalmann. Virtual humans personified. In *Proceedings of the Autonomous Agents Conference (AAMAS)*. Bologna, 2002.
4. R. R. McCrae and O. P. John. An introduction to the five factor model and its applications. *J. of Personality 60*, pp. 175–215, 1992.
5. G. Ritchie. Assessing creativity. Technical Report EDI-INF-RR-0039, School of Informatics, University of Edinburgh, 2001.
6. R. P. Levy. A computational model of poetic creativity with neural network as measure of adaptive fitness. In *Proceedings of the Workshop Program at the Fourth International Conference on Case-Based Reasoning 2001*, 2001.
7. H. M. Manurung. Mcgonagall: The evolutionary poet. 1st prize, Poster competition, the Informatics Jamboree, 23–25 May 2001, 2001.
8. G. Zambetta, F. Catucci. Designing not-so-dull virtual dolls. In *Soft Computing Systems. Design, Management and Applications*, volume 87 of *Frontiers in Artificial Intelligence and Applications*, pp. 513–518, Amsterdam, 2002. IOS Press.
9. W. J. Higginson. *The Haiku Handbook*. Kodansha International, Tokyo, 1989.
10. Fabrizio Sebastiani. Machine learning in automated text categorization. *ACM Computing Surveys*, pp. 1–47, 2002.
11. ALICE AI Foundation. http://www.alicebot.org.
12. The Tomcat Servlet Container. http://jakarta.apache.org/tomcat/index.html.
13. B. Fleming and D. Dobbs. *Animating Facial Features and Expressions*. Charles River Media, Hingham, 1998.
14. The Shout3D API. http://www.eyematic.com/products_shout3d.html.
15. MPEG-4 standard specification. http://mpeg.telecomitalialab.com/standards/mpeg-4/mpeg-4.htm.
16. M. Lamarca. YODA: un prototipo di sistema creativo letterario. Graduation thesis, Università degli Studi di Bari, Dipartimento di Informatica, 2003.
17. The WordNet Website. http://www.cogsci.princeton.edu/~wn.
18. The Java WordNet Library. http://jwordnet.sourceforge.net.
19. P. Ekman and V. W. Friesen. *Facial Action Coding System*. Consulting Psychologists Press, Palo Alto, CA, 1978.
20. R. S. Sutton and A. G. Barto. *Reinforcement learning: an introduction*. MIT Press, 1998.
21. M. Mitchell. *An introduction to genetic algorithms*. MIT Press, 1996.

A Study on the Combination of Evolutionary Algorithms and Stratified Strategies for Training Set Selection in Data Mining

Jose Ramon Cano[1], Francisco Herrera[2] and Manuel Lozano[2]

[1] Dept. of Computer Science, Escuela Politécnica Superior of Linares, University of Jaen, 23700, Linares, Jaen, Spain
jrcano@ujaen.es
[2] Dept. of Computer Science and Artificial Intelligence, University of Granada, 18071, Granada, Spain
herrera@decsai.ugr.es, lozano@decsai.ugr.es

Evolutionary algorithms are adaptive methods based on natural evolution that may be used for search and optimization. As Training Set Selection can be viewed as a search problem, it could be solved using evolutionary algorithms.

In this paper, we have carried out an empirical study of the performance of *CHC* as representative evolutionary algorithm model. This study includes a comparison between this algorithm and other non-evolutionary instance selection algorithms applied in large size data sets for Training Set Selection. The results show that the stratified evolutionary instance selection algorithms consistently outperform the non-evolutionary ones, the main advantages being: better instance reduction rates, higher classification accuracy and models that are easier to interpret.

1 Introduction

Advances in digital and computer technology that have led to the huge expansion of the Internet means that massive amounts of information and collection of data have to be processed. Due to the enormous amounts of data, much of the current research is based on scaling up [15] Data Mining (*DM*) [1, 2, 3] algorithms. Other research has also tackled scaling down data. The main problem of scaling down data is how to select the relevant data and then apply a *DM* algorithm [1]. This task is carried out in the data preprocessing phase in a Knowledge Discovery in Databases (*KDD*) process.

Our attention is focused on Data Reduction (*DR*) [9], which can be achieved in many ways: by selecting features [4], by making the feature-values discrete [11], and by selecting instances [12, 16, 17, 18]. We led our study to instance selection (*IS*) as *DR* mechanism , where we reduce the number of rows in a data set (each

Jose Ramon Cano et al.: *A Study on the Combination of Evolutionary Algorithms and Stratified Strategies for Training Set Selection in Data Mining*, Advances in Soft Computing **1**, 271–284 (2005)
www.springerlink.com

row represents and instance). *IS* can follow different strategies: sampling, boosting, prototype selection (*PS*), and active learning. We are going to study the *IS* from the *PS* perspective.

Instance selection mechanisms have been proposed to choose the most suitable points in the data set to become instances for the training data set used by a learning algorithm. This is our Training Set Selection (*TSS*) problem. *TSS* has been studied previously in the literature, for example, in [10], a Genetic Algorithm [5] (*GA*) is used for training data selection in radial based function networks.

Evolutionary Algorithms (*EAs*) [6, 7] are general-purpose search algorithms that use principles inspired by natural genetic populations to evolve solutions to problems. The basic idea is to maintain a population of chromosomes, which represent plausible solutions to the problem, which evolves over time through a process of competition and controlled variation. *EAs* have been used to solve the *IS* problem, with promising results [10, 19].

As with any algorithm, the issue of scalability and the effect of increasing the size of data on algorithm behaviour are always present. This scaling up drawback appears in *EAs* in the chromosome's size, which reduces the *EAs* convergence capabilities.

To avoid this drawback we offer a combination of *EAs* and the stratified strategy. In large and huge size we can't evaluate the algorithms over the complete data set so the stratification is the only way to carry out the executions. Combining the subset selected per strata we can obtain the subset selected for the whole initial data set. The stratification reduces the data set size, while *EAs* select the best local *TSS*.

The aim of this paper is to study the application of a representative and efficient *EA* model for data reduction (*CHC* [13]) in *TSS*, and to compare it with non-evolutionary instance selection algorithms (hereafter referred to as classical ones) following a stratified strategy.

To address this, we have carried out a number of experiments with increasing complexity and size of data.

In order to do this, this paper is set out as follows. In Sect. 2, we introduce the main ideas about *IS*, describing the process which *IS* algorithms take part (*TSS*), and we also summarise the classical *IS* algorithms used in this study. In Sect. 3, we introduce the foundations of *EAs* and summarise the main features of them, giving details of how *EAs* can be applied to the *TSS* problem in large size data sets. In Sect. 4, we explain the methodology used in the experiments. Sect. 5 deals with the results and the analysis of large and huge data sets. Finally, in Sect. 6, we reach our conclusion.

2 Instance Selection on Data Reduction

In this section we describe the strategy which *IS* takes part in, as a *DR* mechanism, as well as a summary of classical *IS* algorithms.

2.1 Instance Selection

In *IS* we want to isolate the smallest set of instances which enable us to predict the class of a query instance with the same (or higher) accuracy as the original set [9]. By reducing the "useful" data set size, which can reduce both space and time complexities of subsequent processing phases. One can also hope to reduce the size of formulas obtained by a subsequent induction algorithm on the reduced and less noise data sets. This may facilitate interpretation tasks.

IS raises the problem of defining relevance for a prototype subset. From the statistical viewpoint, relevance can be partly understood as the contribution to the overall accuracy, that would be e.g. obtained by a subsequent induction. We emphasize that removing instances does not necessarily lead to a degradation of the results: we have observed experimentally that a little number of instances can have performances comparable to those of the whole sample, and sometimes higher. Two reasons come to mind to explain such an observation. First, some noises or repetitions in data could be deleted by removing instances. Second, each instance can be viewed as a supplementary degree of freedom. If we reduce the number of instances, we can sometimes avoid over-fitting situations.

2.2 Instance Selection for Training Set Selection

There may be situations in which there is too much data and this data in most cases is not equally useful in the training phase of a learning algorithm. Instance selection mechanisms have been proposed to choose the most suitable points in the data set to become instances for the training data set used by a learning algorithm. For example, in [10], a Genetic Algorithm (*GA*) [5] is used for training data selection in radial based function networks.

Figure 1 shows a general framework for the application of an *IS* algorithm for *TSS*. Starting from the data set, *TR*, the *IS* algorithm finds a suitable set, *SS*, then a

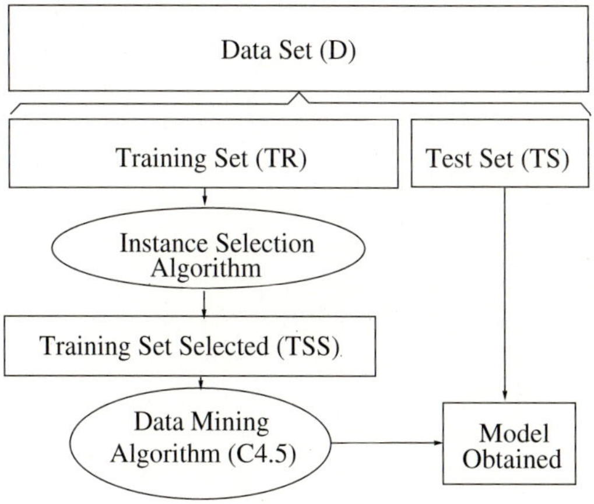

Fig. 1. *TSS* Strategy

learning or *DM* algorithm is applied to evaluate each subset selected ($C4.5$ [8] in our case) to obtain a model from the data set. This model is assessed using the test data set, TS.

2.3 Overview of Instance Selection Algorithms

Historically, *IS* has been mainly aimed at improving the efficiency of the Nearest Neighbour (*NN*) classifier. The *NN* algorithm is one of the most venerable algorithms in machine learning. This algorithm calculates the Euclidean distance (possibly weighted) between an instance to be classified and each training-neighbouring instance. The new instance obtained is assigned to the class of the nearest neighbouring one. More generally, the k-nearest neighbours (*k-NN*) are computed, and the new instance is assigned to the most frequent class among these k neighbours. The *k-NN* classifier was also widely used and encouraged by early theoretical results related to its Bayes error generalization.

However, from a practical point of view, the *k-NN* algorithm is not suitable for dealing with very large sets of data due to the storage requirements it demands and the computational costs involved. In fact, this approach requires the storage of all the instances in memory. Early research in instance selection firstly tried to reduce storage size. Taking as reference our study in [20] we select the most effective classic algorithms to evaluate them.

The algorithms used in this study will be:

Methods based on Nearest Neighbour Rules

- *Cnn* [21] – It tries to find a consistent subset, which correctly classifies all of the remaining points in the sample set. However, this algorithm will not find a minimal consistent subset.
- *Ib*2 [12] – It is similar to *Cnn* but using a different selection strategy.
- *Ib*3 [12] – It outperforms *Ib*2 introducing the acceptable instance concept to carry out the selection.
- Icf [16] – It tries to select the instances which classify more prototypes correctly. Icf uses coverage and reachable concepts to carry out the selection.

Methods Based on Ordered Removal

- *Drop*1 [14] – Essentially, this rule tests to see if removing an instance would degrade leave-one-out cross-validation generalization accuracy, which is an estimate of the true generalization ability of the resulting classifier.
- *Drop*2 [14] – *Drop*2 changes the order of removal of instances. It initially sorts the instances in TR by the distance to their nearest enemy (nearest instance belonging to another class). Instances are then checked for removal beginning at the instance furthest from its nearest enemy. This tends to remove instances furthest from the decision boundary first, which in turn increases the chance of retaining border points.

- *Drop*3 [14] – *Drop*3 uses a noise filtering pass before sorting the instances in *TR*. This is done using the rule: Any instance not classified by its *k*-nearest neighbours is removed.

3 Evolutionary Instance Selection Algorithms

Most of the success of *EAs* is due to their ability to exploit the information accumulated about an initially unknown search space. This is their key feature, particularly in large, complex, and poorly understood search spaces, where classical search tools (enumerative, heuristic, etc.) are inappropriate. In such cases they offer a valid approach to problems requiring efficient and effective search techniques.

In this section we firstly present the key-points of their application to our problem as well as the representation and the fitness function, and then describe the *EA* (*CHC* [13]) used in this study.

3.1 Evolutionary Algorithms

EAs [6, 7] are stochastic search methods that mimic the metaphor of natural biological evolution. All *EAs* rely on the concept of a population of individuals (representing search points in the space of potential solutions to a given problem), which undergo probabilistic operators such as mutation, selection, and (sometimes) recombination to evolve towards increasingly better fitness values of the individuals. The fitness of an individual reflects its objective function value with respect to a particular objective function to be optimized. The mutation operator introduces innovation into the population by generating variations of individuals, and the recombination operator typically performs an information exchange between different individuals from a population. The selection operator imposes a driving force on the process of evolution by preferring better individuals to survive and reproduce when the members of the next generation are selected.

In the appendix we describe the model of *EA* that will be used in this paper as evolutionary instance selection algorithm. *CHC* is a classical model that introduces different features to obtain a trade-off between exploration and exploitation.

3.2 Evolutionary Instance Selection

EAs may be applied to the *IS* problem, because it can be considered as a search problem.

The objective of this paper is to study the performance of the *EAs* as *IS* algorithm applied to *TSS* for data reduction in *KDD*, comparing its results with the ones obtained by the algorithms introduced in Sect. 2.3.

The application of *EAs* to these approach is accomplished by tackling two important issues: the specification of the representation of the solutions and the definition of the fitness function.

Representation

Let's assume a data set denoted TR with n instances. The search space associated with the instance selection of TR is constituted by all the subsets of TR. Then, the chromosomes should represent subsets of TR. This is accomplished by using a binary representation. A chromosome consists of n genes (one for each instance in TR) with two possible states: 0 and 1. If the gene is 1, then its associated instance is included in the subset of TR represented by the chromosome. If it is 0, then this does not occur.

Fitness Function

Let *SS* be a subset of instances of TR to evaluate and be coded by a chromosome. We define a fitness function that combines two values: the classification performance ($clas\_per$) associated with *SS* and the percentage of reduction ($perc\_red$) of instances of *SS* with regards to TR:

$$Fitness(SS) = \alpha \cdot clas\_rat + (1 - \alpha) \cdot perc\_red\,. \tag{1}$$

The 1-*NN* classifier (Sect. 2.3) is used for measuring the classification rate, $clas\_rat$, associated with *SS*. It denotes the percentage of correctly classified objects from TR using only *SS* to find the nearest neighbour. For each object y in *SS*, the nearest neighbour is searched for amongst those in the set $SS \setminus \{y\}$. Whereas, *perc_red* is defined as:

$$perc\_red = 100 \cdot (|TR| - |SS|)/|TR|\,. \tag{2}$$

The objective of the *EAs* is to maximize the fitness function defined, i.e., maximize the classification performance and minimize the number of instances obtained. In the experiments presented in this paper, we have considered the value $\alpha = 0.5$ in the fitness function, as per a previous experiment in which we found the best trade-off between precision and reduction with this value.

Stratification Strategy

The algorithms we have studied, both classical and evolutionary, are affected when the size of the data set increases. The main difficulties they have to face are as follows:

- Efficiency. The efficiency of *IS* algorithms is at least $O(n^2)$, where n is the size of the data set. Most of them present an efficiency order greater than $O(n^2)$. When the size increases, the time needed by each algorithm also increases.
- Resources. Most of the algorithms assessed need to have the complete data set stored in memory to carry out their execution. If the size of the problem were too big, the computer would need to use the disk as swap memory. This loss of resources has an adverse effect on efficiency due to the increased access to the disk.

- Representation. *EAs* are also affected by representation, due to the size of their chromosomes. When the size of these chromosomes is too big, the algorithms experience convergence difficulties.

To avoid these drawback we led our experiments towards a stratified strategy. This strategy divides the initial data set in strata. The strata are disjoints sets with equal class distribution. We evaluate the algorithm over each strata to carry out the selection and finally we reunite the subresult to conform the final one, where we use a *DM* algorithm to evaluate the quality of the subset selected.

In the following section (Fig. 2) we describe the use of stratified strategy combined with *EA*.

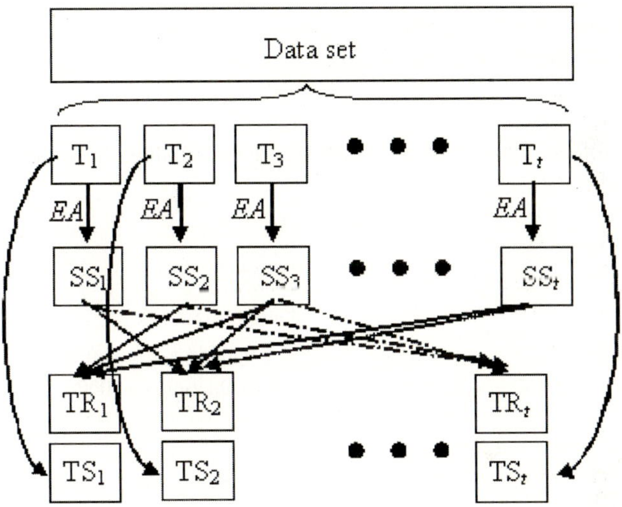

Fig. 2. Combination of *EAs* and Stratified Strategy

Evolutive Algorithms and Stratification Strategy

Following this strategy, data sets are divided into disjoint sets of equal size, $T_1, T_2, \ldots,$ and T_t. We maintain class distribution within each set in the partitioning process. *EA* (*CHC*) is applied to each T_i obtaining a subset selected SS_i. We then conduct t pairs of training and test sets, (TR_i, TS_i), $i = 1, \ldots, t$. For each pair i, the test set, TS_i, is T_i, and the training set, TR_i, is the union of all of the other SS_j, $j \neq i$. We evaluate each pair (TR_i, TS_i) using $C4.5$ for TSS.

4 Experimental Methodology

We have carried out our study of TSS problem using two size problems: large and huge. We intend to study the behavior of the algorithms when the size of the problem increases. Section 4.1 describes the data sets used (large and huge) and introduces the parameters associated with the algorithms. Finally, in Sect. 4.2 we present the tables where we show the results.

4.1 Data Sets and Parameters

The data sets used in the experimentation are the following:

Table 1. Large and Huge size data sets

Data Set	Num. Instances	Num. Features	Num. Classes
Adult	30132	14	2
Kdd Cup'99	494022	41	23

Whether either small or medium data sets are evaluated, the parameters used are the same:

Table 2. Parameters

Algorithm	Parameters
Ib3	Acept. Level = 0.9, Drop Level = 0.7
CHC	Population = 50, Evaluations = 10000

4.2 Table of Results

In the following section we will present the structure of tables where we present the results.

Our table shows the results obtained in TSS by the classical and evolutionary instance selection algorithms, respectively. In order to observe the level of robustness achieved by all algorithms, the table presents the average (for $(TR_i, TS_i)\,, i = 1,\ldots,t$) of the results offered by each algorithm in the data sets evaluated. Each column shows:

- The first column shows the name of the algorithm
- The second column contains the average execution time associated to each algorithm. The algorithms have been run in a *Pentium* 4, 2.4 Gh, 256 RAM, 40 Gb HD.
- The third column shows the average reduction percentage from the initial training sets.

- The fourth column contains the accuracy when *C*4.5 is applied using the training set selected. This column presents two sub-columns: the first one shows the training accuracy when using the training set selected from the initial set; the second sub-column shows the test accuracy of over the test data set. Each sub-column presents the accuracy mean. We study both sub-columns in an effort to evaluate the existence of over-fitting in the instance selection phase.

5 Experimental Study

5.1 Large Size Data Sets: Adult

Table 3 shows the results obtained in *TSS* by instance selection algorithms.

Table 3. *TSS* for Adult data set

	Exec. Time (sec)	% Reduction	*C*4.5 %Ac. Trn	*C*4.5 %Ac. Test
*C*4.5	2		88.72%	85.40%
Cnn Strat	1	97.34%	88.02%	36.15%
*Drop*1 *Strat*	44	95.09%	100.00%	26.31%
*Drop*2 *Strat*	48	70.33%	71.00%	83.09%
*Drop*3 *Strat*	41	95.57%	85.34%	77.29%
*Ib*2 *Strat*	1	99.57%	95.81%	36.37%
*Ib*3 *Strat*	3	76.69%	76.72%	82.73%
Icf Strat	33	85.62%	88.75%	82.21%
CHC Strat	20172	99.38%	97.43%	82.76%

The number of stratus used in Adult data set is 10.

The following conclusions about the instance selection algorithms for *TSS* studying Table 3 can be made:

- In Table 3 we can see that *CHC* has not been improved in their test accuracy by classic methods which offer small reduction rates. *EAs* maintain their behaviour in test and training rates, avoiding over-fitting.
- Classic algorithms produce over-fitting in most of the cases, or offer small reduction rates.
- *CHC* offers the best balance between reduction and accuracy rates.

The main conclusion that can be drawn when using large data sets is that *CHC* is a very good algorithm for data reduction having both high reduction rates and accuracy, and avoiding over-fitting.

Table 4. *TSS* for Kdd Cup'99 data set

	Exec. Time(sec)	% Reduction	C4.5 %Ac. Trn	C4.5 %Ac. Test
C4.5	265		99.97%	99.94%
Cnn Strat	8	81.61%	99.92%	96.43%
Drop1 Strat	111	99.97%	92.33%	34.97%
Drop2 Strat	105	76.66%	99.17%	76.58%
Drop3 Strat	131	56.74%	99.99%	75.38%
Ib2 Strat	7	82.01%	99.40%	95.05%
Ib3 Strat	3	78.92%	98.13%	96.77%
Icf Strat	242	23.62%	99.98%	99.53%
CHC Strat	1960	99.68%	98.97%	97.53%

5.2 Huge Size Data Sets: Kdd Cup'99

The number of stratus used in Kdd Cup'99 data set is 100.

We would make the following conclusions about the instance selection algorithms for *TSS* studying Table 4:

- *CHC* offers the best behaviour in *TSS*. It presents the best reduction rates and accuracy rates. The fourth column in Table 4 shows that *CHC* is the best algorithm offering optimal models that can be applied to *DM*, as per our study.
- The classical algorithms present high accuracy rates but smaller reduction rates than *CHC*.

If we pay attention to execution time we can detect that all classic algoritms increase it when the data set size grows, while the *CHC* one suffer a important reduction. This situation is due to in Kdd Cup'99 data set, *CHC* can develop a higher reduction in the first generations of the algorithm so the number of instances selected in each chromosome are reduced and the time for evaluate each solution is reduced too. So its difficult to talk about execution time for *CHC* as drawback because it depends of the problem where we apply it.

Clearly, when we manage data sets which are bigger than the previous ones, the *CHC* algorithm improves its behaviour giving the best results for scaling down data.

5.3 On the Simplicity Obtained for Training a Decision Tree Algorithm with a Small Instance Selection Set by *EAs* in *TSS*

The presence of noise in data sets is a widely extended pathology in the design of decision trees [22]. Sources of noise can be error in measurements, error in data encoding, error in examples, missing values, etc. Another problem that appears in the design of decision trees is the presence of clashes between examples (same attribute vector but different class). Both problems produce complex hypothesis, poor comprehensibility, over-fitting (hypothesis over-fits the data), and low classification accuracy of new data.

Larger decision trees introduce a loss of human interpretability of the models.

Pruning is the common framework to avoid the problem of over-fitting noisy data. The basic idea is to incorporate a bias towards more general and simpler theories in order to avoid overly specific theories that try to find explanations for noisy examples. The two different ways [23] to deal with this are:

- Pre-pruning: Stop growing the tree when there is not enough data to make reliable decisions, or when the examples are acceptably homogeneous.
- Post-pruning: Grow the full decision tree and then remove nodes for which there is not sufficient evidence.

The drawback of pre-pruning is that it is very difficult to know when to stop growing the tree.

Classical strategies based on post-pruning tend to under-prune or over-prune the decision tree.

We have applied *CHC* in the *TSS* problem as seen in Sects. 5.1 and 5.2, obtaining greater reduction and accuracy rates. In Tables 5 and 6 we show the number of rules offered by the decision tree obtained after *CHC* execution over the data sets. This is compared with the decision tree obtained after applying *C*4.5 to the complete data set.

Table 5. Comparison of the number of rules in the Decision tree for Adult data set

	Number of Rules
*C*4.5	327.4
CHC Strat	16.3

Table 6. Comparison of the number of rules in the Decision tree for Kdd Cup'99 data set

	Number of Rules
*C*4.5	132.7
CHC Strat	9.1

In the first column of tables the name of the algorithm is stated, and the remaining ones give the average number of rules after the evaluation of both algorithms.

An analysis of Tables 5 and 6 give the following conclusions:

- *CHC* offers representative models that maintain the balance between accuracy and reduction.

- Tables 5 and 6 show the reduction that *CHC* gives in the model of a decision tree generated by the subset selected. As we can see, the number of rules associated to decision trees are significantly reduced, with a more interpretable model.

Briefly summarizing this subsection, we conclude that when the *CHC* is applied to *TSS* problems it produces a small model with a high accuracy rate. This small model size increases its speed in classification, reducing its storage necessities and increasing its human interpretability.

6 Conclusion

This paper addressed the analysis of the evolutionary instance selection algorithms (*CHC*) and their use in data reduction for large data sets in *KDD*.

An experimental study was carried out to compare the results of a *EA* model with the classical *IS* ones, over large and huge size data sets. The main conclusions reached are as follows:

- *Stratified CHC* outperform the classical algorithms, simultaneously offering two main advantages: better data reduction percentages and higher classification accuracy.
- *Stratified CHC* avoid over-fitting in the learning phase, maintaining the accuracy rates in the training and the test sets.
- *Stratified CHC* is the most appropriate algorithm evaluated, according to the algorithms that we have compared. It offers the best when we increase the size of the data set.
- *Stratified CHC* algorithm significantly reduces the size of the decision tree associated to the model studied. This characteristic produces decision trees that are easier to interpret.
- In large and huge size data sets, classical algorithms do not present balanced behaviour. If the algorithm reduces the size then its accuracy rate is poor. When accuracy increases there is no reduction. The stratified version of *CHC* offers the best results when the data set size increases.

Therefore, as a final concluding remark, we consider Stratified strategy combined with *CHC* to be a good mechanism for data reduction in *KDD*. It has become a powerful tool to obtain small selected training sets and therefore scaling down data. *CHC* can select the most representative instances, satisfying both the objectives of high accuracy and reduction rates. Stratified strategy permits a reduction of the search space so we can carry out the evaluation of the algorithms with acceptable execution time.

Finally, we would say that future research could be directed towards the study of hybrid strategies between classical and evolutionary instance selection algorithms.

References

1. Adriaans P, Zantinge D (1996). Data Mining. Addison-Wesley.
2. Shanahan J G (2000). Soft Computing for Knowledge Discovery. Kluwer Academic Publishers.
3. Witten I H, Frank E (2000). Data mining: Practical Machine Learning Tools and Techniques with Java implementations. Morgan Kaufmann.
4. Liu H, Motoda H (1998). Feature Selection for Knowledge Discovery and Data Mining. Kluwer Academic Publishers.
5. Goldberg D E (1989). Genetic Algorithms in Search, Optimization, and Machine Learning. Addison-Wesley.
6. Back T, Fogel D, Michalewicz Z (1997). Handbook of Evolutionary Computation. Oxford University Press.
7. Holland J H (1975). Adaptation in Natural and Artificial Systems. The University of Michigan Press.
8. Quinlan J R (1993). C4.5: Programs for Machine Learning. Morgan Kaufmann.
9. Liu H, Motoda H (2001). Data Reduction via Instance Selection. In: Instance Selection and Construction for Data Mining, H. Liu, H. Motoda (Eds.), (pp. 3-20). Kluwer Academic Publishers.
10. Reeves C R, Bush D R (2001). Using Genetic Algorithms for Training Data Selection in RBF Networks. In: Instance Selection and Construction for Data Mining, H. Liu, H. Motoda (Eds.), (pp. 339-356). Kluwer Academic Publishers.
11. Frank E, Witten I H (1999). Making Better Use of Global Discretization. In: Proc. Sixteenth International Conference on Machine Learning , I. Bratko, S. Dzeroski (Eds.), (pp. 115-123). Morgan Kaufmann.
12. Kibbler D, Aha D W (1987). Learning Representative Exemplars of Concepts: An initial Case of Study. In: Proc. of the Fourth International Workshop on Machine Learning, (pp. 24-30). Morgan Kaufmann.
13. Eshelman L J (1991). The CHC Adaptive Search Algorithm: How to Have Safe Search when Engaging in Nontraditional Genetic Recombination. In: Foundations of Genetic Algorithms-1, Rawlins, G.J.E. (Eds.), (pp. 265-283). Morgan Kauffman.
14. Wilson D R, Martinez T R (1997). Instance Pruning Techniques. In: Proceedings of the International Conference, (pp. 403-411). Morgan Kaufmann.
15. Domingo C, Gavalda R, Watanabe O (2002). Adaptative Sampling Methods for Scaling Up Knowledge Discovery Algorithms. Data Mining and Knowledge Discovery 6: 131–152.
16. Brighton H, Mellish C (2002). Advances in Instance Selection for Instance-Based Learning Algorithms. Data Mining and Knowledge Discovery 6: 153–172.
17. Liu H, Motoda H (2002). On Issues of Instance Selection. Data Mining and Knowledge Discovery 6: 115–130.
18. Reinartz T (2002). A Unifying View on Instance Selection. Data Mining and Knowledge Discovery 6: 191–210.
19. Kuncheva L (1995). Editing for the k-Nearest Neighbors Rule by a Genetic Algorithm. Pattern Recognition Letters 16: 809–814.
20. Cano J R, Herrera F, Lozano M (2003). Using Evolutionary Algorithms as Instance Selection for Data Reduction in KDD: An Experimental Study, IEEE Transaction on Evolutionary Computation 7 6: 561–575.
21. Hart P E (1968). The Condensed Nearest Neighbor Rule. IEEE Transaction on Information Theory 18 3: 431–433.

22. Safavian S R, Landgrebe D (1991). A Survey of Decision Tree Classifier Methodology. IEEE Transaction on Systems, Man. and Cybernetics 21 3: 660–674.
23. Esposito F, Malerba D and Semeraro G (1997). A Comparative Analysis of Methods for Pruning Decision Trees. IEEE Transactions on Pattern Analysis and Machine Intelligence 19 5: 476–491.

Part VII

Soft Computing for Modeling, Optimization and Information Processing

Predictive Controller Tuning Using Modified Particle Swarm Optimization Based on Cauchy and Gaussian Distributions

Leandro dos Santos Coelho[1] and Renato A. Krohling[2]

[1] Pontifícia Universidade Católica do Paraná, Grupo Produtrônica, Programa de Pós-Graduação em Engenharia de Produção e Sistemas, Rua Imaculada Conceição, 1155, CEP 80215-901, Curitiba, PR – Brazil
leandro.coelho@pucpr.br

[2] Lehrstuhl Elektrische Steuerung und Regelung, Fakultät für Elektrotechnik und Informationstechnik, Universität Dortmund, D-44221 Dortmund, Germany
krohling@esr.e-technik.uni-dortmund.de

Abstract. Model based predictive control has become an important form of advanced control. Generalised predictive controllers (*GPCs*) have been successfully applied to control complex processes in the industry during the last decade. In this paper, we present a novel method named modified particle swarm optimization, which is based on Gaussian and Cauchy distributions for the optimization of an adaptive *GPC* using an *ARMA* (Auto Regressive Moving Average) mathematical model. The simulation results demonstrated the successful *GPC* tuning and the suitability of the modified particle swarm optimization for the control of a non-linear process of Hammerstein type.

1 Introduction

Model based predictive controllers (*MBPCs*) have been well accepted in the industry, basically due to the development of identification techniques and the use of digital computers [1, 2, 3]. *MBPCs* are a conception of internal model control, whereby a mathematical model is utilised for the prediction of a process output over a finite horizon. At each time step, the *MBPC* calculates a set of optimum process control moves, the first of which is applied to the process in a recent horizon manner [4].

In the last few years some researches have been carried out on the development of design methods and optimization of *MBPCs* using genetic algorithms [5, 6, 7]. In this paper, the design, optimization, and performance of a *MBPC* – an adaptive generalised predictive controller (*GPC*) [8, 9] using a new approach based on Particle Swarm Optimization (*PSO*) [10, 11, 12], are proposed and discussed. The new approach uses Cauchy and Gaussian probability distributions to *PSO*. Particle swarm optimization is a kind of evolutionary algorithm based on a population of individuals and motivated by the simulation of social behaviour instead of the survival of the

Leandro dos Santos Coelho and Renato A. Krohling: *Predictive Controller Tuning Using Modified Particle Swarm Optimization Based on Cauchy and Gaussian Distributions*, Advances in Soft Computing **1**, 287–298 (2005)
www.springerlink.com

fittest individual. In *PSO*, each solution candidate (called particle) has a speed associated to it. The speed is adjusted through an update equation that considers the experience of the corresponding particle and the experience of other particles in the population.

A case study for the optimization of the *GPC* using *PSO* with Cauchy and Gaussian distributions will be examined. The non-linear process has characteristics of the Hammerstein type. In the next sections the design of the adaptive *GPC* and the description of *PSO* are presented. The design and the simulation results showing the behaviour of a control system in closed loop are discussed.

2 Predictive Controller

The *MPBCs* present a series of advantages over the other control methods. The *MPBC* is particularly attractive in the case of complex processes with transport delay, non-minimum phase, instability in open loop, restrictions and uncertainties [1, 8]. The methodologies for *MBPC* use the process model incorporating the foreseen future behaviour of the process in the controller design procedure. Several methodologies for *MBPC* have been studied, but all of them have in common the basic design structure. The main differences among them are how the future forecasts are calculated, the type of mathematical model used, and how the cost function is defined. In this paper, it is described as the design of a *GPC* with the classic algorithm of the recursive squared minimum for identification of the mathematical model of the process.

2.1 Design of Generalized Predictive Control

The *GPC* consists of a generalisation of predictive algorithms of multiple steps (multi-step-ahead). Moreover, the GPC is also a natural extension with extended horizon of single-step-ahead control, named generalised minimum variance controller considered by Clarke & Gawthrop [13]. The *GPC* carries out the forecast of the process output in a horizon of larger time than the transport delay of the process. The control signal is calculated by means of the minimisation of a cost function (quadratic) that it is obtained by the difference between the foreseen output and the reference signal, with restrictions on the future control actions.

Most of strategies in predictive control are basically a combination of: (i) a model, that describes the process, which is necessary to predict the future outputs of the process; (ii) a cost function, usually, quadratic, which takes into account the control and the future errors of the process output. This function is calculated for a fixed number of steps ahead to predict the behaviour of the process; (iii) a known trajectory of the outputs of future references; (iv) restrictions of the system and control variables (optional); and (v) an optimization algorithm to find the control sequence that minimises the cost function.

In the present study, the *GPC* with incremental model is considered in order to prevent the error in steady state, which occurs in the positional model many times. The multi-step-ahead predictive controllers use a process model of the kind *CARIMA* (Controlled Auto-Regressive Integrated Moving Average) as described by the following difference equation [14]:

$$A(q^{-1})y(t) = q^{-d}B(q^{-1})u(t) + C(q^{-1})\xi(t)/\Delta \tag{1}$$

which can be rewritten in the following way:

$$A(q^{-1})\Delta y(t) = q^{-d}B(q^{-1})\Delta u(t) + C(q^{-1})\xi(t) \tag{2}$$

where

$$\begin{aligned} A(q^{-1}) &= 1 + a_1 q^{-1} + \cdots + a_{n_a} q^{-n_a} , \\ B(q^{-1}) &= b_0 + b_1 q^{-1} + \cdots + b_{n_b} q^{-n_b} , \\ C(q^{-1}) &= c_0 + c_1 q^{-1} + \cdots + c_{n_c} q^{-n_c} , \end{aligned}$$

where $y(t)$ is the system output, $u(t)$ is the *control signal*, $\xi(t)$ is a random sequence of the kind white noise and $\Delta = (1 - q^{-1})$.

The design adopted for the adaptive *GPC* is based on the approach proposed by Clarke et al. [8, 9]. In that case, the cost function is given by the equation:

$$J_{GPC} = E\left\{ \sum_{j=N_1}^{N_2} [y(t+j) - y_r(t+j)]^2 + \sum_{j=1}^{N_u} \Gamma \Delta u^2(t+j-1) \right\} \tag{3}$$

where Γ is a weighting factor of the control signal, N_1 and N_2 are the horizon of forecast of the initial and the final output respectively, and N_u is the control horizon. The terms $y(t+j)$ and $y_r(t+j)$ represent the output signal and the reference signal j steps ahead and $\Delta u(t+j-1)$ is the increment of the control signal in the time $(t+j-1)$.

The prediction horizon and the weighting of the control are the main tuning parameters of the *GPC*. From the selection of these parameters, it is possible to obtain different kinds of predictive controllers and to adjust the desired performance of the controlled system.

In the stochastic case, some suitably defined expectation of J_{GPC} should be minimized. In this case, using the following polynomial equations:

$$1 = F_j \bar{A} + z^{-j} G_j \tag{4}$$

$$BF_j = E_j + z^{-j} H_j \tag{5}$$

with $j = 1, 2, \ldots, N_2$. Equations (4) and (5) can be written as:

$$y(t+j) = E_j \Delta u(t+j-1) + G_j y(t) + H_j \Delta u(t-1) . \tag{6}$$

In the vector form, the equation (6) is described by:

$$y = \boldsymbol{E u} + Gy(t) + H\Delta u(t-1) \tag{7}$$

$$\boldsymbol{y}^T = [y(t+1), \ldots, y(t+N_2)] \tag{8}$$

$$\boldsymbol{u}^T = [\Delta u(t), \ldots, \Delta u(t+N_2-1)] \tag{9}$$

$$G^T = [G_1, \ldots, G_{N2}] \; H^T = [H_1, \ldots, H_{N2}] \tag{10}$$

where E is an inferior triangular matrix with size $(N_2 x N_2)$ given by:

$$E = \begin{bmatrix} e_0 & & & \\ e_1 & e_0 & & \\ \cdots & & & \\ e_{N_2-1} & e_{N_2-2} & \cdots & e_0 \end{bmatrix} \tag{11}$$

From the definitions of y, $\boldsymbol{u}$, and using

$$\boldsymbol{y}_r^T = [y_r(t+1), \ldots, y_r(t+N_2)] \tag{12}$$

the cost function can be described by:

$$J_{GPC} = (\boldsymbol{y} - \boldsymbol{y}_r)^T (\boldsymbol{y} - \boldsymbol{y}_r) + \Gamma \boldsymbol{u}^T \boldsymbol{u} \,. \tag{13}$$

Substituting (7) into (13) and let J_{GPC} minimised in relation to $\boldsymbol{u}$, results the following control law:

$$\boldsymbol{u} = (E^T E + \Gamma I)^{-1} E^T [\boldsymbol{y}_r - Gy(t) - H\Delta u(t-1)] \tag{14}$$

In order to avoid the singularity of $(E^T E + \Gamma I)$ and reduce the computational effort for the horizon of output with a large value, it was used a control horizon N_u $(N_u < N_2)$, i.e., for $j \geq N_u$, the incremental control is $\Delta u(t+j) = 0$ (a constant control after the time N_u).

The control vector, $\boldsymbol{u}$, and the matrix E with size $(N_2 \times N_u)$ have the form:

$$\boldsymbol{u}^T = [\Delta u(t), \ldots, \Delta u(t+N_u-1)] \tag{15}$$

$$E = \begin{bmatrix} e_0 & & & \\ e_1 & e_0 & & \\ \cdots & & & \\ e_{Nu-1} & e_{Nu-2} & \cdots & e_0 \\ \cdots & & & \\ e_{N2-1} & e_{N2-2} & \cdots & e_{N2-Nu} \end{bmatrix} \tag{16}$$

2.2 Adaptive Generalized Predictive Control

Adaptive *GPC*, or even the MBPCs in its general form, can be obtained by means of two design procedure: direct and indirect. The direct procedure uses a recursive estimator to directly obtain the parameters of the control law trough measures from the input and output of the process. The indirect procedure combines a recursive estimator to obtain the parameters of the process model and a procedure to design the controller.

In summary, for the adaptive *GPC* design, the estimator of parameters must be iterative, and the model of the system is updated in each sampling period when new measured are available. In this paper, the recursive squared minimum (*RSM*) algorithm was used [15] which is summarised in the following steps: (i) obtain the inputs and outputs of the process to be identified; (ii) update the vector of measures; (iii) calculate the prediction error; (iv) calculate the gain of the estimator; (v) update the vector of estimated parameters; (vi) update the covariance matrix; and (vii) return to the step (i).

2.3 Particle Swarm Optimization

2.3.1 Standard Particle Swarm Optimization

Particle Swarm Optimization (*PSO*) is an evolutionary computation technique originally developed by Kennedy and Eberhart in 1995 [10, 11]. *PSO* is initialised with a population of random solutions. Unlike the most evolutionary algorithms, each potential solution (individual) in *PSO* is also associated with a randomised velocity, and the potential solutions, called *particles*, are then "flown" through the problem space.

Each particle keeps track of its coordinates in the problem space, which are associated with the best solution (fitness) it has achieved so far. This value is called *pbest*. Another "*best*" value that is tracked by the *global* version of the particle swarm optimiser is the overall best value, and its location, obtained so far by any particle in the population. This location is called *gbest*.

The particle swarm optimization concept consists of, at each time step, changing the velocity (accelerating) of each particle flying toward its *pbest* and *gbest* locations (global version of *PSO*). Acceleration is weighted by random terms, with separate random numbers being generated for acceleration toward *pbest* and *gbest* locations, respectively. The procedure for implementing the global version of *PSO* is given by the following steps [16]:

(i) Initialise a population (array) of particles with random positions and velocities in the n dimensional problem space.
(ii) For each particle, evaluate its fitness value.
(iii) Compare each particle's fitness with the particle's *pbest*. If current value is better than *pbest*, then set *pbest* value equal to the current value and the *pbest* location equal to the current location in n-dimensional space.
(iv) Compare fitness with the population's overall previous best. If current value is better than *gbest*, then reset *gbest* to the current particle's array index and value.
(v) Change the velocity and position of the particle according to (17) and (18), respectively [17, 18]:

$$\boldsymbol{v}_i = w \cdot \boldsymbol{v}_i + c_1 \cdot ud() \cdot (\boldsymbol{p}_i - \boldsymbol{x}_i) + c_2 \cdot Ud() \cdot (\boldsymbol{p}_g - \boldsymbol{x}_i) \tag{17}$$

$$\boldsymbol{x}_i = (\boldsymbol{x}_i + \boldsymbol{v}_i) \; . \tag{18}$$

(iv) Loop to step (ii) until a stop criterion is met, usually a sufficiently good fitness or a maximum number of iterations (generations).

where $x_i = [x_{i1}, x_{i2}, \ldots, x_{in}]^T$ stands for the position of the i-th particle, $v_i = [v_{i1}, v_{i2}, \ldots, v_{in}]^T$ stands for the velocity of the i-th particle and $p_i = [p_{i1}, p_{i2}, \ldots, p_{in}]^T$ represents the best previous position (the position giving the best fitness value) of the i-th particle. The index g represents the index of the best particle among all the particles in the group. Variable w is the inertia weight, $c_1$and c_2 are positive constants; *ud*() and *Ud*() are two random functions in the range [0, 1]. Particles' velocities on each dimension are clamped to a maximum velocity *Vmax*. If the sum of accelerations would cause the velocity on that dimension to exceed *Vmax*, which

20. Chellapilla, K. (1998) Combining mutation operators in evolutionary programming, *IEEE Trans. on Evolutionary Computation* 2(3), pp. 91–96
21. Rudolph, G. (1997) Local convergence rates of simple evolutionary algorithms with Cauchy mutations, *IEEE Trans. on Evolutionary Computation* 1, pp. 249–258
22. Katende, E.; Jutan, A.; Corless, R. (1998) Quadratic nonlinear predictive control, *Industrial and Engineering Chemistry Research* 37(7), pp. 2721–2728
23. Drogies, S.; De Geest, D. (1999) Predictive control: propositions for the design methodology, *American Control Conference*, San Diego, CA, pp. 647–651
24. Rani, K. Y.; Unbehauen, H. (1997) Study of predictive controller tuning methods, *Automatica* 33(12), 2243–2248

A Realistic Information Retrieval Environment to Validate a Multiobjective GA-P Algorithm for Learning Fuzzy Queries*

Oscar Cordón[1], Enrique Herrera-Viedma[1], María Luque[1], Felix Moya[2] and Carmen Zarco[3]

[1] Dept. of Computer Science and A.I. E.T.S. de Ingeniería Informática, University of Granada. 18071 – Granada (Spain)
{ocordon,viedma,mluque}@decsai.ugr.es

[2] Dept. of Information Sciences. Faculty of Information Sciences, University of Granada. 18071 – Granada (Spain)
felix@ugr.es

[3] PULEVA S.A., Camino de Purchil, 66. 18004 – Granada (Spain)
czarco@puleva.es

Summary. IQBE has been shown as a promising technique to assist the users in the query formulation process. In this framework, queries are automatically derived from sets of documents provided by them. However, the different proposals found in the specialized literature are usually validated in non realistic information retrieval environments. In this work, we design several experimental setups to create real-like retrieval environments and validate the applicability of a previously proposed multiobjective evolutionary IQBE technique for fuzzy queries on them.

1 Introduction

Information retrieval (IR) may be defined as the problem of the selection of documentary information from storage in response to search questions provided by a user [2]. Information retrieval systems (IRSs) deal with documentary bases containing textual, pictorial or vocal information and process user queries trying to allow the user to access to relevant information in an appropriate time interval.

The paradigm of Inductive Query by Example (IQBE) [4], where queries describing the information contents of a set of documents provided by a user are automatically derived, has proven to be useful to assist the user in the query formulation process. This is especially useful for fuzzy IRSs [3], as they consider

* This work was supported by the Spanish Ministerio de Ciencia y Tecnología under projects TIC2003-07977 and TIC2003-00877, including FEDER fundings.

Oscar Cordón et al.: *A Realistic Information Retrieval Environment to Validate a Multiobjective GA-P Algorithm for Learning Fuzzy Queries*, Advances in Soft Computing **1**, 299–309 (2005)
www.springerlink.com

complex queries composed of weighted query terms joined by the logical operators AND and OR, which are difficult to be formulated by non expert users.

The most known existing approach is that of Kraft et al. [15], based on genetic programming (GP) [13]. Several other approaches have been proposed based on more advanced evolutionary algorithms (EAs) [1], such as genetic algorithm-programming (GA-P) [12] or simulated annealing-programming, to improve Kraft et al.'s [7, 8].

In view of the latter, the IQBE paradigm seems to perform properly but most of the existing proposals are validated by only running the algorithms on a whole document collection selected from those typical in IR, as Cranfield, CISI, etc. This does not show the real environment in which an IRS will be used.

In this paper, we will design real-like environments to test these kinds of algorithms by: i) dividing the documentary base (Cranfield, in our case) into a training document set, in which the queries will be learned, and a test document set, against to which the obtained queries will be tested; and ii) using training collections with a number of irrelevant documents that shows the real behaviour of the users in an IQBE or a user profile [13] environment. Then, we will validate a specific algorithm able to generate several queries with a different trade-off between precision and recall in a single run, the multiobjective GA-P IQBE technique proposed in [9], in the designed IR environments.

The paper is structured as follows. Section 2 is devoted to the preliminaries, the basis of FIRSs and a short review of IQBE. Then, Sect. 3 describes the design of realistic IR test environments. The multiobjective GA-P proposal is reviewed in Sect. 4. Section 5 presents the experiments developed and the analysis of results. Finally, the conclusions are pointed out in Sect. 6.

2 Preliminaries

2.1 Fuzzy Information Retrieval Systems

FIRSs are constituted of the following three main components:

The Documentary Data Base, that stores the documents and their representations (typically based on index terms in the case of textual documents).

Let D be a set of documents and T be a set of unique and significant terms existing in them. An indexing function $F : D \times T \rightarrow [0,1]$ is defined as a fuzzy relation mapping the degree to which document d belongs to the set of documents "about" the concept(s) represented by term t. By projecting it, a fuzzy set is associated to each document ($d_i = \{< t, \mu_{d_i}(t) > | t \in T\}$; $\mu_{d_i}(t) = F(d_i, t)$) and term ($t_j = \{< d, \mu_{t_j}(d) > | d \in D\}$; $\mu_{t_j}(d) = F(d, t_j)$).

In this paper, we will work with Salton's normalized *inverted document frequency* (IDF) [2]: $w_{d,t} = f_{d,t} \cdot log(N/N_t); F(d,t) = \frac{w_{d,t}}{Max_d w_{d,t}}$, where $f_{d,t}$ is the frequency of term t in document d, N is the total number of documents and N_t is the number of documents where t appears at least once.

The Query Subsystem, allowing the users to formulate their queries and presenting the retrieved documents to them. Fuzzy queries are expressed using a query language that is based on weighted terms, where the numerical or linguistic weights represent the "subjective importance" of the selection requirements.

In FIRSs, the query subsystem affords a fuzzy set q defined on the document domain specifying the degree of relevance (the so called *retrieval status value* (RSV)) of each document in the data base with respect to the processed query: $q = \{< d, \mu_q(d) > | d \in D\}$; $\mu_q(d) = RSV_q(d)$.

The Matching Mechanism, that evaluates the degree to which the document representations satisfy the requirements expressed in the query (i.e., the RSV) and retrieves those documents that are judged to be relevant to it.

When using the *importance* interpretation [3], the query weights represent the relative importance of each term in the query. The RSV of each document to a fuzzy query q is then computed as follows [17]. When a single term query is logically connected to another by the AND or OR operators, the relative importance of the single term in the compound query is taken into account by associating a weight to it. To maintain the semantics of the query, this weighting has to take a different form according as the single term queries are ANDed or ORed. Therefore, assuming that A is a fuzzy term with assigned weight w, the following expressions are applied to obtain the fuzzy set associated to the weighted single term queries A_w (*disjunctive queries*) and A^w (*conjunctive ones*):

$$A_w = \{< d, \mu_{A_w}(d) > | d \in D\}\ ; \quad \mu_{A_w}(d) = Min\ (w, \mu_A(d))$$
$$A^w = \{< d, \mu_{A^w}(d) > | d \in D\}\ ; \quad \mu_{A^w}(d) = Max\ (1 - w, \mu_A(d))$$

If the term is negated in the query, a negation function is applied to obtain the corresponding fuzzy set: $\overline{A} = \{< d, \mu_{\overline{A}}(d) > | d \in D\}; \mu_{\overline{A}}(d) = 1 - \mu_A(d)$.

Finally, the RSV of the compound query is obtained by combining the single weighted term evaluations into a unique fuzzy set as follows:

$$A\ AND\ B = \{< d, \mu_{A\ AND\ B}(d) > | d \in D\}\ ; \quad \mu_{A\ AND\ B}(d) = Min(\mu_A(d), \mu_B(d))$$
$$A\ OR\ B = \{< d, \mu_{A\ OR\ B}(d) > | d \in D\}\ ; \quad \mu_{A\ OR\ B}(d) = Max(\mu_A(d), \mu_B(d))$$

2.2 Inductive Query by Example

IQBE was proposed in [4] as "a process in which searchers provide sample documents (examples) and the algorithms induce (or learn) the key concepts in order to find other relevant documents". This way, IQBE is a technique for assisting the users in the query formulation process performed by machine learning methods. It works by taking a set of relevant (and optionally, non relevant documents) provided by a user and applying an off-line learning process to automatically generate a query describing the user's needs from that set. The obtained query can then be run in other IRSs to obtain more relevant documents.

3 Real-like IR Environments to Test IQBE Algorithms

As said, the experimental studies developed in most IQBE contributions [4, 6, 7, 8, 15] do not represent a real environment where an IRS will be used. In this work, we aim at designing realistic retrieval environments to test IQBE algorithms.

Two problems are found in the experimental setups usually considered. On the one hand, the document set provided to the IQBE algorithm is the whole documentary collection. In this set, the relevant documents are those which are relevant to the selected query, while the irrelevant documents are the rest of them. Hence, the amount of irrelevant documents is very high (for example, in the first Cranfield query, 1369 of the 1398 documents are irrelevant). This does not represent a realistic environment as when the user provides a set of (relevant and irrelevant) documents, for which he wants to learn the best possible query retrieving them, the amount of irrelevant documents provided uses to be significantly smaller.

On the other hand, the real goal of the query learning system (the derivation of queries modeling the information needs represented by the set of documents provided by the user, which are able to retrieve new relevant documents when applied to a different documentary collection) is not actually tested.

Considering the previous aspects, and following the usual machine learning operation mode, we propose to divide the documentary base into a training document set from which the queries will be learned, and a test document set against to which those queries will be tested. Besides, in order to confront the former aspect, we use training collections with a realistic number of irrelevant documents that matches with the real behaviour of the users. In short, we will implement four different environments – two corresponding to a usual IQBE framework and other two coming from the user profile field –, by considering a different number of irrelevant documents for the training set in each case[1].

These four proposals are analyzed as follows and their characteristics are summarized in the left side of Table 1, where $\%non-rel$ and $\%rel$ refer to the percentages of the whole irrelevant and relevant documents, respectively, included in the training document set.

Table 1. Characteristics of the IR environments designed

	A1/A2	B1/B2	C1/C2	D1/D2
% non-rel	0	2	10	50
% rel	50,25	50,25	50,25	50,25

	A	B	C	D
Pob-Size	200	400	400	800
#Eval	1000	25000	25000	50000

3.1 Classic IQBE Test Environments

We include within this group those environments where the training document set is assigned a number of irrelevant documents similar to that a user would provide in a

[1] Notice that two different variants are obtained from each of these four environments by taking two different values for the number of relevant documents considered.

real IQBE case. It is expected that a normal user can afford up to 30 or 40 irrelevant documents to represent an information need. On the other hand, other real (and very usual) case is that where the user does only provide relevant documents and does not give any irrelevant one at all.

As said, we are working with Cranfield, which has a total of 1398 documents. We have decided to design two different classic IQBE environments: one of them where the training set has a 2% of the whole irrelevant documents (randomly selected) and another where no irrelevant document is included on that set.

Both variants will allow us to check if the amount of documents normally provided by a user to a IQBE process is enough to derive queries satisfying the user's needs or if there is a need of using any additional assistance mechanism.

3.2 User Profile-Based IQBE Test Environments

These two environments are based on incorporating a large number of irrelevant documents to the training documentary set, a 10% and a 50% of the overall number existing in the whole collection. At first sight, we could think there is a discrepancy with a realistic IR environment, since a user is not able to provide so many irrelevant documents (around 690 when working with Cranfield). However, this operation mode is clearly justified when considering user profiles.

User profile derivation [13] is based on a relevance feedback framework where a user runs queries on an IRS and judges the relevance of the retrieved documents to his informations needs. Then, the system makes use of this information to build a *user profile*, representing the user information needs, considered to enhance the retrieval efficacy of future queries of that user. Hence, the system can store user relevance judgements from different queries in an automatic way. As said in [11], these techniques will be useful for users having a persistent need for the same type of information in order to increase the retrieval effectiveness.

4 A Multiobjective GA-P Algorithm for Automatically Learning Fuzzy Queries

The components of our multiobjective IQBE algorithm to learn fuzzy queries based on the GA-P paradigm [9] are described next.

Coding Scheme: The expressional part (GP part) encodes the query composition – terms and logical operators – and the coefficient string (GA part) represents the term weights, as shown in Fig. 1. A real coding scheme is considered for the GA part.

Fitness Function: The multiobjective GA-P (MOGA-P) algorithm is aimed at jointly optimizing the precision and recall criteria [2], as follows:

$$Max\, P = \frac{\sum_d r_d \cdot f_d}{\sum_d f_d}\ ; \quad Max\, R = \frac{\sum_d r_d \cdot f_d}{\sum_d r_d}$$

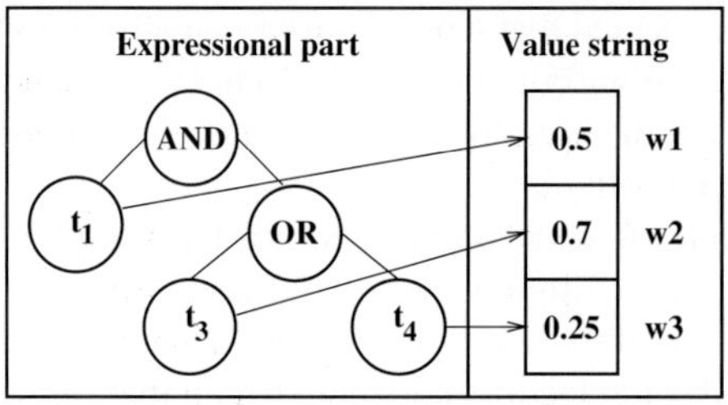

Fig. 1. GA-P individual representing the fuzzy query $0.5\ t_1\ AND\ (0.7\ t_3\ OR\ 0.25\ t_4)$

with $r_d \in \{0,1\}$ being the relevance of document d for the user and $f_d \in \{0,1\}$ being the retrieval of document d in the processing of the current query.

Pareto-Based Multiobjective Selection and Niching Scheme: The Pareto-based multiobjective EA considered is Fonseca and Fleming's Pareto-based MOGA [5]. Therefore, the selection scheme of our MOGA-P algorithm involves the following four steps:

1. Each individual is assigned a rank equal to the number of individuals dominating it plus one (non-dominated individuals receive rank 1).
2. The population is increasingly sorted according to that rank.
3. Each individual is assigned a fitness value according to its ranking in the population: $f(C_i) = \frac{1}{rank(C_i)}$.
4. The fitness assignment of each group of individuals with the same rank is averaged among them.

Then, a niching scheme is applied in the objective space to obtain a well-distributed set of queries with a different trade-off between precision and recall (see [9] for details). Finally, the intermediate population is obtained by Tournament selection [16].

Genetic Operators: The BLX-α crossover operator [10] is applied twice on the GA part to obtain two offsprings; it ramdoly generates offsprings's genes from the associated neighborhood of the genes in the parents, allowing a suitable balance between exploration and exploitation. Michalewicz's non-uniform mutation operator [16] is considered to perform mutation on that part.

The usual GP crossover [14] is considered for the GP part. Two different mutation operators are applied: random generation of a new subtree, and random change of a query term by another not present in the encoded query.

5 Experiments Developed and Analysis of Results

As said, the documentary set used to design our IR frameworks has been the *Cranfield* collection, composed of 1398 documents about Aeronautics. It has been automatically indexed by first extracting the non-stop words, applying a stemming algorithm, thus obtaining a total number of 3857 different indexing terms, and then using the normalized IDF scheme (see Sect. 2.1) to generate the term weights in the document representations.

Among the 225 queries associated to the Cranfield collection, we have selected those presenting 20 or more relevant documents (queries 1, 2, 23, 73, 157, 220 and 225). The number of relevant documents associated to each of these seven queries are 29, 25, 33, 21, 40, 20 and 25, respectively.

For each one of these queries and each retrieval environment, the documentary collection has been divided into two different, non overlapped, document sets, training and test, each of them composed of the percentage of relevant and irrelevant documents showed in the left side of Table 1.

MOGA-P has been run ten different times on the training document set associated to each query during a fixed number of evaluations (see the right side of Table 1[2]). The common parameter values considered are a maximum of 20 nodes for the expression parts, a Tournament size t of 1% of the population size, 0.8 and 0.2 for the crossover and mutation probabilities in both the GA and the GP parts.

The Pareto sets obtained in the ten runs performed for each query have been put together, and the dominated solutions removed from the unified set. Then, five queries well distributed on the Pareto front were selected from each of the seven unified Pareto sets and run on the corresponding test set once preprocessed[3].

Table 2. Statistics of the Pareto sets obtained by the MOGA-P algorithm (option C1)

$\#q$	$\#p$	$\sigma_{\#p}$	$\#dp$	$\sigma_{\#dp}$	M_2^*	$\sigma_{M_2^*}$	M_3^*	$\sigma_{M_3^*}$
1	30.7	4.318	2.9	0.263	6.455	1.305	0.601	0.025
2	45.1	5.098	1.8	0.190	8.421	1.192	0.448	0.052
23	34.0	4.825	3.9	0.411	9.216	2.061	0.74	0.042
73	44.3	3.699	1.7	0.145	6.735	1.714	0.338	0.071
157	33.5	5.301	4.8	0.395	8.944	2.046	0.803	0.035
220	36.6	2.021	1.1	0.095	1.308	1.241	0.044	0.041
225	36.9	6.082	2.2	0.19	8.899	2.605	0.485	0.056

As there is not enough space in the contribution to report every experiment developed, several illustrative results have been selected to be showed. Tables 2 and 3 collect several data about the composition of the ten Pareto sets generated for each query in environments C1 and D1, always showing the averaged value and its standard deviation. From left to right, the columns contain the number of

[2] We have tried with different parameter values choosing those for which the systems behaves better.

[3] As the index terms of the training and test documentary bases can be different, there is a need to translate training queries into test ones, removing those terms without a correspondence in the test set.

Table 3. Statistics of the Pareto sets obtained by the MOGA-P algorithm (option D1)

$\#q$	$\#p$	$\sigma_{\#p}$	$\#dp$	$\sigma_{\#dp}$	M_2^*	$\sigma_{M_2^*}$	M_3^*	$\sigma_{M_3^*}$
1	110.0	10.5	5.3	0.348	39.901	4.574	0.918	0.044
2	127.4	7.462	4.3	0.202	47.023	3.235	0.895	0.033
23	133.8	5.805	6.9	0.170	52.156	3.004	1.042	0.015
73	93.0	12.435	2.6	0.210	24.893	5.133	0.730	0.041
157	118.9	7.886	7.8	0.310	45.264	3.943	1.066	0.006
220	91.1	6.897	1.9	0.221	18.987	4.395	0.437	0.094
225	98.1	6.243	2.3	0.202	22.931	4.266	0.626	0.083

non-dominated solutions obtained ($\#p$), the number of different objective vectors (i.e., precision-recall pairs) existing among them ($\#dp$), and the values of two of the usual multiobjective metrics $\mathcal{M}_2^*$ and $\mathcal{M}_3^*$ [5].

On the other hand, Table 4 shows the retrieval efficacy of the five queries selected from the unified Pareto sets for several Cranfield queries in three of the retrieval environments (B2, C1 and D1). In that table, Sz stands for the query size, P and R for the precision and recall values, and $\#rr/\#rt$ for the number of relevant and retrieved documents, respectively. Finally, the following subsections summarize the conclusions drawn in the different experiments developed.

5.1 Classic IQBE *Versus* User Profile Test Environments

In the two classic IQBE environments, A and B, both the precision and the recall of every learned query is always equal to 1 in the training set, regardless the number of relevant documents. Besides, both values are also very close to 1 in most of the cases in the user profile-based evironment C (see B2 and C1 in Table 4). However, in the other user profile-based framework D, it is very difficult to find a query with both recall and precision equal to 1 (see D2 in Table 4). Hence, as the number of irrelevant documents increases, it is more difficult for the learned query to only retrieve relevant documents.

On the other hand, in the test results, as the number of irrelevant documents in the training set increases, the precision values also increase, whereas the recall values diminish. The reason is that when there are a lot of irrelevant documents in the test set (a few of them in the training set), the queries get all the relevant documents by retrieving a very large number of documents, thus obtaining a very low precision. However, there are cases in option D where the query does not retrieve any document or just one, showing a usual machine learning phenomenon called *over-learning* (see the results for query 225 in C1 and D1 in Table 4).

5.2 50% Relevant Documents *Versus* 25% Relevant Documents

In both classic IQBE environments (A and B), there is no significant differences between option 1 (50% of the relevant documents in the training set) and 2 (25%).

Table 4. Retrieval efficacy of the selected queries on the training and test collections

	#q		Training set Sz	P	R	#rr/#rt	Test set Sz	P	R	#rr/#rt
B2	1	1	19	1.0	1.0	7/7	19	0.036	0.364	8/221
		2	19	1.0	1.0	7/7	19	0.063	0.455	10/158
		3	19	1.0	1.0	7/7	19	0.057	0.500	11/192
		4	19	1.0	1.0	7/7	19	0.062	0.500	11/176
		5	15	1.0	1.0	7/7	15	0.094	0.636	14/149
	225	1	19	1.0	1.0	6/6	17	0.026	0.211	4/151
		2	17	1.0	1.0	6/6	13	0.043	0.263	5/117
		3	17	1.0	1.0	6/6	17	0.021	0.211	4/193
		4	19	1.0	1.0	6/6	19	0.050	0.579	11/220
		5	15	1.0	1.0	6/6	15	0.045	0.579	11/244
C1	2	1	19	0.923	1.0	12/13	11	0.096	0.769	10/104
		2	19	0.923	1.0	12/13	17	0.131	0.846	11/84
		3	19	0.923	1.0	12/13	19	0.133	0.769	10/75
		4	19	0.923	1.0	12/13	19	0.110	0.846	11/100
		5	19	1.0	0.833	10/10	17	0.046	0.231	3/65
	225	1	19	0.923	1.000	12/13	17	0.016	0.077	1/61
		2	19	0.923	1.000	12/13	19	0.000	0.000	0/52
		3	19	1.000	0.917	11/11	13	0.017	0.077	1/60
		4	19	0.800	1.000	12/15	17	0.016	0.077	1/61
		5	17	1.000	0.833	10/10	11	0.021	0.077	1/48
D1	157	1	19	0.299	1.0	20/67	15	0.195	0.8	16/82
		2	19	0.39	0.8	16/41	19	0.119	0.25	5/42
		3	19	0.593	0.8	16/27	17	0.3	0.3	6/20
		4	19	0.789	0.75	15/19	15	0.25	0.15	3/12
		5	19	1.0	0.5	10/10	15	0.375	0.15	3/8
	225	1	17	0.324	1.0	12/37	15	0.0	0.0	0/33
		2	17	0.324	1.0	12/37	15	0.0	0.0	0/33
		3	19	0.579	0.917	11/19	11	0.0	0.0	0/15
		4	19	0.688	0.917	11/16	15	0.0	0.0	0/8
		5	19	1.0	0.917	11/11	15	1.0	0.077	1/1

However, in those based on user profiles (C and D), the differences are more significant. In both cases, option 1 with more relevant documents in the training set performs better than the other in the test results. This shows that the larger the number of irrelevant documents in the training set, the more relevant documents are needed to get good queries for the test one.

In addition, option C2 usually presents more queries with precision and recall values equal to 0, having a stronger *over-learning* than C1.

6 Concluding Remarks

Several real-like retrieval environments with different characteristics have been proposed to test IQBE algorithms. The Cranfield collection has been considered to define several retrieval needs and, for each of them, relevant and irrelevant documents have been divided into several training-test partitions with a different number of documents. Then, a previous multiobjective evolutionary IQBE proposal for learning fuzzy queries has been tested in the designed environments analyzing the retrieval efficacy obtained in each case.

As future works, we will use retrieval measures considering not only the absolute number of relevant and non relevant documents retrieved, but also their relevance order in the retrieved document list, which is a fuzzy IR ability very useful for the user.

References

1. T. Bäck, D.B. Fogel, and Z. Michalewicz. *Handbook of Evolutionary Computation*. IOP Publishing and Oxford University Press, 1997.
2. R. Baeza-Yates and B. Ribeiro-Neto. *Modern Information Retrieval*. Adisson, 1999.
3. G. Bordogna, P. Carrara, and G. Pasi. Fuzzy Approaches to Extend Boolean Information Retrieval. In P. Bosc and J. Kacprzyk, editors, *Fuzziness in Database Management Systems*, pp. 231–274. 1995.
4. H. Chen and et al. A Machine Learning Approach to Inductive Query by Examples: An Experiment Using Relevance Feedback, ID3, Genetic Algoritms, and Simulated Annealing. *Journal of the American Society for Information Science*, 49(8):693–705, 1998.
5. C. A. Coello, D. A. Van Veldhuizen, and G. B. Lamant. *Evolutionary Algorithms for Solving Multi-Objective Problems*. Kluwer Academy Publisher, 2002.
6. O. Cordón, E. Herrera-Viedma, and M. Luque. Evolutionary Learning of Boolean Queries by Multiobjective Genetic Programming. In *Proc. PPSN-VII*, pp. 710–719, Granada (Spain), 2002. LNCS 2439.
7. O. Cordón, F. Moya, and C. Zarco. A GA-P Algorithm to Automatically Formulate Extended Boolean Queries for a Fuzzy Information Retrieval System. *Mathware & Soft Computing*, 7(2-3):309–322, 2000.
8. O. Cordón, F. Moya, and C. Zarco. A new Evolutionary Algorithm combining Simulated Annealing and Genetic Programming for Relevance Feedback in Fuzzy Information Retrieval Systems. *Soft Computing*, 6(5):308–319, 2002.
9. O. Cordón, F. Moya, and C. Zarco. Automatic Learning of Multiple Extended Boolean Queries by Multiobjective GA-P Algorithms. In V. Loia, M. Nikravesh, and L. A. Zadeh, editors, *Fuzzy Logic and the Internet*. Springer, 2003. In press.
10. L. J. Eshelman and J. D. Schaffer. Real-coded Genetic Algorithms and Interval-Schemata. In L. D. Whitley, editor, *Foundations of Genetic Algorithms* 2, pp. 187–202. 1993.
11. W. Fan, M. D. Gordon, and P. Pathak. Personalization of Search Engine Services for Effective Retrieval and Knowledge Management. In *Proceedings of the 2000 International Conference on Information Systems (ICIS)*, Brisbane, Australia, 2000.
12. L. Howard and D. D'Angelo. The GA-P: A Genetic Algorithm and Genetic Programming Hybrid. *IEEE Expert*, 3(10):11–15, 1995.

13. R.R. Korfhage. *Information Storage and Retrieval*. Wiley, 1997.
14. J. Koza. *Genetic Programming. On the Programming of Computers by Means of Natural Selection*. The MIT Press, 1992.
15. D.H. Kraft, F.E. Petry, B.P. Buckes, and T. Sadasivan. Genetic Algorithms for Query Optimization in Information Retrieval: Relevance Feedback. In E. Sanchez, T. Shibata, and L.A Zadeh, editors, *Genetic Algorithms and Fuzzy Logic Systems*, pp. 155–173. 1997.
16. Z. Michalewicz. *Genetic Algorithms + Data Structures = Evolution Programs*. Springer-Verlag, 1996.
17. E. Sanchez. Importance in Knowledge Systems. *Information Systems*, 6(14):455–464.

Evolutionary Learning of a Fuzzy Controller for Mobile Robotics

M. Mucientes, D.L. Moreno, A. Bugarín, and S. Barro

Dept. of Electronics and Computer Science, University of Santiago de Compostela, 15782 Santiago de Compostela, Spain
{manuel, dave, alberto, senen}@dec.usc.es

1 Introduction

Fuzzy control has shown to be a very useful tool in the field of autonomous mobile robotics, characterized by a high uncertainty in the knowledge about the environment where robot evolves.

The design of a fuzzy controller is generally made using expert knowledge about the task to be controlled. Expert knowledge is applied in order to decide the number of linguistic labels for each variable, to tune the membership functions, to select the most adequate linguistic values for the consequents, and to define the rules in the fuzzy knowledge base. This process is tedious and highly time-consuming. For this reason, automated learning techniques, such as evolutionary algorithms, have been employed for helping in some, or in all, of the tasks involved in the design process.

A knowledge base is formed of a data base (number and definition of the linguistic values, and universe of discourse of each of the variables) and the rule base. Some authors have utilized evolutionary algorithms for learning or tuning fuzzy controllers in robotics [1, 2, 3]. In some of the approaches evolutionary algorithms are used just for tuning the membership functions. In others, the complete rule base is learned, starting from a hand designed data base. But only in a few of them both the data base and the rule base are learned.

In this paper we describe the learning of a fuzzy controller for the wall-following behaviour in a mobile robot. No restrictions are placed neither in the number of linguistic labels, nor in the values of the membership functions. The methodology is based on the Iterative Rule Learning (IRL) approach [4]. The paper is organized as follows: in Sect. 2 some general comments about the evolutionary learning of knowledge bases are made, whilst in Sect. 3 the genetic learning methodology employed is explained. Section 4 describes the application of the proposed algorithm to the wall-following behaviour, and Sect. 5 presents the results we obtained. Finally, conclusions and future work are pointed out in Sect. 6.

M. Mucientes et al.: *Evolutionary Learning of a Fuzzy Controller for Mobile Robotics*, Advances in Soft Computing **1**, 311–321 (2005)
www.springerlink.com

2 Evolutionary Learning of Knowledge Bases

Learning of knowledge bases using evolutionary algorithms has three main approaches: Michigan, Pittsburgh and IRL [5]. In the Michigan approach [6], each chromosome represents an individual rule, and the entire population is the rule base. Rules evolve along time due to their interaction with the environment. The major problem of this approach is that of resolving the conflict between the performance of individual rules and that of the rule base. The objective is to obtain a good rule base, which means to obtain good individual rules, but also rules that cooperate between each other to get adequate outputs. This could be sometimes conflicting, for example when an individual rule that receives a high payoff is not adequately cooperating with other rules. This problem is addressed in [7].

This conflict is overcome by the Pittsburgh approach [8], where each chromosome represents a full knowledge base. Length of the chromosomes can be variable, which permits dealing with knowledge bases with a variable number of rules. This approach has a higher computational cost, because several knowledge bases have to be evaluated, while for the Michigan approach a single rule base is evaluated.

In the third approach (IRL [4]), each chromosome represents an individual rule, but contrary to the Michigan approach, a single rule is learned by the evolutionary algorithm and not the whole rule base. After each sequence of iterations, the best rule is selected and added to the final rule base. The selected rule must be penalised in order to induce niche formation in the search space. Niching is necessary for solving multimodal problems, as occurs with knowledge bases learning. In this case, each of the rules of the knowledge base is a solution (highly multimodal problem), and all the solutions must be taken into account to get the complete knowledge base. A common way to penalize the rules that have been obtained is to delete those training examples that have been covered by the set of rules that integrate the final rule base. The final step of the IRL approach is to check whether the obtained set of rules is a solution to the problem. In the case it is not, the process is repeated. A weak point of this approach is that the cooperation between rules is not taken into account when a rule is evaluated.

When learning knowledge bases, many approaches use a predefined number of linguistic labels for each variable, or the shape of the membership functions is constrained, facilitating the learning, but also limiting the solutions space and the strength of the method. The learning of both the data and rule bases can be done simultaneously, or in different stages. Some approaches firstly learn the data base and finally generate the rule base using the learned data base. Other approaches obtain the rule base using a predefined data base, and then tuning the shape of the membership functions to improve the learned knowledge base.

3 Learning of Fuzzy-Rule Based Controllers

Our proposal consists on a learning method based on the IRL approach in which both the data and rule bases are simultaneously learned. The only predefined parameters are the universe of discourse and the granularity of each variable. Both the number of linguistic labels, the shape of the membership functions and the rules' structure (a variable could not be considered in a rule) will be learned. The algorithm has the following steps:

1. Obtain a rule for the system.
   (a) Initialise population.
   (b) Evaluate population.
   (c) Eliminate bad rules and fill up population.
   (d) Scale the fitness values.
   (e) While the maximum number of iterations is not exceeded.
      i. Select the individuals of the population.
      ii. Crossover and mutate the individuals.
      iii. Evaluate population.
      iv. Eliminate bad rules and fill up population.
      v. Scale the fitness values.
2. Add the best rule to the final rule set.
3. Penalize the selected rule.
4. If the knowledge base does not solve the problem, return to step 1.

The rules that are going to be learned are conventional fuzzy rules like:

$$\begin{aligned} R^i &: \text{If } X_1^i \text{ is } A_1^i \text{ and } \ldots \text{ and } X_{NA}^i \text{ is } A_{NA}^i \\ &\text{Then } Y_1^i \text{ is } B_1^i \text{ and } \ldots \text{ and } Y_{NC}^i \text{ is } B_{NC}^i \end{aligned} \tag{1}$$

where $R^i, i = 1,\ldots,NR$, is the i-th rule, $X_j^i, j = 1,\ldots,NA$, and Y_k^i, $k = 1,\ldots,NC$, are linguistic variables of the antecedent and consequent parts, respectively. NR is the number of rules, NA the number of antecedents in a rule, NC the number of consequents, and A_j^i and B_k^i are linguistic values (labels) of these variables.

A set of examples has been chosen for learning the knowledge base. These examples cover the universe of discourse of all the variables in the antecedent part of the rule. The universes of discourse have been discretised, in order to minimize the search space, with a step or granularity $g_n, n = 1,\ldots,NV$, where $NV = NA + NC$ is the number of variables. Prior to the application of the learning method, the best action for each one of the examples is determined and saved in the variables in the consequent part of that example. The function, SF, that scores the action of a rule over an example (not the fitness function) is application dependent.

An example, e^l, is covered by rule R^i if it complies with the following two conditions:

$$A_1^i\left(e_1^l\right) \wedge \cdots \wedge A_{NA}^i\left(e_{NA}^l\right) > 0 \tag{2}$$

where $A_j^i(e_j^l)$ represents the membership degree of the value of variable j in the example e^l to A_j^i. A new parameter, δ, is defined with the aim of selecting the relation between the number of rules and the quality and accuracy of the controller. In that way, a second condition is imposed:

$$\frac{SF\left(R^i\left(e^l\right)\right)}{\max\left(SF\left(e^l\right)\right)} > \delta \tag{3}$$

where $SF(R^i(e^l))$ is the score assigned to the state reached by the system after applying rule R^i over example e^l, and $\max(SF(e^l))$ is the maximum score that an action can obtain for example e^l. The value of parameter δ can be adjusted in a range between 0 and 1. A low value of δ produces a lower number of rules in the final knowledge base, but the quality and the accuracy of the controller decreases. On the contrary, a high value of δ increases the quality of the controller, but also the number of rules.

The use of δ can be clarified by means of the following example. Let us suppose a robot must reach a point by turning 30°. Although this may be labelled as the best control action, also a rule proposing a turning of 20° should be considered as a good rule, even though the goal point is not fully reached. Parameter δ indicates the minimum quality a rule must have in order to be a valid rule for being added to the final knowledge base.

The shape of the membership functions that are going to be learned is shown in Fig. 1. Parameters b_n^i and c_n^i are learned, but points a_n^i and d_n^i are calculated as:

$$a_n^i = \max\left\{b_n^i - g_n, lower_n\right\} \tag{4}$$

$$d_n^i = \min\left\{c_n^i + g_n, upper_n\right\} \tag{5}$$

where $lower_n$ and $upper_n$ are the extreme points of the universe of discourse of variable n. In that way, it is not possible to exceed the range of the universe of discourse of the variables during the learning process. For example, for a variable n having the following values: $lower_n = 0, upper_n = 60, g_n = 10, b_n^i = 40$ and $c_n^i = 60$, then $a_n^i = \max\{40 - 10, 0\} = 30$ and $d_n^i = \min\{60 + 10, 60\} = 60$.

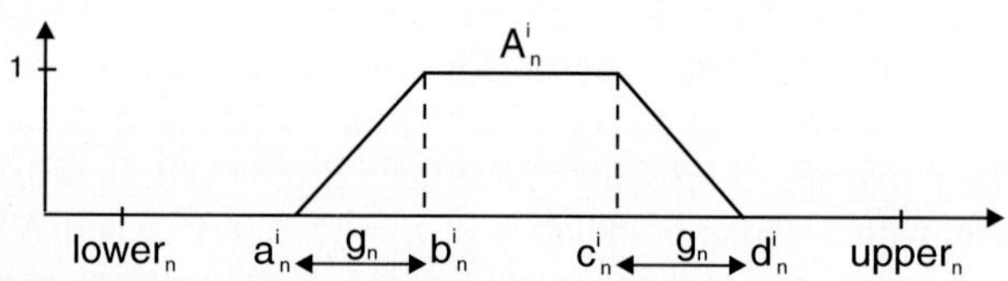

Fig. 1. Shape of the membership function for the linguistic value A_n^i

Chromosomes are real coded, and a rule as the one shown in (1) is encoded into a chromosome C^i as:

$$C^i = \left(a_1^i, b_1^i, c_1^i, d_1^i, \cdots, a_{NV}^i, b_{NV}^i, c_{NV}^i, d_{NV}^i\right) \tag{6}$$

The first step of the genetic algorithm consists in initialising the population. Rules in the initial population are created in the following way. An example currently not covered by any rule in the final knowledge base is randomly selected. The created rule is going to cover that single example, named e^l, at this initial generation. The membership functions for this rule are constructed as: $b_n^i = c_n^i = e_n^l$, whilst a_n^i and d_n^i are calculated using (4) and (5) respectively.

The evaluation of each individual of the population (each rule) is done with a two level fitness function. The first level (FF) consists on counting the number of examples that are covered by this rule, i.e., that fulfil (2) and (3), and that are not yet covered by a rule of the final knowledge base. If an example is covered by a rule of the final knowledge base, it will not contribute to the fitness value of any rule in the population. A second level for the fitness function is added in order to distinguish between rules with the same antecedent part but different consequents. It is the average value of the scoring function, SF, for all the examples that verify (2):

$$ASF^i = \frac{\sum_{l=1}^{NE} SF\left(R^i\left(e^l\right)\right)}{NEC^i} \tag{7}$$

where NE is the number of examples, and NEC^i is the number of examples that verify (2) for rule i. The second level of the fitness function is only used (in conjunction with the first level) when the final generation has been reached and the best rule of the population has to be added to the final knowledge base.

If for any example a rule verifies (2) but not (3), then this rule is deleted from the population. After the deletion of all the bad rules, the population must be filled up, until its size reaches NR. The rules that will be added are the best rules in the population. Finally the fitness values of the individuals of the population must be linearly scaled in order to prevent premature convergence of the population.

The selection procedure that has been employed is the stochastic remainder without replacement. An individual i will be selected $int(\frac{FF^i}{AFF})$ times, where FF^i is the value of the fitness function (first level) for individual i, and AFF is the average value of all FF^i. Taking into account $frac(\frac{FF^i}{AFF})$ the population is randomly filled up. After selection, the individuals are crossed (one-point crossover), mutated, and finally added to the new population. Elitism has been applied to avoid the loss of the best individuals due to crossover and mutation.

Crossover is done taken into account that the combination of points a_n^i and b_n^i cannot be truncated, and the same occurs to points c_n^i and d_n^i. This means that the slope of the sides of a membership function cannot be modified. After crossover, each chromosome is reordered for repairing bad definitions of the membership functions (e.g. $b_n^i > c_n^i$).

The mutation operator has three equally probable options to operate on a gene. It will only modify genes of type b_n^i or c_n^i: increasing or decreasing the value of the gene in an amount of g_n, or leaving the gene unchanged. This will provoke the extension or contraction of the membership function in a quantity equal to the granularity of each variable, implementing a local search in that way. After mutation, each chromosome will be reordered, and values of a_n^i and d_n^i will be calculated using (4) and (5), respectively.

Once the maximum number of iterations has been reached, the best rule of the population is added to the final knowledge base, and all the examples covered by this rule are marked. In that way these examples will not contribute to the fitness value of the individuals in the next sequence of iterations. If all the examples are covered by the rules of the final knowledge base, then this is a solution to the problem and the algorithm ends.

4 Learning the Wall-Following Behaviour

The methodology presented in the previous section is applied here for the design of a fuzzy controller for the wall-following behaviour in mobile robotics.

The wall-following behaviour is usually implemented when the robot is exploring an unknown area, or when it is moving between two points in a map. A good wall-following controller is characterized by three features: to maintain a suitable distance from the wall that is being followed, to move at a high velocity whenever the layout of the environment is permitting, and finally to avoid sharp movements, making smooth and progressive turns and changes in velocity. The controller can be configured modifying the values of two parameters: the reference distance, which is the desired distance between the robot and the selected wall, and the maximum velocity attainable by the robot. In what follows we assume that the robot is going to follow a contour that is on its right side. Of course, the robot could also follow the left-hand wall, but this can be easily dealt with by simply interchanging the sensorial inputs.

The input variables of the control system are the right-hand distance (*RD*), the distances quotient (*DQ*), which is calculated as:

$$DQ = \frac{\textit{left-hand distance}}{RD} \tag{8}$$

As it can be seen (Fig. 2), *DQ* shows the relative position of the robot inside a corridor, which provides with information that is more relevant to the problem than simply using the left-hand distance. A high value for *DQ* means that the robot is closer to the right-hand wall, whilst a low value indicates that the closer wall is the left-hand one. The other input variables are the linear velocity of the robot (LV), and the orientation of the robot with respect to the wall it is following. A positive value of the orientation indicates that the robot is approaching to the wall, whilst a negative value means the robot is moving away from the wall. The output variables are the linear acceleration and the angular velocity.

All the information used to calculate distances and orientations comes from the ultrasound sensors of a Nomad 200 robot. The distances and the orientation are obtained in two ways: if any of the walls (left or right) can be modelled with a straight

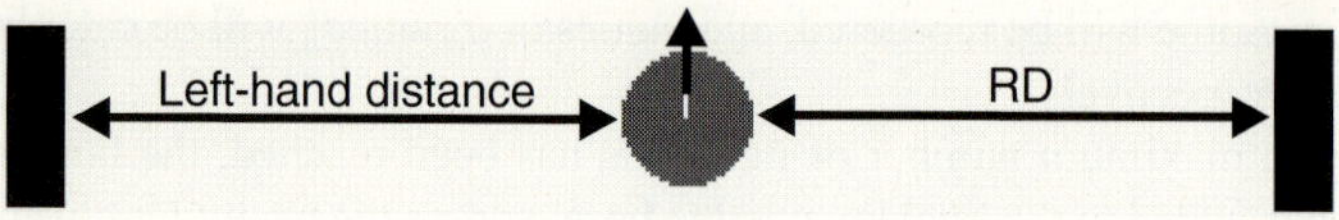

Fig. 2. Description of some of the distances used for the calculation of the input variables

line using a least square mean of the raw sensor data, then the corresponding distance and orientation are measured from that line. Otherwise, distance is measured as the minimum distance of a set of sensors, and the orientation will be the orientation of that sensor with respect to the advance direction.

The function *SF* used at (3) is defined for this application as:

$$SF(R^i(e^l)) = \frac{1}{\alpha_1 + \alpha_2 + \alpha_3 + 1} \tag{9}$$

where α_1, α_2, and α_3 are respectively:

$$\alpha_1 = 100\frac{|RD\text{-}reference\ distance|}{g_{RD}} \tag{10}$$

$$\alpha_2 = 10\frac{|maximum\ velocity - LV|}{g_{LV}} \tag{11}$$

$$\alpha_3 = \frac{|orientation|}{g_{orientation}} \tag{12}$$

and g_{RD}, g_{LV}, and $g_{orientation}$ are the granularities of the respective input variables. The granularities are used in these equations in order to evaluate the deviations of the values of the variables from the desired ones in a relative manner (the deviation of the value of variable n from the desired one is measured in units of g_n). This makes the comparison of the deviations of different variables possible and, as a consequence, the assignment of the weights for each one of the variables. These weights (100, 10 and 1 for (10), (11), and (12) respectively) have been heuristically determined, and indicate how much important the deviation in the value of a variable is with respect to the deviation of other variables.

SF takes values in $[0, 1]$. The highest weight has been assigned to the distance, as small variations of *RD* with respect to the reference distance should be highly penalised. An intermediate weight is associated to velocity and, finally, the least important contribution to function *SF* is for the orientation of the robot.

The defuzzification method that has been used for the learned fuzzy controller is the height defuzzifier [9]:

$$O_k = \frac{\sum_{i=1}^{NR} \overline{Y}^i_k B^i_k(\overline{Y}^i_k)}{\sum_{i=1}^{NR} B^i_k(\overline{Y}^i_k)} \tag{13}$$

where $\overline{Y}^i_k$ is the centre of gravity of the linguistic label B^i_k, and O_k is the defuzzified value for variable k. This method does not take into account the width of the membership functions from the consequent part of the rule. For this reason, and in order to simplify the learning process, the membership functions of the variables of the consequent part are crisp. This simplification does not affect the quality of the obtained controller.

5 Results

The system described in the previous sections has been implemented with a crossover probability of 0.2 and a mutation probability (per gene) of 0.4 (remember that mutation can increase, decrease or maintain unchanged a gene with equal probability). These values have been selected in order to focus search in a few promising areas (low crossover probability), but exploiting those areas doing a local search due to the high mutation probability, in a similar way evolution strategies, for example $(1+1)$-ES, work. We noticed that a high crossover probability distracted the search due to the combination of rules from quite different areas of the search space, but a low crossover probability is still necessary to discover new promising areas.

The population size is 300 individuals, and the maximum number of generations is 50. Once that value is reached, the best rule of the population is added to the final knowledge base. The process is repeated until the final knowledge base covers all the examples. Different values for parameter δ (3) have been tried. All the parameters of evolutionary learning have been heuristically obtained.

Figure 3 shows the variation of the number of rules and the average velocity change of the final knowledge base with δ. As δ increases the number of rules rises. Values of $\delta < 0.06$ have been discarded since they did not produce valid controllers. The average velocity change of the robot was measured for the environment shown in Fig. 4(b). This variable evaluates the average change of the robot's velocity between two consecutive control iterations. Low values of the average velocity change indicate smooth and progressive changes in velocity, reflecting more accuracy and quality in the control actions. As can be seen in Fig. 3, as δ increases the average velocity change decreases, and consequently the accuracy and quality of the controller rises.

All the controllers that have been obtained were tested in three simulated environments using the Nomad 200 robot simulation software. We have to emphasize

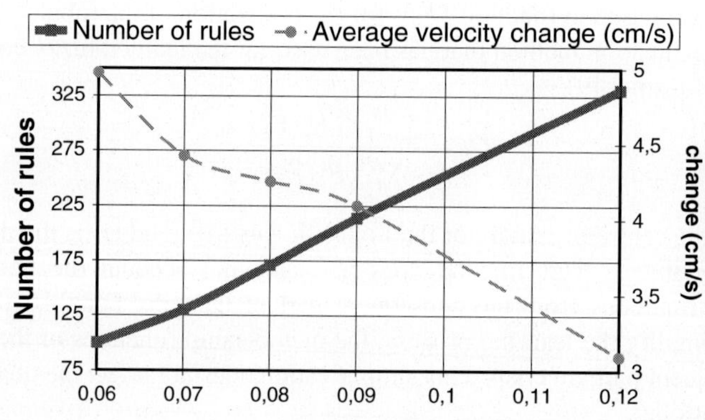

Fig. 3. Variation of the number of rules and the average velocity change of the final knowledge base with δ

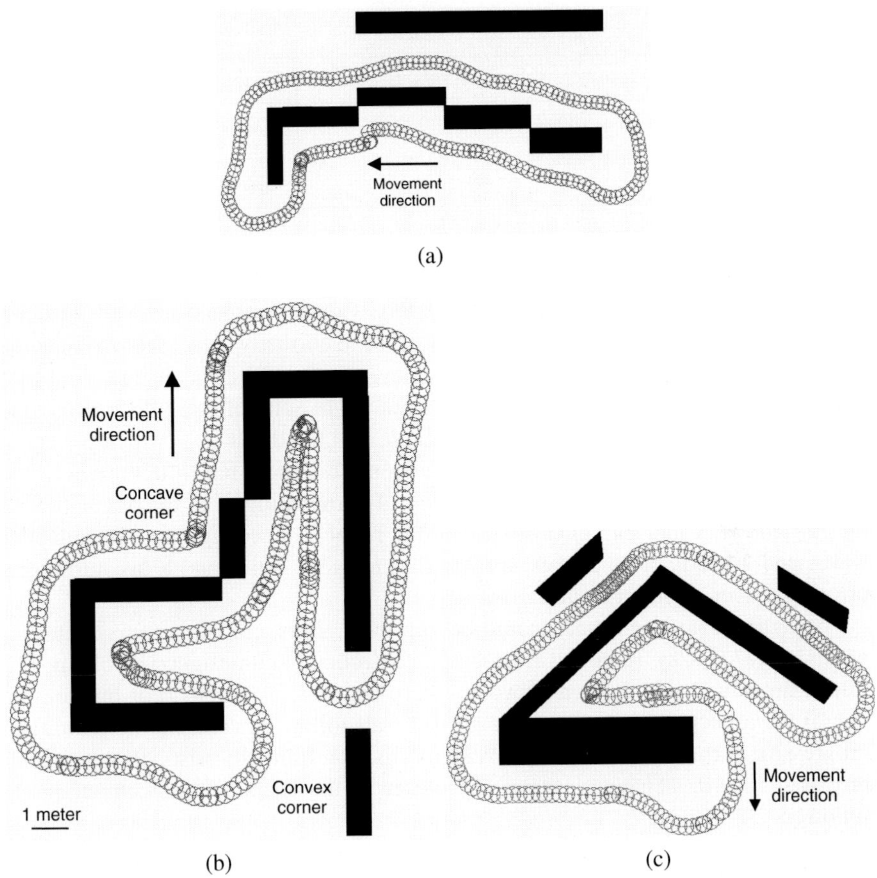

Fig. 4. Path of the robot in different simulated environments for $\delta = 0.06$

that the environments training set is different from the testing set. Learning only depends on function *SF*, which has to be carefully selected, and also on the examples, that must be chosen covering the input space with an adequate granularity (selection of the granularity of the variables is also of high importance). None of the environments shown in this paper (Fig. 4) have been used during the learning process.

Figure 4 shows as an example the robot path in three simulated environments for $\delta = 0.06$. The robot trajectory is represented by circular marks. A higher concentration of marks indicates lower velocity. The learned controller has 100 rules, the maximum velocity the robot can reach is 61 cm/s, and the reference distance at which the robot should follow the right wall is 51 cm. Ten tests have been done for each one of the environments. The average values measured for some parameters that reflect the controller performance are shown in Table 1.

Table 1. Average values of some parameters for the environments of Fig. 4

Environment	RD (cm)	Velocity (cm/s)	Velocity Change (cm/s)	Time (s)
4(a)	70	57	4.23	62
4(b)	65	51	4.59	106
4(c)	63	48	5.48	87

Environment 4(b) is quite complex, with three concave corners and seven convex corners in a circuit of a length of 54 meters. Convex corners are truly difficult situations, because the robot's sensors may cease to correctly detect the wall at some given moments, even though some of them may occasionally detect it. The controller must also significantly reduce velocity at corners. In spite of these difficulties, the obtained average velocity has been quite high, and the distance at which the robot should follow the wall is near the desired reference distance. The difference between both distances is caused by the high number of corners, in which the orientation of the robot is very bad (at concave corners the robot is detecting two perpendicular walls, and sometimes at convex corners it detects no wall), and a fast turning is prioritised over a correct distance.

Nevertheless, some aspects could be improved. Thus, comparing this controller with [10, 11] (Fuzzy Temporal Rule-based controller, hand-designed involving 313 rules), the obtained behaviour provokes sharp changes in velocity. For this reason, one of the aspects that characterizes a good wall-following controller (smooth and progressive turns and changes in velocity) is not fulfilled. Also, the controller proposed in [10, 11] gets a trajectory closer to the shape of the contour being followed.

6 Conclusions and Future Work

A genetic algorithm based on the IRL approach for the learning of fuzzy controllers has been described. Learning has no restrictions neither in the number of linguistic values for each variable, nor in the values that define the membership functions. The process only depends on function *SF*, and on the selected granularities of the variables, which are application dependent and must be carefully chosen. The algorithm has been applied to the learning of the wall-following behaviour. The learned control systems (for different values of δ) have been tested in a simulated environment with a high number of corners, showing a good performance both in the distance the wall was followed and in the average velocity.

Some aspects must be improved in the future, in order to get a better cooperation among rules of the final knowledge base, and enhance the interpretability of the rules. Another challenging work will be the learning of Fuzzy Temporal Rule-based controllers [10, 11, 12], which have a high degree of expressiveness and of analysing the evolution of variables, whilst taking past values into account.

Acknowledgements

Authors wish to acknowledge support from the Spanish Ministry of Education and Culture through grant TIC2000-0873-C02-01.

References

1. Braunstingl, R., Mujika, J., Uribe, J.P.: A wall following robot with a fuzzy logic controller optimized by a genetic algorithm. In: Proceedings of the International Joint Conference of the Fourth IEEE International Conference on Fuzzy Systems and the Second International Fuzzy Engineering Symposium. Volume 5., Yokohama (Japan) (1995) 77–82
2. Leitch, D.: Genetic algorithms for the evolution of behaviours in robotics. In: Genetic Algorithms and Soft Computing. Volume 8 of Studies in fuzziness and soft computing. Physica-Verlag (1996) 306–328
3. Pratihar, D.K., Deb, K., Ghosh, A.: A genetic-fuzzy approach for mobile robot navigation among moving obstacles. International Journal of Approximate Reasoning **20** (1999) 145–172
4. Cordón, O., Herrera, F.: Hybridizing genetic algorithms with sharing scheme and evolution strategies for designing approximate fuzzy rule-based systems. Fuzzy sets and systems **118** (2001) 235–255
5. Cordón, O., Herrera, F., Hoffmann, F., Magdalena, L.: Genetic fuzzy systems: evolutionary tuning and learning of fuzzy knowledge bases. Volume 19 of Advances in Fuzzy Systems – Applications and Theory. World Scientific (2001)
6. Magdalena, L., Velasco, J.R.: Fuzzy Rule-Based Controllers that Learn by Evolving their Knowledge Base. In: Genetic Algorithms and Soft Computing. Volume 8 of Studies in fuzziness and soft computing. Physica-Verlag (1996) 172–201
7. Bonarini, A.: Evolutionary Learning of Fuzzy rules: competition and cooperation. In Pedrycz, W., ed.: Fuzzy Modelling: Paradigms and Practice. Kluwer Academic Press, Norwell (USA) (1996) 265–284
8. Carse, B., Fogarty, T.C., Munro, A.: Evolving fuzzy rule based controllers using genetic algorithms. Fuzzy sets and systems **80** (1996) 273–293
9. Mendel, J.M.: Fuzzy logic systems for engineering: A tutorial. Proceedings of the IEEE **83** (1995) 345–377
10. Mucientes, M., Iglesias, R., Regueiro, C.V., Bugarín, A., Barro, S.: A fuzzy temporal rule-based velocity controller for mobile robotics. Fuzzy Sets and Systems **134** (2003) 83–99
11. Mucientes, M., Iglesias, R., Regueiro, C.V., Bugarín, A., Barro, S.: A fuzzy temporal rule-based approach for the design of behaviors in mobile robotics. In: Intelligent Systems: Technology and Applications. Volume 2. Fuzzy Systems, Neural Networks and Expert Systems of CRC Press International Volumes on Intelligent Systems Techniques and Applications. CRC Press (2003) 373–408
12. Mucientes, M., Iglesias, R., Regueiro, C.V., Bugarín, A., Cariñena, P., Barro, S.: Fuzzy temporal rules for mobile robot guidance in dynamic environments. IEEE Transactions on Systems, Man and Cybernetics-Part C: Applications and Reviews **31** (2001) 391–398

Genetic Algorithm in Process Optimisation Problems

Victor Oduguwa, Ashutosh Tiwari and Rajkumar Roy

Department of Enterprise Integration, School of Industrial and Manufacturing Science, Cranfield University, Cranfield, Bedford, MK43 0AL, UK
{v.oduguwa, a.tiwari and r.roy}@cranfield.ac.uk

Abstract. Genetic Algorithm (GA) is generating considerable interest for solving industrial optimisation problems. It is proving robust in delivering global optimal solutions and helping to resolve limitations encountered in traditional methods. However there are fewer GA applications in the process optimisation. This paper presents an overview of recent GA applications in process optimisation. The paper explores the features of process optimisation and critically evaluates how current GA techniques are suited for such complex problems. The survey outlines the current status and trends of GA applications in process related industries. For each industry, the paper describes the general domain problem, common issues, current trends, and the improvements generated by adopting the GA strategy. The paper concludes with an outline of future research directions.

1 Introduction

Genetic Algorithm (GA) are biologically inspired robust optimisers which in recent years are generating increasing interest for solving industrial optimisation problems. It is proving robust in delivering global optimal solutions and helping to resolve limitations encountered in traditional methods. The application of GA techniques is increasing with successful applications in many areas including but not limited to: engineering design optimisation, manufacturing system, process control, medical diagnosis, finance and simulation and communication systems. Despite the wider applications of GA in various industrial sectors, there are fewer publications in the area of process optimisation. This can be largely attributed to the inherent limitations in GA for large-scale process problems.

This paper attempts to bridge the gap between theory and practice by exploring characteristics of process optimisation and presenting a recent survey of GA applications. The paper begins by presenting definitions and the essential features of process optimisation problems. A survey of recent applications of GA to process optimisation problems is presented and a critical analysis of GAs potential to deal with such problems is outlined. The survey explores the current status and trends of GA applications in process related industries. For each application, the paper describes the problem, common issues, current trends, and the improvements

Victor Oduguwa et al.: *Genetic Algorithm in Process Optimisation Problems*, Advances in Soft Computing **1**, 323–333 (2005)
www.springerlink.com

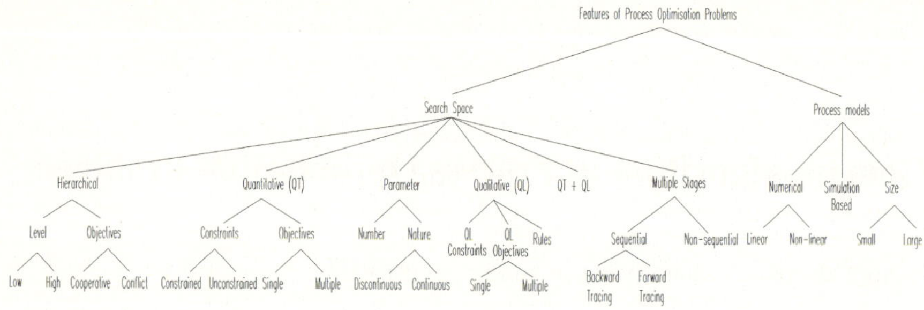

Fig. 1. Features of Process Optimisation Problems

achieved by adopting the GA strategy. The paper concludes with an outline of challenges and future research directions.

2 Definition and Features of Process Optimisation

Process optimisation involves the generation of optimal design solutions for individual units of the process where design information for the unit, are used to meet the desired requirement of the overall process. Various search algorithms have been developed for different types of process optimisation problems. In order to explore the efficacy of such algorithms it is often necessary to examine the features of the process problem. This section introduces a classification scheme shown in Fig. 1 to provide an explanatory shell for exploring the complexity of process optimisation problems.

2.1 Search Space

Process optimisation search space represents how the problem is represented in conventional optimisation theory. The main features of the problem space is summarised in this section using the classification given by Oduguwa [1]. The objective functions are mostly multiple and often conflicting, since a range of criteria is taken into account. Constraints are usually quite large, and could be either hard, soft equality or inequality. The parameter space are highly dimensional and could be discontinuous, continuous or a combination of both. Large number of parameters also implies exponential increase in interactions [2] among those parameters which give result to the "Curse of dimensionality" [3].

The search space can be quantitative (Q^T) or qualitative (Q^L) in nature. While Q^T describes the behaviour of the problem in numerical terms, Q^L is a non-numerical method of reasoning.

Multiple Stages: Individual units of he process systems are connected to form multiple stages of the overall process problem. The nature of this association could sequential or non-sequential. Sequential nature of the process offers the separate subsystems a dependency link where the output relationship of a subsystem becomes the input relationship of the subsequent subsystem in an orderly manner.

Hierarchical nature of process problem implies the "pecking order" of the subsystems problem space. Most process optimisation problems can be hierarchical in nature with multiple level problem space. A two level problem is described here where the subsystems problem is considered the low-level problem and the system level that acts as the coordinator is considered the high level problem. The low-level search provides insight into the intricacies of subsystem while the high-level search aggregates the overall search behaviour with respect to the system requirement.

2.2 Process Models

The process model is related to the search space and characterised in terms of its influence on the complexity of the process problem. Process models can be small or large but most real life process models are large and hence the combinatorial explosion. In these problems objective functions are often analytically unknown and are often substituted by fast-simplified models to gain access to less difficult functional models. In most cases, the function evaluation is mainly achievable through computer simulations.

3 Classical Approaches in Process Optimisation

Several mathematical optimisation algorithms for process optimisation have been used to solve the following process related optimisation problems: scheduling [4], chemical process engineering [5] and heat integration distillation [6]. Many of the solution approaches are based on mixed integer non-linear programme (MINLP) formulations, some of which have been successful in locating optimal and near-optimal solutions. However, these methods have some inherent limitations. The objective functions and the constraint are required to be twice differentiable and initial starting point guesses are required for the search, which may influence the type of solution achieved. This tends to reduce the consistency of the result achievable by this method. Also the user is required to have a detailed knowledge of the search space when applying these optimisation techniques. This is generally difficult to achieve in real-life problems since the search space are too complex to be known. Dynamic Programming (DP) is also mostly used in scheduling and process planning problems. The basic idea behind DP comes from the observation that any partial sub-path that ends at a point along the true optimal must itself be the optimal path leading up to that point. Thus the optimal path can be found by incremental extension of optimal sub-paths. The main limitation of dynamic programming is scalability. As the problem scale increases, the dimensionality explode hence the technique is not suitable for large-scale problems.

3.1 Stochastic Search Techniques

Recently, stochastic search techniques such as Simulated Annealing (SA), Taboo Search (TS) and Genetic algorithms (GA) are emerging for solving this class of problems. This paper focuses on GAs. GA are adaptive methods used to solve search and optimisation problems, based on genetic processes of biological organisms [7]. GA was developed in the 1970's by John Holland and students at the University of Michigan. Their aim was to simulate adaptive process of natural systems and to develop artificial systems that retain features of natural systems. The canonical form of the GA encodes each candidate solution to a given problem as a binary, integer, or real-valued string, referred to as chromosome. GAs simulates the genetic state (chromosomes) of a population of individuals using recombination operator with mutation. Each individual is evaluated and fitness assigned in proportion to the value of the objective function for the individual. New individuals created by these operators are selected on the basis of their fitness for the next generation.

4 GA Industrial Process Applications

This section presents an overview of interesting GA applications from the metal-forming (MF), scheduling, production, network and the chemical process industries. Most of these process optimisation problems have the characteristics described in Sect. 2. For each of the industrial areas explored, this section describes the problem area, traditional technique adopted, common issues, current trends, and the improvements achieved by adopting the GA strategies are discussed. An outline of GA applications in industry is given in Table 1.

4.1 Process Design Problems in the MF Industry

The MF process is a complex operation requiring a simple geometry to be transformed into a complex one. The main goal of optimisation in MF is to produce sound products through optimal process design, since the process material and die parameters significantly influences the process. Classical approaches such as trial and error for arbitrarily assigning reduction are tedious, ill-structured, time consuming and costly. DP even though can handle continuous and discrete variable, it is limited since the MF process normally involves large amount of process parameters with wide range of values, which may be active in the optimisation problem. Also derivative based approaches are not suitable since the objective function may reveal multiple stationary points. Several authors [8] have shown that the GA's based approached can be used to deal with this complex problems.

In rolling system optimisation several studies have been reported using SGA for optimal design [9], for multi-objective optimisation problems [10, 12], and recently for Q^T and Q^L search space [13]. These approaches have been shown to offer a more structured approach to process optimal design problems and the benefits to catalogue optimal design solution for future re-use. However the main

Table 1. Summary of GA Applications in Process Industry

Process Problems	Characteristics	EA Component
Steel Making	• A real coded GA for optimal process parameter setting in CC of steel [8]	GA SGA,
	• Parameter optimisation of re-heat furnaces for bloom delivery [14]	DE, μGA
	• Pareto converging GA for optimal CC velocity [10]	PCGA
	• SGA for optimal process parameter setting in CC of steel [11]	SGA
Metal forming	• Inverse elitist evolutionary search for optimal design of metal forging [15]	GA
	• Inverse elitist evolutionary search for optimal design of extrusion [15]	GA
	• Multi-objective GA for pass schedule optimal design in cold rolling of steel strips [9]	GA NSGAII
	• Multi-objective optimisation of rod product design [12]	NSGAII,
	• A Q^T and Q^L design optimisation approach to minimise both load and shape in rod rolling product design [16]	FL
	• Three objective optimisation of elongation, strain distribution and shape for rod rolling [13, 14, 15, 16, 17]	NSGAII, FL
Manufacturing Process	• SGA for process optimal design of textile manufacturing process [18]	SGA
	• SGA to determine optimal cutting conditions [19]	SGA
	• SGA for optimal set points for leavening process in bread-making plant [20]	SGA
Production Scheduling	• Scheduling complex products using repair strategy to obtain feasible solutions [21]	SGA
	• Optimal operational planning of district heating and planning [23]	GA+NLP
	• Optimal production planning to meet time-varying stochastic demand [24]	GA+DP
	• Global topological design for hierarchical network [25]	GA
Chemical	• Optimal operating conditions for improved unit performance [26]	NSGA
	• Optimisation of process parameter for carbon nanotubes [27]	SGA

problem experienced using GA in this environment is due to the expensive function evaluations. Since objective functions are often analytically unknown, function evaluations can only be achieved through costly computer simulations. The slow convergence criteria to near optimal solution with very small tolerance accompanied with the large population of solutions required for the evolutionary process also result to expensive evaluations. Oduguwa et al. [16, 17] developed meta-models as surrogate to these expensive computational FE simulations. The meta-model were used for function evaluations in the optimisation of rolling design problems. This reduced computational cost significantly. Several other authors have adopted micro GA (μGA) as embedded optimisers within expensive computational simulation softwares to reduce the computational cost. The μGA attempts rapid convergence using a small population size which typically consists of five individuals created using selection and crossover. Several authors have applied μGA optimal design. Deb and Chakraborti [14] conducted an experimental study to identify the optimal setting of the burners in continuous casting using μGA. Roy et al. [28] proposed adaptive μGA for optimal design of process variables in multi-pass wire drawing. In spite of the small population, the authors reported expensive computation runs required for convergence to optimal solutions with very small tolerance.

4.2 Scheduling and Process Planning Problems in Production Industries

Scheduling is an important aspect of the economy defined as the allocation of resources over time to perform a collection of activities. These problems are normally difficult to solve due to various constraints, complex product structures with many levels of assembly and large number of potential sequences. Mathematical programming techniques such as mixed integer programming techniques although have provided near optimal solutions, they tend to be suitable to small scheduling problems [21]. Most GA applications to these problems, encodes the sequence of the entire schedule in the chromosomes. Such a method can render the SGA unusable since the large amounts of offspring can become infeasible after the genetic operations. To overcome these problems, most GA research in this area focuses on developing approaches that improves the feasibility of chromosomes in the population.

Hicks et al. [21] proposed a repair strategy in their GA application for scheduling complex product in Capital Goods Company. Marian et al. [22] developed a more elaborate approach based on Huygens' wave propagation in Principles of Physics for an assembly sequence planning problem. The principle is based on determining the components to be added to the partial assembly at the next stage. Sakawa [23] proposed a GA using chromosome encoded with integer genes and fitness based on linear programming problem to decide feasible solutions for district heating and cooling (DHC) plant. In a similar approach, Yokoyama and Lewis III [24] proposed a two level solution method using GA and DP for production planning problem. At the upper level GA was used to determine decision variables that minimise set-up cost and the production cost while at the lower level, the DP were used to determine the decision variables for the production quantities. In telecommunications network design problem, Pablo et al. [25] guaranteed feasible solutions in GA by using problem specific heuristics. The network problem was augmented with fictitious nodes rules to establish feasible communication links.

Most of the approaches outlined above attempts to solve the feasibility problem caused by concatenating the decision variables into the chromosomes. It is arguable if this approach could sustain the increasing need to incorporate more problem features into the process problem, resulting in the need to develop advanced techniques to encourage feasible solutions.

4.3 Classical GA Applications in Related Process Industries

This section evaluates the success of classical GAs in related process industries. Several authors have applied classical GA, in manufacturing process [19, 20], and chemical industries [26, 27] in chemical industry. The main reason for classical GAs success in such complex environments is due to the size of the problem. Only the necessary part of the solution is coded in the chromosome. The problem size tends range from 5 to about 19. Since the number of parameters is relatively small there is a higher chance of obtaining feasible solutions. This approach can be considered appropriate with respect to the sparsity effects principle [29]. Where it is assumed that the process problem are dominated by the chosen parameters and the interactions are negligible. It is interesting to note that most authors suggested the need to include more parameters for a complete optimisation analysis. Since this approach is only suitable for small sized problems, future research might necessary to determine the maximum number of parameters the classical GA can cope with.

5 Future Directions

Future GA researches are required to address the issues outlined in the previous paragraph. This section briefly describes possible research directions.

Studies are required to explore search behaviours within hierarchical and sequential process environment. Exploiting special features of such problem could result to more realistic solutions.

Decomposition strategies from mathematical programming and super computing technologies can be explored to deal with large scale problems.

Integrated design strategies that deal with both Q^T and Q^L knowledge simultaneously within an optimisation framework would form an interesting area of research.

6 Current Research Efforts

Our research effort focuses on three problems areas regarding GA application to process design problem. These areas are: hierarchical process optimisation, approximate solution models and integrated design strategy for Q^L and Q^T models.

The authors have developed solution approaches suitable for improving GA application in process design. An outline of the proposed approaches is given below.

6.1 Hierarchical Process Optimisation

Most real-life process optimisation problems such as rolling system design can be hierarchical in nature where the subsystems problem is considered as the low-level problem and the system level is considered the high level problem. Most GA based solution approach assumes maximisation of objective function subject to various constraints. While such assumptions may be required at an aggregate level so that the problem is tractable for mathematical analysis, it is less certain that the assumptions will lead to valid conclusions at the low-level.

In addition, while aggregation may even out fluctuation caused by many of the underlying factors influencing the search directions, this may not always be the case and could result to deceptions in the search problem. Including the hierarchical property into search algorithms offers both advantages and disadvantages. The low-level search provides insight into the intricacies of subsystem but their search path is biased to the subsystems and this limits their application at system level. Subsystem search directions are also of limited value at the system level since the coordinator is mainly concerned with aggregating the overall search behaviour with respect to the system requirement. However, outputs from low-level search directions serves as important pre-requisites for guiding the high-level search directions. Since the optimal design path and convergence characteristics are key features for efficient search performance, it is therefore necessary to include the hierarchical behaviour in process optimisation problems.

A number of GA based approaches [30, 31] have been developed for solving this class of problems. Recently, the authors [32] proposed a bi-level genetic algorithm (BiGA) for solving the two level problem within a single framework. BiGA is an elitist optimisation algorithm developed to encourage limited asymmetric cooperation between the two players. Interested researchers can see the paper [32] for more details.

6.2 Integrated Qualitative and Quantitative Design Approaches

Most process optimisation techniques search the problem space using quantitative models (Q^T). Although they provide various detailed information about the process, unfortunately, these approaches do not consider qualitative effect of the problem simultaneously. The Q^T models are often incomplete such that key qualitative (Q^L) characteristics are missed. It is therefore necessary to develop integrated design strategies that incorporate the Q^L characteristics with the Q^T features for realistic optimal solutions.

One of the early works reported that considered Q^L characteristics within engineering search problem was Roy [33]. A GA based approach was developed to identify optimal solutions and then fuzzy logic was used to qualitatively evaluate a selection of solutions separately. Recently, Oduguwa et al. [13, 34] extended this work by developing an integrated Q^L evaluation optimisation approach which combines Q^L evaluation from designers with Q^T formulation of the design problem within an optimisation framework. The elaborate approach adopts the principle of multi-objective optimisation to explore the functional relationship between the Q^T and Q^L knowledge. In the proposed solution strategy, fuzzy logic-based models are developed to account for the Q^L characteristics of the design problem.

7 Conclusion

This paper presents an overview of recent GA in process optimisation. The paper explores the features of process optimisation and critically evaluates how current GA techniques are suited for such complex problems. Based on our observation, GA method offers improved solutions over classical strategies. However, future research work is required to address the underlying issues experienced to broaden the application of GA for the process optimisation problems.

References

1. V. Oduguwa, "Rolling System Design Optimisation using Soft Computing Techniques," EngD Thesis, Cranfield University, Bedford, UK, 2003.
2. H. W. Ray and J. Szekely, *Process Optimization with Applications in Metallurgy and Chemical Engineering*. New York: John Wiley and Sons, 1973.
3. R. Bellman, *Dynamic Programming*: Princeton University Press, 1957.
4. G. V. Reklatis, A. Sunol, K., D. W. T. Rippin, and O. Hortaçsu, "Overview of scheduling and planning operations: batch processing systems engineering," in *Batch Processing Systems: Fundamentals and Applications for Chemical Engineering*, O. Hortaçsu, Ed.: Springer, 1996.
5. C. Floudas, A, *Nonlinear and mixed-integer optimization: fundamentals and applications (topics in chemical engineering)*. New York: Oxford University Press, 1995.
6. R. Yokoyama and K. Ito, A revised decomposition method for MILP problems and its application to operational planning of thermal storage systems, *Journal of Energy Resources Technology*, 118: 277–284, 1996.
7. D. E. Goldberg, *Genetic Algorithm in Search, Optimization and Machine Learning*. Massachusetts: Addison Wesley, 1989.
8. B. Sarler, B. Filipic, M. Raudensky, and J. Horsky, An interdisciplinary approach towards optimal continuous casting of steel. In *Materials Processing in the Computer Age III: Proceedings of TMS*, pp. 27–36, Nashville, Tennessee. Warrendale, Pennsylvania, 2000.
9. D. D. Wang, A. K. Tieu, F. G. de Boer, B. Ma, and W. Y. D. Yuen, Towards a heuristic optimum design of rolling schedules for tandem cold rolling mills, *Engineering Application of Artificial Intelligence*, 13: 397–406, 2000.
10. N. Chakraborti, R. Kumar, and D. Jain, A study of the continuous casting mold using a pareto-converging genetic algorithm, *Applied Mathematical Modelling*, 25: 287–297, 2001.
11. C. A. Santos, J. A. Spim Jr, M. C. F. Ierardi, and A. Garcia, The use of artificial intelligence technique for the optimisation of process parameters used in the continuous casting of steel, *Applied Mathematical Modelling*, 26: 1077–1092, 2002.

12. J. S. Chung, S. M. Byon, H. J. Kim, and S. M. Hwang, Process Optimal Design in Metal Forming by Double-Objective Genetic Algorithm, *Transactions of the NAMRI/SME*, XXVII: 51–56, 2000.
13. V. Oduguwa and R. Roy, An Integrated Design Optimisation approach for Quantitative and Qualitative Search Space, In *Proceedings of ASME: 2003 ASME Design Engineering Technical Conference*, pp., Chicago, IIlinois, 2003 (*accepted for publication*).
14. N. Chakraborti, K. Deb, and A. Jha, A genetic algorithm based heat transfer analysis of a bloom re-heating furnace, *Steel research*, 71 (10): 396–420, 2000.
15. C. A. Conceicao Antonio and N. Magalhaes Dourado, Metal-forming process optimisation by inverse evolutionary search, *Journal of Material Processing Technology*, 121: 403–413, 2002.
16. V. Oduguwa and R. Roy, Multi-Objective Optimisation of Rolling Rod Product Design using Meta-Modelling Approach, In *Proceedings of the Genetic and Evolutionary Computation Conference (GECCO-2002)*, pp. 1164–1171, New York, 2002.
17. R. Roy and V. Oduguwa, Multiobjective Optimisation of Rod Design in Long Product Rolling within a Quantitative and Qualitative Search Space, In *2003 IEEE International Symposium on Computational Intelligence in Robotics and Automation*, Kobe, Japan, 2003 (*accepted for publication*).
18. S. Kim and G. J. Vanchtsevanos, An Intelligent approach to integration and control of textile process, *Information Sciences*, 123: 181–199, 2000.
19. F. Cus and J. Balic, Optimization of cutting process by GA approach, *Robotics and Computer Integrated Manufacturing*, 19: 113–121, 2003.
20. M. L. Fravolini, A. Ficola, and M. La Cava, Optimal operation of the leavening process for a bread-making industrial plant, *Journal of Food Engineering*, 2003 (submitted for publication).
21. P. Pongcharoen, C. Hicks, P. M. Braiden, and D. J. Stewardson, Determining optimum Genetic Algorithm parameters for scheduling the manufacturing and assembly of complex products, *International Journal of Production Economics*, 78: 311–322, 2002.
22. R. M. Marian, L. H. S. Luong, and K. Abhary, Assembly sequence planning and optimisation using genetic algorithm, *Applied Soft Computing*, 2/3F: 223–253, 2003.
23. M. Sakawa, K. Kato, S. Ushiro, and M. Inaoka, Operation planning of district heating and cooling plants using genetic algorithms for mixed integer programming, *Applied Soft Computing (ASC) Journal*, 1: 139–150, 2001.
24. M. Yokoyama and H. W. Lewis III, Optimization of the stochastic dynamic production cycling problem by a genetic algorithm, *Computer & Operations Research*, 30: 1831–1849, 2003.
25. P. Cortes, M. S. Caraballo, J. M. Garcia, J. Larraneta, and L. Onieva, GA for Planning Cable Telecommunication Networks, In *WSC5 Proceedings of the 5th Online World Conference on Soft Computing Methods in Industrial Applications*, pp. 135–141, 2000.
26. J. K. Rajesh, S. K. Gupta, G. P. Rangaiah, and A. K. Ray, Multi-objective optimization of industrial hydrogen plants, *Chemical Engineering Science*, 56: 999–1010, 2001.
27. M. Grujicic, G. Cao, and B. Gersten, Optimization of the chemical vapor deposition process for carbon nanotubes fabrication, *Applied Surface Science*, 199: 90–106, 2002.
28. S. Roy, S. Ghosh, and R. Shivpuri, Optimal Design of Process Variables in Multi-Pass Wire Drawing by Genetic Algorithms, *Journal of Manufacturing Science and Engineering*, 118, 1996.
29. D. C. Montgomery, *Design and Analysis of Experiments*, Fourth ed: John Wiley & Sons, 1997.
30. M. Gendreau, P. Marcotte, and G. Savard, A Hybrid Tabu-Ascent Algorithm for the Linear Bilevel Programming Problem, *Journal of Global Optimization*, 8 (3): 217–233, 1996.

31. Y. Yin, Genetic Algorithm based approach for bilevel programming models, *Journal of Transportation Engineering*, 126 (2): 115–120, 2000.
32. V. Oduguwa and R. Roy, Bilevel Optimisation using Genetic Algorithm, In *2002 IEEE International Conference on Artificial Intelligence Systems, (ICAIS 2002)*, pp. 322–327, Divnomorskoe, Russia, 2002.
33. R. Roy, "Adaptive Search and the Preliminary Design of Gas Turbine Blade Cooling System," PhD Thesis, University of Plymouth, Plymouth, 1997.
34. V. Oduguwa, R. Roy and D. Farrugia, Fuzzy Multi-Objective Optimisation Approach for Rod Shape Design in Long Product Rolling, Fuzzy Sets and Systems IFSA 2003, 10th International Fuzzy Systems Association World Congress. 2003. Istanbul, Turkey: Springer-Verlag. pp. 636–643.

Author Index

Subject Index